Energy Harvesting Wireless Communications

Energy Harvesting Wireless Communications

Chuan Huang
University of Electronic Science and Technology of China, Chengdu, China

Sheng Zhou
Tsinghua University, Beijing, China

Jie Xu
Guangdong University of Technology, Guangzhou, China

Zhisheng Niu
Tsinghua University, Beijing, China

Rui Zhang
National University of Singapore, Singapore

Shuguang Cui
University of California, Davis, USA

WILEY

IEEE PRESS

This edition first published 2019

Registered Offices
John Wiley & Sons, Inc., 111 River Street, Hoboken, NJ 07030, USA
John Wiley & Sons Singapore Pte. Ltd, 1 Fusionopolis Walk, #07-01 Solaris South Tower, Singapore 138628

Editorial Office
The Atrium, Southern Gate, Chichester, West Sussex, PO19 8SQ, UK

For details of our global editorial offices, customer services, and more information about Wiley products visit us at www.wiley.com.

Wiley also publishes its books in a variety of electronic formats and by print-on-demand. Some content that appears in standard print versions of this book may not be available in other formats.

Library of Congress Cataloging-in-Publication Data

Names: Huang, Chuan, 1983- author.
Title: Energy harvesting wireless communications / Dr. Chuan Huang, Chengdu, China, Dr. Sheng Zhou, Beijing, China, Mr. Jie Xu, Guangzhou, China, Mr. Zhisheng Niu, Beijing, China, Mr. Rui Zhang, Singapore, Mr. Shuguang Cui, Davis, USA.
Description: First edition. | Hoboken, NJ : John Wiley & Sons, Inc., [2019] | Includes bibliographical references and index. |
Identifiers: LCCN 2018047797 (print) | LCCN 2018048603 (ebook) | ISBN 9781119295976 (Adobe PDF) | ISBN 9781119295969 (ePub) | ISBN 9781119295945 (hardcover)
Subjects: LCSH: Energy harvesting. | Wireless communication systems–Power supply.
Classification: LCC TK2896 (ebook) | LCC TK2896 .H83 2019 (print) | DDC 621.382/32–dc23
LC record available at https://lccn.loc.gov/2018047797

Cover Design: Wiley
Cover Image: © Adyna / Getty Images

Set in 10/12pt WarnockPro by SPi Global, Chennai, India
Printed in Singapore by C.O.S. Printers Pte Ltd

10 9 8 7 6 5 4 3 2 1

Contents

1
Introduction

Energy harvesting (EH) is expected to have abundant applications in future wireless communication systems to power transceivers by utilizing environmental energy such as solar, thermal, wind, and kinetic energy. It becomes a promising technology that enables sensor networks, cellular networks, and wireless communications in wide rural areas.

Since renewable energy is generally clean and cheap, EH offers various benefits compared with conventional energy supplies such as batteries and fossil-fuel-based generators. For example, in cellular networks, solar panels and wind farms have been deployed to power base stations, thus lowering the expenses of energy bills, reducing the level of carbon dioxide emissions, and improving the flexibility of deployment. Besides, in wireless sensor networks, EH has been considered as a good substitute for the traditional battery, which in principle prolongs the network operation time to almost infinity. In short, EH in turn leads to a promising future for wireless networks: green and self-sustainable.

1.1 Energy Harvesting Models and Constraints

Despite many advantages, the use of EH also imposes new challenges on the design of wireless communication. Obviously, the harvested energy from solar, thermal, wind, and kinetic energy sources is not stable and might change randomly over time. Therefore, besides the randomness of the channel fading, there is another dimension of stochastic resource to be dealt with, and it brings new constraints in the optimization of EH wireless communication systems.

Wireless communication channels often fluctuate more substantially and dynamically than practical EH rates (e.g. the channel changes on the order of milliseconds, while the EH rate changes on the order of seconds or minutes), while channel fading is the main challenge faced in the design of reliable wireless communications. To illustrate this issue, we adopt a point-to-point wireless communication system, which consists of one transmitter powered by an energy harvester and one receiver with a reliable power supply, to show the phenomenon of the multi-time-scale channel/EH rate variations. In practice, the coherence time of EH processes is often much larger than that of wireless channels, as previously mentioned. Therefore, a block-based quasi-static EH model is practically valid, where the EH rate remains constant within each EH coherent block and may change from one block to another, and at the same time each EH block spans over many communication channel coherent blocks, as shown in Figure 1.1. For the

Energy Harvesting Wireless Communications, First Edition.
Chuan Huang, Sheng Zhou, Jie Xu, Zhisheng Niu, Rui Zhang and Shuguang Cui.

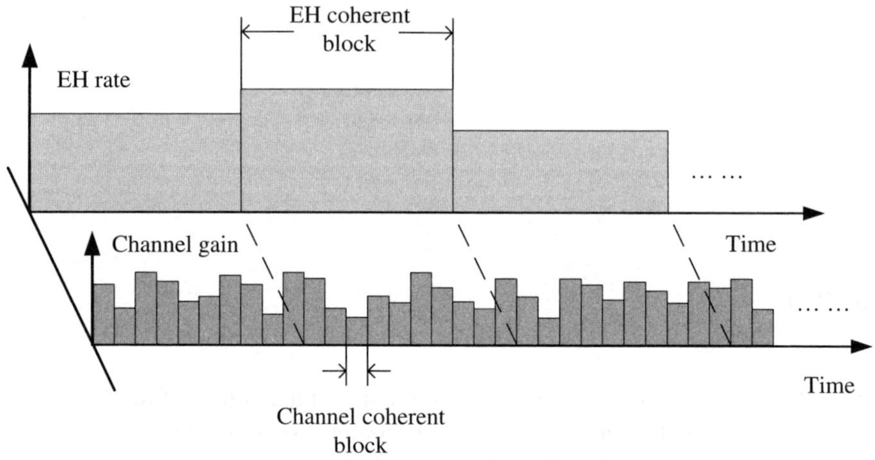

Figure 1.1 Time variations in EH process versus wireless channel.

purpose of exposition, we consider wireless data transmissions over a finite horizon of $M \leq 1$ EH blocks. Each EH block is further divided into $N \leq 1$ communication blocks each of one unit time and a constant channel gain.

Moreover, the random and intermittent characteristics of renewable energy impose a new type of EH constraint: the available energy at an EH communication node up to any time is bounded by its accumulatively harvested energy at that time. This is in contrast to conventional communication systems with stable energy sources, in which the available energy at any time is either unbounded or only limited by the remaining energy in the storage device (e.g. battery).

Mathematically, let $E_m \geq 0$ denote the EH rate in the mth EH block and $h_{n,m} \geq 0$ the channel power gain of the (n, m)th communication block (i.e. the nth communication block of the mth EH block) with $n = 1, \dots, N, m = 1, \dots, M$. Furthermore, we use $P_{n,m} \geq 0$ to denote the power consumption at the transmitter in the (n, m)th communication block. Unless otherwise stated, we consider that $P_{n,m}$ represents the transmit power at the transmitter and ignore the power consumption by circuit, signal processing, etc. Assuming an ideal energy storage device (i.e. with infinite capacity and no energy leakage) employed at the transmitter, we have the *EH constraints* on the scheduled power consumptions $\{P_{n,m}\}$; that is, the energy accumulatively consumed up to any communication block (n, m), i.e. $\sum_{j=1}^{m-1}\left(\sum_{i=1}^{N} P_{i,j}\right) + \sum_{i=1}^{n} P_{i,m}$, should be no larger than the energy accumulatively harvested by then, i.e. $N\sum_{j=1}^{m-1} E_j + nE_m$. In other words, we have the EH constraints as

$$\sum_{j=1}^{m-1}\sum_{i=1}^{N} P_{i,j} + \sum_{i=1}^{n} P_{i,m} \leq N\sum_{j=1}^{m-1} E_j + nE_m, \quad n = 1, \dots, N, m = 1, \dots, M. \tag{1.1}$$

Due to both the new EH constraints and the multi-time-scale channel/EH rate variations, it is a challenging problem to jointly optimize the communication scheduling and energy management in EH-based wireless communications.

Finally, the availabilities of the channel state information (CSI) $\{h(n, m)\}$ and the energy state information (ESI) $\{E(m)\}$ at the transmitter, respectively, can significantly affect the performance of EH communication systems. Among all different assumptions about the channel state information at the transmitter (CSIT) and energy state information at the transmitter (ESIT), there are four cases of primary interest in this book, listed below:

(1) *Case 1*: Noncausal CSIT and ESIT. At the beginning of the transmission, the transmitter perfectly knows the past, current, and future CSI and ESI. This case approximates the practical scenario when the transmitter can accurately predict the future CSI (e.g. slowly varying channels in low-mobility applications) and the future ESI (e.g. based on historical data in a periodically varying energy environment). The optimal solution in this case provides a performance upper bound for all other CSIT/ESIT availability cases.
(2) *Case 2*: Causal CSIT and ESIT. At the beginning of each EH/communication block, the transmitter knows the past and current CSI/ESI, as well as the statistical information (e.g. distributions) of future CSI/ESI. In general, the solution of this case achieves the lowest utility among the first three cases considered herein.
(3) *Case 3*: Causal CSIT and noncausal ESIT. This is a hybrid model based on cases 1 and 2, in which all ESI is perfectly known at the beginning of the transmission while only the past and current CSI is known.
(4) *Case 4*: No CSIT and noncausal/causal ESIT. During the transmission, the transmitter does not have any CSI and only has statistical information on the CSI. The noncausal or causal ESIT is defined as that in Case 1 or 2 above. Note that in all the above cases, we assume that at each communication block, the receiver perfectly knows the CSI in that block.

1.2 Structure of the Book

Based on the previous section, it is observed that EH brings a new dimension to the wireless communication problems, in the form of intermittency and randomness of the available energy, as well as the possibility of the energy cooperations among the transmission nodes in wireless networks. In this book, we summarize the progresses taken in the past few years in this new research field. This book is divided into three parts:

(1) In part I, we focus on the optimal transmission design for EH wireless communication systems. In particular, Chapter 2 addresses the optimal power allocation problems for the point-to-point EH channels to maximize the system throughput or minimize the average outage probability and also considers the effects of imperfect circuits and limited battery storage. Chapter 3 examines the power allocation for various multi-node wireless channels powered by energy harvesters, including the multiple-access channels (MACs), relay channels, and large relay networks. Chapter 4 studies the cross-layer design for EH communications, considering some upper layer issues such as transmission delay and traffic variations over time.
(2) In part II, we focus on the design and optimization of some EH networks. Chapter 5 considers the *ad hoc* networks, where there is no central control of the whole network, and studies the opportunistic access control schemes and the corresponding throughput scaling behavior. Chapter 6 considers a standard cellular network with multiple base stations powered by energy harvesters and studies the energy and

communication cooperations among them. Chapter 7 considers several new issues in the next-generation cellular networks and studies EH-based hyper-cellular networks with control and traffic separation and proactive content caching and push for better utilization of the renewable energy on small-cell base stations.

(3) Part III includes three appendices about the basic tools widely used in this book.

Part I

Energy Harvesting Wireless Transmission

2

Power Allocation for Point-to-Point Energy Harvesting Channels

To start with, in this chapter we consider the simplest point-to-point energy harvesting (EH) channel with a single EH-powered wireless transmitter communicating with a single wireless receiver. As shown in Figure 2.1, the transmitter can use an energy harvester to harvest the renewable energy from the environment and then store the harvested energy in a rechargeable storage device, to be used for sending information to the receiver. On the other hand, the receiver can either be powered by fixed energy supply (such as battery or grid) or EH, which will be specified later. The aim of this chapter is to characterize the fundamental limits of the communication performance of the point-to-point EH channel via designing the power allocation at the wireless transmitter. As this channel can be viewed as a building block of more general multiuser EH channels and EH-powered wireless networks, the results obtained in this chapter will provide key design insights for general EH networks.

In the point-to-point EH channel, the introduction of renewable energy imposes new design challenges on the power allocation at the wireless transmitter. In particular, as the renewable energy arrival rates or EH rates are generally random and intermittent, a new type of EH constraints is introduced, i.e. the available energy at the EH wireless transmitter up to any time is bounded by its accumulatively harvested energy by that time. This is in sharp contrast to conventional wireless communication systems with fixed energy sources, in which the available energy at any time is either unbounded or only limited by the remaining energy in the storage device (e.g. battery). In addition, wireless communication channels and practical EH rates often fluctuate at different time scales, while channel fading is the main challenge faced in the design of reliable wireless communications. Due to both the new EH constraints and the multi-time scale channel/EH rate variations, it is a challenging problem to design the power allocation at the EH wireless transmitter for communication performance optimization [1–3].

In this chapter, we first build a general utility optimization framework to reveal the key design principles of the point-to-point EH channel by considering the scenario with the receiver powered by a fixed energy supply and an ideal transmitter with an infinite energy storage and perfect circuits. Then, we apply this framework to solve two specific problems including the throughput maximization and the outage probability minimization, which are commonly adopted in the wireless literature as the communication performance metrics. Next, we extend the power allocation strategies to more practical setups by considering the limited energy storage and imperfect circuits at the wireless transmitter. In addition, we discuss the power allocation when both the transmitter and the receiver are EH powered.

Energy Harvesting Wireless Communications, First Edition.
Chuan Huang, Sheng Zhou, Jie Xu, Zhisheng Niu, Rui Zhang and Shuguang Cui.

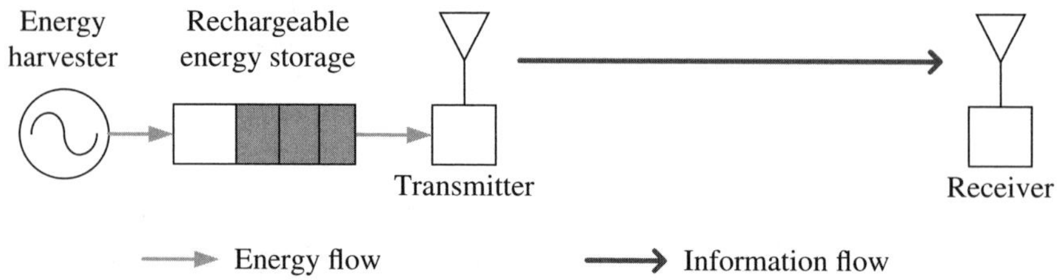

Figure 2.1 A point-to-point EH channel with an EH transmitter using the harvested power to send information to a receiver.

2.1 A General Utility Optimization Framework for Point-to-Point EH Channels

In the following sections, we focus on the case when the receiver has a fixed energy supply, unless otherwise stated. We consider block-based quasi-static models for the EH process at the wireless transmitter as well as the wireless communication channel from the transmitter to the receiver, where both the EH rate and the wireless channel remain constant over each block and may change from one block to another. Since the coherence time of the EH process is often much larger than that of the wireless channel as previously mentioned, we consider that each EH coherent block spans over many communication channel coherent blocks, as shown in Figure 2.1. For the purpose of exposition, we consider the wireless data transmission over a finite horizon of length T (in, e.g. seconds) that consists of $M \geq 1$ EH blocks. Each EH block is further divided into $N \geq 1$ communication blocks each of one unit time and a constant channel gain.

Let $E_m \geq 0$ denote the EH rate in the mth EH block and $h_{n,m} \geq 0$ the channel power gain of the (n, m)th communication block (i.e. the nth communication block of the mth EH block) with $n = 1, \dots, N, m = 1, \dots, M$. Furthermore, we use $P_{n,m} \geq 0$ to denote the power consumption at the transmitter in the (n, m)th communication block. Unless otherwise stated, we consider that $P_{n,m}$ represents the transmit power at the transmitter and ignore the power consumption by circuit, signal processing, etc. Assuming an ideal energy storage device (i.e. with infinite capacity and no energy leakage) employed at the transmitter, we have the *EH constraints* on the scheduled power consumptions $\{P_{n,m}\}$; that is, the energy accumulatively consumed up to any communication block (n, m), i.e. $\sum_{j=1}^{m-1}\left(\sum_{i=1}^{N} P_{i,j}\right) + \sum_{i=1}^{n} P_{i,m}$, should be no larger than the energy accumulatively harvested by then, i.e. $N\sum_{j=1}^{m-1} E_j + nE_m$ [4, 5]. In other words, we have the EH constraints as

$$\sum_{j=1}^{m-1}\sum_{i=1}^{N} P_{i,j} + \sum_{i=1}^{n} P_{i,m} \leq N\sum_{j=1}^{m-1} E_j + nE_m,$$
$$n = 1, \dots, N, \ m = 1, \dots, M. \tag{2.1}$$

Over the (n, m)th communication block, we denote the input–output relationship from the transmitter to the receiver in the point-to-point wireless channel as

$$y_{n,m} = \sqrt{P_{n,m}h_{n,m}}x_{n,m} + v_{n,m}, \quad n = 1, \dots, N, \ m = 1, \dots, M, \tag{2.2}$$

where $y_{n,m}$ is the channel output or the received signal at the receiver, $x_{n,m}$ is the corresponding channel input or the transmitted signal by the transmitter with zero mean and unit average power, and $v_{n,m}$ is the independent and identically distributed (i.i.d.) circularly symmetric complex Gaussian (CSCG) noise at the receiver with zero mean and unit variance. For each communication block, the achievable data rate (in bps Hz^{-1}) corresponds to the instantaneous mutual information of the channel, assuming the optimal Gaussian codebook for the transmitted signals, which is given by

$$I_{n,m}(h_{n,m}, P_{n,m}) = \log_2(1 + h_{n,m}P_{n,m}). \tag{2.3}$$

To characterize the communication quality of service (QoS) measured at the receiver, we define a general utility function $U_{n,m}(P_{n,m})$ for the (n, m)th block, which is dependent on the instantaneous mutual information $I_{n,m}(h_{n,m}, P_{n,m})$ at that block. As a result, the general utility maximization problem over the M EH blocks could be formulated as

$$\max_{\{P_{n,m}\geq 0\}} \sum_{m=1}^{M} \sum_{n=1}^{N} U_{n,m}(P_{n,m}) \tag{2.4}$$

$$\text{s.t.} \sum_{j=1}^{m-1} \sum_{i=1}^{N} P_{i,j} + \sum_{i=1}^{n} P_{i,m} \leq N \sum_{j=1}^{m-1} E_j + nE_m, \quad n = 1, \ldots, N, \; m = 1, \ldots, M. \tag{2.5}$$

In practice, the utility function $U_{n,m}(P_{n,m})$ can be defined more explicitly as throughput [4–10], non-outage probability [11, 12], or other performance metrics such as end-to-end distortion in an EH-based estimation system [13], which could be either deterministic or statistical average based on the availabilities of the channel state information (CSI), i.e. $\{h_{n,m}\}$, and the energy state information (ESI), i.e. $\{E_m\}$, at the transmitter, namely, CSIT and ESIT, respectively. In the following sections, we will particularly focus on the throughput maximization and outage probability minimization by considering the throughput and the non-outage probability as the utility $U_{n,m}(P_{n,m})$. Moreover, among all different assumptions about the channel state information at the transmitter (CSIT) and energy state information at the transmitter (ESIT), there are four cases of our primary interest including:

- *Case 1*: Noncausal CSIT and ESIT.
- *Case 2*: Causal CSIT and ESIT.
- *Case 3*: Causal CSIT and noncausal ESIT.
- *Case 4*: No CSIT and noncausal/causal ESIT.

Note that in all the above cases, we consider the case that at each communication block, the receiver perfectly knows the CSI in that block.

2.2 Throughput Maximization for Gaussian Channel with EH Transmitter

First, we consider the throughput maximization problem in the EH channel by considering the instantaneous mutual information $I_{n,m}(h_{n,m}, P_{n,m})$ in (2.3) as the utility function. Specifically, we consider the Gaussian channel case with the channel power gains remaining constant over the whole time horizon. As a result, we have $h_{n,m} = h, n = 1, \ldots, N, m = 1, \ldots, M$. In this case, the utility function is expressed as

$$U_{n,m}(P_{n,m}) = I_{n,m}(h, P_{n,m}) = \log_2(1 + hP_{n,m}) \tag{2.6}$$

in bps Hz^{-1} with normalized noise power. Furthermore, with the Gaussian channel, we have the number of communication blocks in each EH block as $N = 1$ and define $P_m \triangleq P_{1,m}$ without loss of generality. Accordingly, the general utility maximization problem in (2.4) can be expressed as the following throughput maximization problem:

$$\max_{\{P_m \geq 0\}} \sum_{m=1}^{M} \log_2(1 + hP_m) \tag{2.7}$$

$$\text{s.t.} \sum_{j=1}^{m} P_j \leq \sum_{j=1}^{m} E_j, \quad m = 1, \ldots, M. \tag{2.8}$$

The optimal power allocation solution to the problem (2.7) depends on the availability of the ESIT. In the following, we consider two cases with noncausal and causal ESIT, respectively.

2.2.1 The Case with Noncausal ESIT

When the transmitter noncausally knows the current and future ESI, we refer to the solution to problem (2.7) as an *offline* optimization. In the following, we first obtain the optimal solution structure to problem (2.7) and then discuss its implementation in practice.

2.2.1.1 Staircase Power Allocation to Problem (2.7)

It is easy to show that problem (2.7) is a convex optimization problem, since the objective function is concave and the EH constraints are linear. As a result, this problem can be efficiently solved via standard convex optimization techniques [14]. Furthermore, since problem (2.7) satisfies Slater's condition, strong duality holds between (2.7) and its dual problem. In this case, we can apply the Karush–Kuhn–Tucker (KKT) conditions to reveal its optimal solution.

Let $\lambda_m \geq 0$ denote the Lagrange multiplier associated with the mth constraint in (2.8) and $\mu_m \geq 0$ denote the Lagrange multiplier associated with the power constraint $P_m \geq 0, m = 1, \ldots, M$. The Lagrangian associated with problem (2.7) is

$$\mathcal{L}_{2.2}(\{P_m\}, \{\lambda_m\}, \{\mu_m\}) = \sum_{m=1}^{M} \log_2(1 + hP_m) - \sum_{m=1}^{M} \lambda_m \left(\sum_{j=1}^{m} P_j - \sum_{j=1}^{m} E_j \right) + \sum_{m=1}^{M} \mu_m P_m. \tag{2.9}$$

Then the necessary and sufficient conditions for $\{P_m^\star\}$, $\{\lambda_m^\star\}$, and $\{\mu_m^\star\}$ to be the primal and dual optimal solutions to problem (2.7) are given by the following KKT conditions:

$$0 \leq P_m^\star, \tag{2.10}$$

$$0 \geq \sum_{j=1}^{m} P_j^\star - \sum_{j=1}^{m} E_j, \tag{2.11}$$

$$0 \leq \lambda_m^\star, \tag{2.12}$$

$$0 \leq \mu_m^\star, \tag{2.13}$$

$$0 = \lambda_m^\star \left(\sum_{j=1}^{m} P_j^\star - \sum_{j=1}^{m} E_j \right), \tag{2.14}$$

$$0 = \mu_m^\star P_m^\star, \tag{2.15}$$

$$0 = \frac{\partial \mathcal{L}_{2.2}(\{P_m^\star\}, \{\lambda_m^\star\}, \{\mu_m^\star\})}{\partial P_m^\star}, \tag{2.16}$$

for all $m = 1, \ldots, M$. From (2.10), (2.13), (2.15), and (2.16), we obtain the optimal power allocation as

$$P_m^\star = \left[v_m - \frac{1}{h} \right]^+, \quad m = 1, \ldots, M, \tag{2.17}$$

where $[x]^+ \triangleq \max(x, 0)$, $v_m \triangleq \left(\ln 2 \sum_{j=m}^{M} \lambda_j^\star \right)^{-1} \geq 0$, and the $\lambda_j^\star$'s satisfy the above KKT conditions.

For the ease of description, we define an EH block $t \in \{1, \ldots, M\}$ as a *transition block* if the transmit power changes after EH block t, i.e. $P_t^\star \neq P_{t+1}^\star$. We define the last EH block M also as a transition block (say, by defining $P_{M+1}^\star$ to be infinity); hence there is at least one transition block. We collect all transition blocks as the set $\mathcal{S} = \{t_1, \ldots, t_{|\mathcal{S}|}\}$, where $t_i < t_j$ for $i < j$ and $t_{|\mathcal{S}|} = M$. Then, we have the following structural properties for the optimal power allocation $\{P_m^\star\}$, as shown in Figure 2.2.

Proposition 2.1 *The optimal solution $\{P_m^\star\}$ in (2.17) to problem (2.7) satisfies the following properties:*

- *The optimal transmit power monotonically increases over time, i.e. $P_1^\star \leq P_2^\star \leq \cdots \leq P_M^\star$. We say that the optimal solution performs staircase power allocation over blocks, since the transmit power $\{P_m^\star\}$ in (2.17) is a staircase-like function.*
- *If EH block t is a transition block, then the battery storage is empty after this block, or equivalently, the accumulative energy consumed up to EH block t equals that harvested up to then, i.e. (2.11) holds with equality for $t \in \mathcal{S}$.*

Proof: Since $\lambda_j^\star \geq 0$, it follows that $v_1 \leq v_2 \leq \cdots \leq v_M$. Accordingly, based on (2.17), we have $P_1^\star \leq P_2^\star \leq \cdots \leq P_M^\star$. Therefore, the first property is verified.

Suppose that the transmit power changes after EH block t, and thus we have $v_t \neq v_{t+1}$. Since by definition $v_t = \left(\ln 2 \sum_{j=t}^{M} \lambda_j^\star \right)^{-1}$, we get $\lambda_t^\star \neq 0$. From (2.13), we get $\lambda_t^\star > 0$. It then follows from the complementary slackness condition (2.14) that (2.11) holds with equality for EH block t. This proves the second property. □

2.2.1.2 Efficient Algorithm to Solve Problem (12.7)

Based on the structural properties of the optimal solution, we develop an efficient algorithm to implement the staircase power allocation to optimally solve problem (2.7). Such an algorithm was proposed initially in [4] and in [6] in parallel.

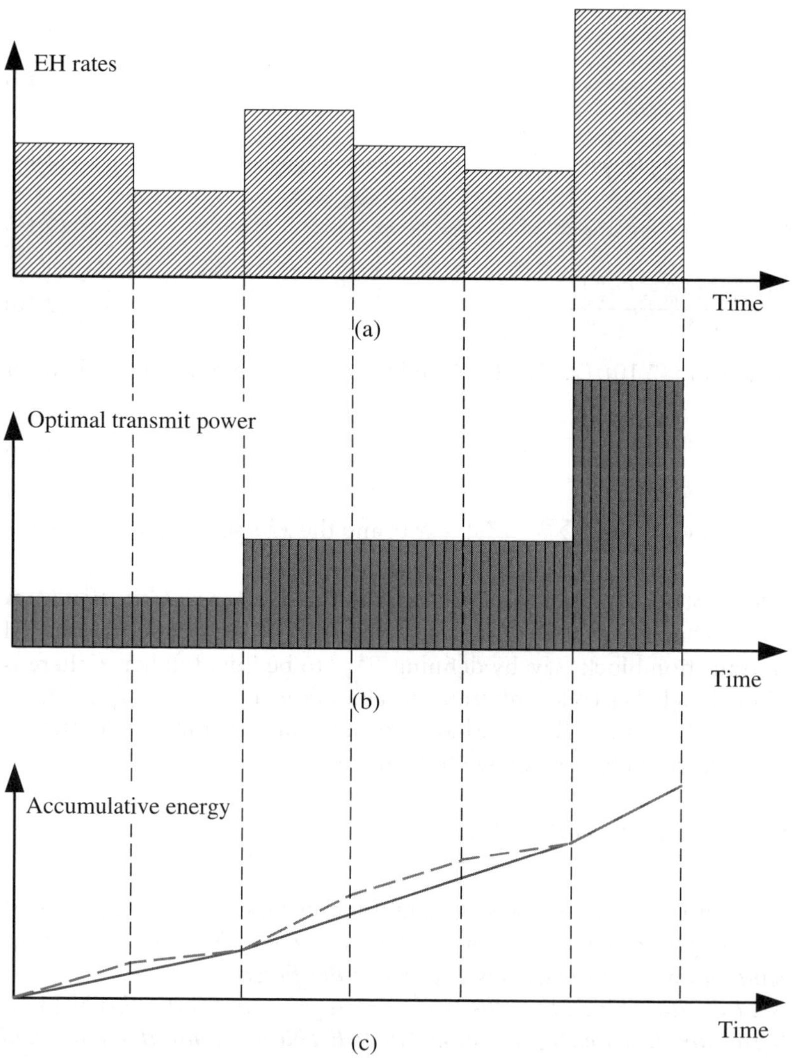

Figure 2.2 Staircase structure of the optimal transmit power allocation. (a) The EH rates over time. (b) The optimal staircase transmit power over time. (c) The accumulatively harvested energy (upper curve) and accumulatively consumed energy (lower curve) over time.

Some definitions are in order. For convenience, let $t_0 = 0$. We define the ith *block interval* $\tilde{S}_i$, where $i = 1, \ldots, |S|$, as the EH blocks between the ith and the $(i+1)$th transition block, i.e. $\tilde{S}_i \triangleq \{t_{i-1}+1, \ldots, t_i\}$. Thus, $\cup_i \tilde{S}_i = \{1, \ldots, M\}$ and $\tilde{S}_i \cap \tilde{S}_j = \phi$ for $i \neq j$. Let $S^\star = \{t_1^\star, \ldots, t_{|S^\star|}^\star\}$ denote the optimal set of transition blocks corresponding to an optimal power allocation.

Then the optimal staircase power allocation is performed as follows. From the second property in Proposition 2.1, all the harvested energy available in the ith block interval (i.e. $\sum_{m=t_{i-1}^\star+1}^{t_i^\star} E_m$) needs to be used during the ith block interval. Moreover, the optimal transmit power should remain constant over this block interval. In other words, the optimal power allocation at the ith block interval is given as

$$P_m^\star = \frac{\sum_{m=t_{i-1}^\star+1}^{t_i^\star} E_m}{t_i^\star - t_{i-1}^\star}, \quad m = t_{i-1}^\star + 1, \dots, t_i^\star. \tag{2.18}$$

As a result, the original optimization problem (2.7) can thus be reduced to a search for the optimal transition block set $S^\star$ that has a size from 1 to at most M:

$$\max_{1\le|S|\le M} \max_{S} \sum_{i=1}^{|S|} (t_i - t_{i-1}) \log\left(1 + h\frac{\sum_{m=t_{i-1}+1}^{t_i} E_m}{t_i - t_{i-1}}\right) \tag{2.19}$$

subject to the power allocation $\left\{P_m = \frac{\sum_{m=t_{i-1}+1}^{t_i} E_m}{t_i - t_{i-1}}\right\}$, satisfying the EH constraints in (2.8). A brute force search based on (2.19) is of a high computational complexity. Nevertheless, it turns out that it is optimal to simply employ a forward-search procedure, starting with the search of the optimal $t_1^\star$, then of the optimal $t_2^\star$, and so on until the last optimal transition block $t_{|S^\star|}^\star$ equals M, at which point the optimal size $|S^\star|$ is also obtained.

The first optimal $t_1^\star$ can be found in Lemma 2.1 given below; by induction, the search of the subsequent optimal transition blocks will follow similarly. Lemma 2.1 requires the following *feasible-search procedure*:

(1) Initialize S_1 as an empty set.
(2) For $t_1 = 1, \dots, M$, obtain the optimal power allocation from block 1 to block t_1 as

$$P_{t_1} = \frac{\sum_{m=1}^{t_1} E_m}{t_1}, \quad t_1 = 1, \dots, M. \tag{2.20}$$

(3) Admit t_1 in the set S_1 if the corresponding optimal power allocation satisfies the EH constraints in (2.8).

Note that the set S_1 is non-empty, as it contains at least the element $t_1 = 1$. Moreover, the set S_1 includes all possible candidates for the optimal $t_1^\star$.

Lemma 2.1 *Let S_1 be the feasible set of t_1 obtained by the feasible-search procedure. Then the optimal transition block is given by the largest element in S_1, which corresponds to the element with the smallest value of $\left\{\frac{\sum_{m=1}^{t_1} E_m}{t_1}\right\}$, i.e.*

$$t_1^\star = \max_{t_1\in S_1} t_1 = \arg\min_{1\le t_1\le M} \frac{\sum_{m=1}^{t_1} E_m}{t_1}. \tag{2.21}$$

Proof: If $|S_1| = 1$, then the only t_1 must be optimal. Henceforth, assume $|S_1| \ge 2$. Consider two blocks $t', t'' \in S_1$, where $t' < t''$. Denote their respective optimal transmit power obtained from (2.20) as $P_{t'} = \frac{\sum_{m=1}^{t'} E_m}{t'}$ and $P_{t''} = \frac{\sum_{m=1}^{t''} E_m}{t''}$. Then $P_{t'} \ge P_{t''}$. Otherwise if $P_{t'} < P_{t''}$, then more power is allocated for each time block $k = 1, \dots, t'$, with the transmit power $P_{t''}$ used, as compared with the case with $P_{t'}$ used. But since the transmit power $P_{t'}$ has used all available power at block t' due to second property in Proposition 2.1, the transmit power $P_{t''}$ is infeasible and thus cannot be optimal.

We now show that t' cannot be the optimal $t_1^{\star}$ by contradiction. Suppose that $t_1^{\star} = t'$, i.e. the transmit power $P_{t'}$ is used from block 1 to block t'. The transmit power must then subsequently decrease at some block $t' < k \leq t''$; otherwise the sum power allocated from block 1 to block t'' will be more than the sum power allocated with the (constant) transmit power $P_{t''}$, which violates the sum power constraint. But from the first property in Proposition 2.1, the optimal transmit power is non-decreasing. Thus, $t_1^{\star} \neq t'$ by contradiction. By induction, all elements in S_1, except for the largest one, are suboptimal. The only candidate left, namely, the largest element, must then be optimal.

Based on the above discussion, $t_1^{\star}$ should correspond to the element with smallest value of $\left\{ P_{t_1} = \frac{\sum_{m=1}^{t_1} E_m}{t_1} \right\}$ from (2.20). As a result, it must follow that $t_1^{\star} = \arg\min_{1 \leq t_1 \leq M} \frac{\sum_{m=1}^{t_1} E_m}{t_1}$, and this lemma is proved. □

We now propose Algorithm 2.1 to find the optimal transition blocks for solving problem (2.19) or equivalently problem (2.7), for which the optimality is shown in Proposition 2.2. Briefly, Algorithm 2.1 computes $t_1^{\star}, t_2^{\star}, \dots, t_{|S^{\star}|}^{\star}$ in each of the iteration until $t_{|S^{\star}|}^{\star} = M$. Given that $t_1^{\star}, \dots, t_{i-1}^{\star}$ is found, $t_i^{\star}$ is obtained based on (2.21) in Lemma 2.1.

Proposition 2.2 *Algorithm 2.1 obtains the optimal $S^{\star}$ that solves the optimization problem in (2.19) or equivalently problem (2.7).*

Proof: The ith iteration of Algorithm 2.1 finds the optimal $t_i^{\star}$ based on (2.21) in Lemma 2.1. Next, from the second property in Proposition 2.1, in the first block interval, all the power available would be used. Since no power is available for subsequent block intervals given $t_1^{\star}$, the power allocation for subsequent block intervals can be optimized independent of the actual power allocated in the first block interval. The throughput maximization problem from block $t_1^{\star} + 1$ onward can be solved similarly as before (after removing time blocks $1, \dots, t_1^{\star}$). Thus, we determine $t_2^{\star}$ based on (2.21), similarly for $t_3^{\star}$ and so on, as reflected in the iteration of Algorithm 2.1. The iteration ends when the optimal transition block equals M, which is the largest possible value as stated in the optimization problem in (2.19). □

Algorithm 2.1: Finding the optimal transition blocks.

Input: $t_0^{\star} = 0$

Output: $t_i^{\star}$

```
for i = 1, 2, ..., M do
    P_{t_i} ← (Σ_{j=t*_{i-1}+1}^{m} E_j) / (m − t*_{i-1})
    t_i ← t*_{i-1} + 1
    t*_i ← arg min_{t*_{i-1}+1 ≤ t_i ≤ M} P_{t_i}
    if t*_i − M then
        Exit the algorithm
    end if
end for
```

To better illustrate the optimal power allocation structure at the EH transmitter, Figure 2.2c shows the accumulatively harvested energy (the upper curve) and accumulatively consumed energy (the lower curve) over time. Here, both the upper and the lower curves increase monotonically over time, and their slopes represent the EH rate and the optimal transmit power, respectively. First, the accumulatively consumed energy curve is observed to always lie below the accumulatively harvested energy curve, which is in order to satisfy the EH constraints. Next, the slope of the accumulatively consumed energy curve is observed to monotonically increase over time, which is expected as in the first property of Proposition 2.1. Furthermore, it is observed that when the slope of the accumulatively harvested energy curve changes, it touches the accumulatively harvested energy curve at that time instant. This is consistent with the second property of Proposition 2.1, showing that the harvested energy must be used up at that time instant. This also implies that the first line segment of the accumulatively consumed energy curve should be the line connecting the original and the corner point of the accumulatively harvested energy curve and with the minimum slope, for which the idea is used to develop Algorithm 2.1 to obtain the optimal staircase power allocation solution.

2.2.2 The Case with Causal ESIT

Next, we consider the other case with only causal ESIT, i.e. at each EH block m, the transmitter only knows the knowledge of past and present E_j's, $j = 1, \dots, m$, but is not aware of future E_j's, $j = m + 1, \dots, M$. In this case, the transmitter cannot solve problem (2.7) via the offline optimization approach with convex optimization techniques due to the unawareness of E_j's, $j = m + 1, \dots, M$. Alternatively, we need to use an *online* optimization for problem (2.7) in the following.

2.2.2.1 Dynamic Programming

Dynamic programming is known as the optimal online approach to solve problem (2.7), provided that the harvested energy E_m's follow a stochastic process with certain distributions, and the transmitter knows such distribution information. In this case, the transmitter aims to maximize the expected throughput over the finite horizon of M EH blocks, i.e. $\mathbb{E}\left(\sum_{m=1}^{M}\log_2(1+hP_m)\right)$, subject to the EH constraints in (2.8). The policy can be optimized offline and implemented in real time via a look-up table that is stored at the transmitter.

At each EH block m, we denote the state of the system as the harvested energy E_m and the energy storage level, denoted by B_m, at that block. Here, B_m's are given as

$$B_m = \sum_{i=1}^{m-1}(E_i - P_i), \quad m = 1, \dots, M, \tag{2.22}$$

where $B_1 = 0$. Note that with B_m at hand, the EH constraints in (2.8) can be re-expressed as

$$P_m \le B_m + E_m, \quad m = 1, \dots, M. \tag{2.23}$$

Then we have the following proposition.

Proposition 2.3 *Given E_1 and B_1, the optimal value achieved by maximizing $\mathbb{E}\left(\sum_{m=1}^{M}\log_2(1+hP_m)\right)$ subject to the EH constraints in (2.8) is given by $J_1(E_1, B_1)$, which can be computed recursively based on the following Bellman equations, starting from $J_M(E_M, B_M)$, $J_{M-1}(E_{M-1}, B_{M-1})$, and so on until $J_1(E_1, B_1)$:*

$$J_M(E_M, B_M) = \max_{P_M \geq 0} \ \log_2(1+hP_M) \quad \text{s.t. } P_M \leq B_M + E_M, \tag{2.24}$$

$$J_m(E_m, B_m) = \max_{P_m \geq 0} \ \log_2(1+hP_m) + \bar{J}_{m+1}(B_m - P_m) \quad \text{s.t. } P_m \leq B_m + E_m, \tag{2.25}$$

for $m = 1, \ldots, M-1$, where

$$\bar{J}_{m+1}(B_m - P_m) = \mathbb{E}_{E_{m+1}}(J_{m+1}(E_{m+1}, B_m - P_m)), \tag{2.26}$$

where $\mathbb{E}_{E_{m+1}}(\cdot)$ denotes the expectation over E_{m+1}. An optimal policy is accordingly given by $\pi^\star = \{P_m^{\text{DP}}(E_m, B_m)\}$, where $P_m^{\text{DP}}(E_m, B_m)$ is the optimal solutions to problem (2.24) for $m = M$ and (2.25) for $m = 1, \ldots, M-1$.

Proof: The proof follows directly by applying Bellman equations [15] and thus is omitted here for brevity. □

For problem (2.24), the optimal solution is trivial, i.e. $P_M^{\text{DP}}(E_M, B_M) = B_M + E_M$, which means that the transmitter uses all the available energy for transmission in the last EH block M. We can interpret the maximization in (2.25) as a tradeoff between the present and future rewards. This is because the throughput $\log_2(1+hP_m)$ represents the present reward, while $\bar{J}_{m+1}(B_m - P_m)$, commonly known as the value function, is the expected future throughput accumulated from block $m+1$ until block M.

Next, we present structural properties of the maximum throughput $J_m(E_m, B_m)$'s and the corresponding optimal policy $\pi^\star$ in Propositions 2.4 and 2.5. The two propositions are proved based on the concavity of the throughput function $\log_2(1+P_m)$, for which the details can be found in [4].

Proposition 2.4 *We have the following properties for $J_m(E_m, B_m)$'s:*

(1) $J_m(E_m, B_m)$'s in (2.24) and (2.25) are concave in B_m for $m = 1, \ldots, M$.
(2) $\bar{J}_{m+1}(B_m - P_m)$ in (2.26) is concave in B_m for $m = 1, \ldots, M$.

Proposition 2.5 *Given E_m, the optimal power allocation $P_m^{\text{DP}}(E_m, B_m)$ that solves (2.24) and (2.25) is non-decreasing in B_m, where $m = 1, \ldots, M$.*

The structural properties in Propositions 2.4 and 2.5 simplify the numerical computation of the optimal power allocation solution in Proposition 2.3. In particular, from (2.24), we get the optimal solution for block M as $P_M^{\text{DP}}(E_M, B_M) = B_M + E_M$. Now, consider the problem of finding the optimal $P_m^{\text{DP}}(E_m, B_m)$ to obtain $J_m(E_m, B_m)$, $m = 1, \ldots, M-1$. As $\bar{J}_{m+1}(B_m - P_m)$ in (2.26) is concave as shown in Proposition 2.4, the objective function in (2.25) is concave. Thus, problem (2.25) is a convex optimization

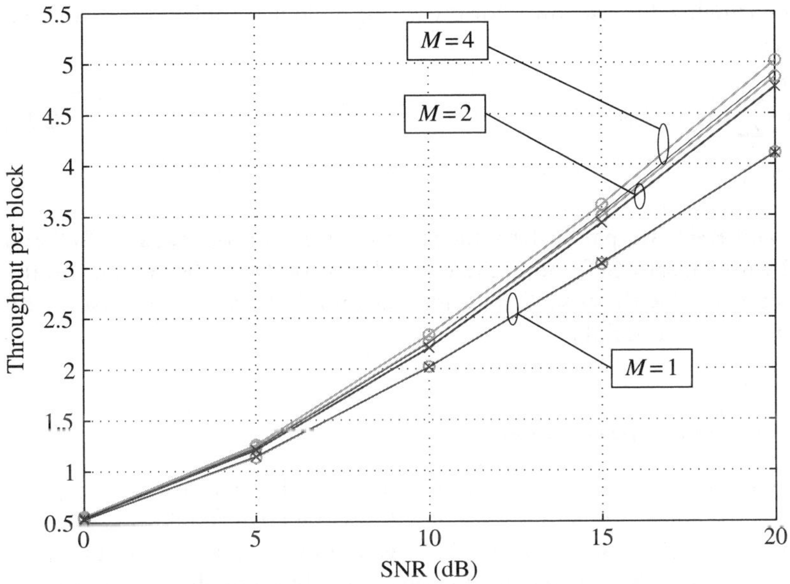

Figure 2.3 Optimal throughput with noncausal ESI (light gray with "o" markers) or causal ESI (dark gray with "×" markers) at the transmitter for $M = 1, 2, 4$.

problem and has a unique solution, which can thus be easily solved using numerical techniques such as a bisection search [14].

Figure 2.3 compares the throughput per EH block, i.e. the sum throughput divided by the number of EH blocks M, with noncausal ESI versus that with causal ESI. In this simulation, the harvested energy E_m's are i.i.d. over EH blocks m. We assume the harvested energy E_m takes a value in $\{0, 0.5, 1\}$ with equal probability. It is observed that for $M = 1$, the throughput in both cases with noncausal ESI and causal ESI is the same, because any ESI cannot be exploited for future EH blocks. However, in both cases the throughput per block increases as M increases. The increment is more substantial when noncausal ESI is available, intuitively because the ESI can then be much better exploited. The incremental improvement as M increases is significant when M is small but becomes less significant when M is large. The throughput with either noncausal ESI or causal ESI does not differ significantly, possibly because the ESI that can be further exploited from noncausal ESI is limited in our i.i.d. scenario.

2.3 Throughput Maximization for Fading Channel with EH Transmitter

Now, we move to the case with fading channels, where the transmitter aims to maximize the throughput over the finite horizon with M EH blocks each consisting of N communication blocks. In this case, the utility function is given as

$$U_{n,m}(P_{n,m}) = \log_2(1 + h_{n,m}P_{n,m}) \tag{2.27}$$

in bps Hz^{-1} with normalized noise power for the (n, m)th block. Problem (2.4) can then be reformulated as

$$\max_{\{P_{n,m}\geq 0\}} \sum_{m=1}^{M}\sum_{n=1}^{N} \log_2(1 + h_{n,m}P_{n,m}) \tag{2.28}$$
$$\text{s.t. (2.5).}$$

As for the CSIT and ESIT, we particularly focus on Case 1 with noncausal CSIT and ESIT and Case 2 with causal CSIT and ESIT. We will provide some discussions on the other two cases, i.e. Case 3 with causal CSIT and noncausal ESIT, as well as Case 4 with no CSIT and noncausal/causal ESIT.

2.3.1 The Case with Noncausal CSIT and ESIT

First, we consider the case with noncausal CSIT and ESIT. In this case, we adopt an offline optimization to solve problem (2.28). Since the objective function is concave and the constraints are all linear, problem (2.28) is a convex optimization problem. Furthermore, as problem (2.28) satisfies Slater's condition, strong duality holds between problem (2.28) and its dual problem. In this case, we use the KKT condition to obtain the optimal solution.

2.3.1.1 Water-Filling Power Allocation

Before proceeding to consider the general case where the EH constraint (2.5) is imposed for all $m = 1, \ldots, M$ and $n = 1, \ldots, N$, we impose the constraint (2.5) only for the last communication block, i.e. only for $(n, m) = (N, M)$. This then corresponds to the conventional problem of maximizing the sum throughput with a *sum* energy constraint of $P_{\max} = N\sum_{m=1}^{M} E_m$:

$$\mathcal{T}^{\mathrm{WF}}(\{h_{n,m}\}_{(1,1)}^{(N,M)}, P_{\max}) = \max_{\{P_{n,m}\geq 0\}} \sum_{m=1}^{M}\sum_{n=1}^{N} \log_2(1 + h_{n,m}P_{n,m}) \tag{2.29}$$

$$\text{s.t.} \sum_{m=1}^{M}\sum_{n=1}^{N} P_{n,m} \leq P_{\max}. \tag{2.30}$$

Since less constraints are imposed, the maximum throughput in (2.29) is no smaller than that of problem (2.28). It is well known that the optimal solution to problem (2.29) is given by (see, e.g. [14, 16])

$$P_{n,m}^{\mathrm{WF}} = \left[\nu - \frac{1}{h_{n,m}}\right]^{+}. \tag{2.31}$$

This optimal solution is implemented by the *water-filling power allocation algorithm*, where the *water-level* (WL) $\nu \geq 0$ is chosen such that (2.30) holds with equality by using the optimal power allocation in (2.31). For completeness, an implementation of the *water-filling power allocation algorithm* is given in Algorithm 2.2.

Algorithm 2.2: Water-filling power allocation algorithm.

Input: $P = 0$, error tolerance ϵ, $\nu^{\text{low}} = 0, \nu^{\text{up}} = \infty$ (a large number), $P^{\theta}_{n,m} := [\nu^{\theta} - 1/h_{n,m}]^{+}, \theta \in \{\text{low}, \text{up}\}, n = 1, \dots, N, m = 1, \dots, M$

Output: $P^{\text{WF}}_{n,m} = \left[\nu - \frac{1}{h_{n,m}}\right]^{+}, n = 1, \dots, N, m = 1, \dots, M$

while $|P_{\max} - P| > \epsilon$ or $P > P_{\max}$ **do**
  $\nu \leftarrow (\nu^{\text{low}} + \nu^{\text{up}})/2$
  $P_{n,m} \leftarrow [\nu - 1/h_{n,m}]^{+}, n = 1, \dots, N, m = 1, \dots, M$
  $P \leftarrow \sum_{m=1}^{M} \sum_{n=1}^{N} P_{n,m}$
  if $P > P_{\max}$ **then**
    $\nu^{\text{up}} = \nu$
  else
    $\nu^{\text{low}} = \nu$
  end if
end while

2.3.1.2 Staircase Water-Filling Power Allocation

Now, we solve our original problem (2.28) with the EH constraints in (2.5). It turns out that the conventional water-filling algorithm is no longer optimal. Instead, it is necessary to use a generalized type of water filling where the WL is a staircase-like function.

With noncausal ESIT and CSIT, we solve problem (2.28) via the KKT condition [14]. Let $\lambda_{n,m} \geq 0$ denote the Lagrange multiplier associated with the (n, m)th constraint in (2.5) and $\mu_{n,m} \geq 0$ denote the Lagrange multiplier associated with the power constraint $P_{n,m} \geq 0$. Then the Lagrangian associated with problem (2.28) is expressed as

$$\begin{aligned}\mathcal{L}_{2.3}(\{P_{n,m}\}, \{\lambda_{n,m}\}, \{\mu_{n,m}\}) &= \sum_{m=1}^{M} \sum_{n=1}^{N} \log(1 + h_{n,m} P_{n,m}) \\ &- \sum_{m=1}^{M} \sum_{n=1}^{N} \lambda_{n,m} \left(\sum_{j=1}^{m-1} \sum_{i=1}^{N} P_{i,j} + \sum_{i=1}^{n} P_{i,m} - N \sum_{j=1}^{m-1} E_j - nE_m \right) \\ &+ \sum_{m=1}^{M} \sum_{n=1}^{N} \mu_{n,m} P_{n,m}. \end{aligned} \tag{2.32}$$

The necessary and sufficient conditions for $\{P^{\star}_{n,m}\}$, $\{\lambda^{\star}_{n,m}\}$, and $\{\mu^{\star}_{n,m}\}$ to be the primal and dual optimal solutions to problem (2.28) are given by the following KKT conditions:

$$0 \leq P^{\star}_{n,m}, \tag{2.33}$$

$$0 \geq \sum_{j=1}^{m-1} \sum_{i=1}^{N} P^{\star}_{i,j} + \sum_{i=1}^{n} P^{\star}_{i,m} - N \sum_{j=1}^{m-1} E_j - nE_m, \tag{2.34}$$

$$0 \leq \lambda^{\star}_{n,m}, \tag{2.35}$$

$$0 \leq \mu^{\star}_{n,m}, \tag{2.36}$$

$$0 = \lambda^{\star}_{n,m} \left(\sum_{j=1}^{m-1} \sum_{i=1}^{N} P^{\star}_{i,j} + \sum_{i=1}^{n} P^{\star}_{i,m} - N \sum_{j=1}^{m-1} E_j - nE_m \right), \tag{2.37}$$

$$0 = \mu^{\star}_{n,m} P^{\star}_{n,m}, \tag{2.38}$$

$$0 = \frac{\partial \mathcal{L}_{2.3}(\{P^{\star}_{n,m}\}, \{\lambda^{\star}_{n,m}\}, \{\mu^{\star}_{n,m}\})}{\partial P^{\star}_{n,m}}, \tag{2.39}$$

for all $n = 1, \dots, N$ and $m = 1, \dots, M$. From (2.33), (2.36), (2.38), and (2.39), we obtain the optimal power allocation as

$$P^{\star}_{n,m} = \left[\nu_{n,m} - \frac{1}{h_{n,m}} \right]^{+}, \quad n = 1, \dots, N, \;\; m = 1, \dots, M, \tag{2.40}$$

where $\nu_{n,m} \triangleq \left(\ln 2 \left(\sum_{j=m+1}^{M} \sum_{i=1}^{N} \lambda^{\star}_{i,j} + \sum_{i=n}^{N} \lambda^{\star}_{i,m} \right) \right)^{-1} \geq 0$ and the $\lambda^{\star}_{i,j}$'s satisfy the above KKT conditions. Analogous to the problem in (2.29) with only a single power constraint, we say $\nu_{n,m}$ is the WL for the (n, m)th block. It is interesting to compare the optimal power allocation in (2.40) for the fading channel versus the staircase power allocation in (2.17) for the Gaussian channel. It is observed that in the fading channel case, the optimal transmit power is adaptively adjusted based on the wireless channel gains. In particular, under the same WL, more power is allocated to the blocks with stronger channel power gains.

For notational convenience, for any block $t = (n, m)$, we define the subsequent block as block $t + 1 = (n + 1, m)$, where block $(N + 1, m)$ is defined to be equivalent to block $(1, m + 1)$. Then, similar as in the previous section with Gaussian channels, we define the (n, m)th block as a *transition block* if the transmit power changes after the (n, m)th block, i.e. $P^{\star}_{n,m} \neq P^{\star}_{n+1,m}$. We define the last block $(n, m) = (N, M)$ also as a transition block (say, by defining $\nu_{N+1,M}$ to be infinity); hence there is at least one transition block. We collect all transition blocks as the set $\mathcal{S} = \{t_1, \dots, t_{|\mathcal{S}|}\}$, where $t_i < t_j$ for $i < j$ and $t_{|\mathcal{S}|} = (N, M)$. Then, we have the following structural properties for the optimal power allocation $\{P^{\star}_{n,m}\}$, as shown in Figure 2.4.

Proposition 2.6 *The optimal solution $\{P^{\star}_{n,m}\}$ in (2.40) to problem (2.28) satisfies the following properties:*

- *The WL monotonically increases over time, i.e. $\nu_{1,1} \leq \cdots \leq \nu_{N,M}$. We say that the optimal solution performs staircase water filling over blocks, since the WL is a staircase-like function.*
- *If block (n, m) is a transition block, then the energy storage is empty after this block, or equivalently, the accumulative energy consumed up to block (n, m) equals that harvested up to then, i.e. (2.34) holds with equality for $(n, m) \in \mathcal{S}$.*

Proof: The proof of this proposition is similar as that for Proposition 2.1 and thus is omitted here for brevity. □

From Proposition 2.6, we have the following additional structural properties in Corollaries 2.1 and 2.2.

Corollary 2.1 *If the channel power gain is non-decreasing over blocks, then the optimal power allocation is non-decreasing over blocks.*

Proof: This follows immediately from (2.40) and the first property in Proposition 2.6, which implies that $P^{\star}_{n,m} \leq P^{\star}_{i,j}$ if $h_{n,m} \leq h_{i,j}$ for $(n, m) \leq (i, j)$. □

An example that illustrates Corollary 2.1 is given in Figure 2.4c, where we see that the inverse of the channel power gain is non-increasing. It is easy to see that the converse of Corollary 2.1 is not true in general. That is, if the channel power gains are non-increasing, then the optimal power allocation may not be non-increasing over blocks (in particular for the block immediately after the transitions block). In

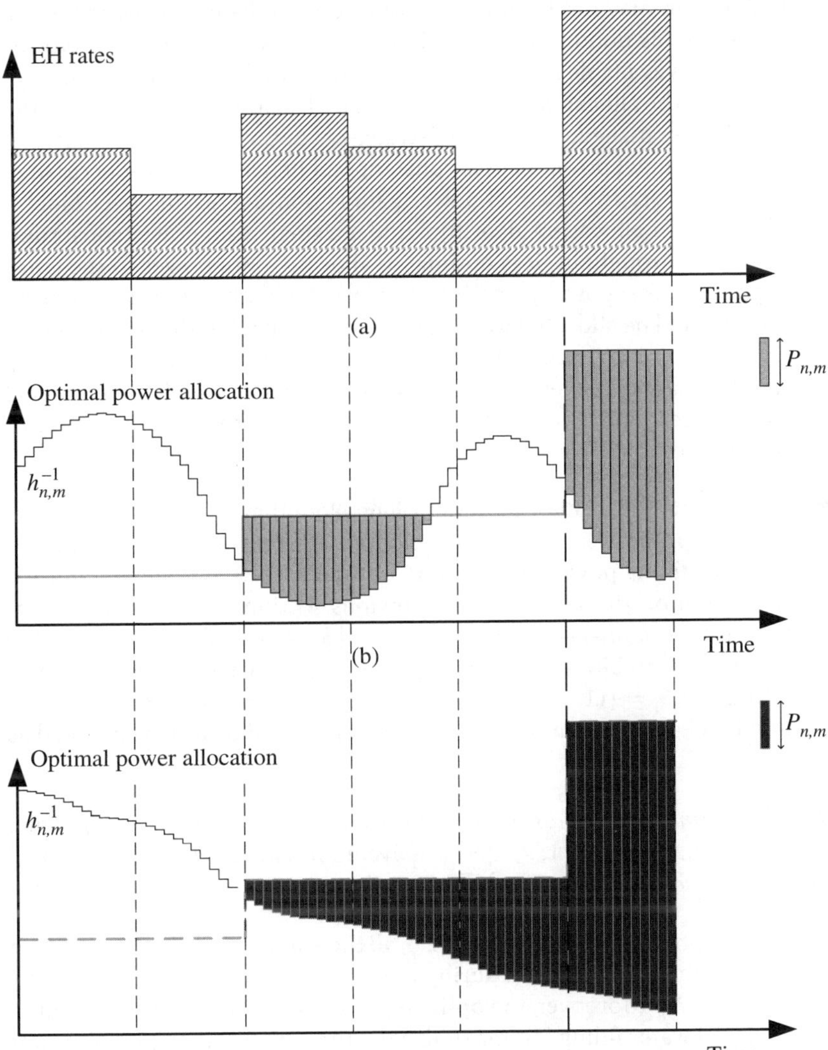

Figure 2.4 Staircase water-filling power allocation. (a) The EH rates over time. (b) The optimal staircase water filling over time, where there are two transition blocks and hence three different WLs. (c) The optimal staircase water filling over time with monotonically increasing wireless channel gains, where there are two transition blocks and hence three different WLs.

conventional water filling, however, both Corollary 2.1 and its converse hold, i.e. the optimal power allocation is non-decreasing over blocks *if and only if* the channel power gains are non-decreasing over blocks.

We give an intuitive understanding of Corollary 2.1, and why the converse does not hold, to shed light on how the EH constraints lead to a different optimal power allocation. First, let us consider the Gaussian channel where the channel power gain is constant over blocks (see Section 2.2). If all the harvested energy is already available in the first block, i.e. there is only a single sum power constraint, a uniform power allocation is optimal for the Gaussian channel. However, in an EH communication system, maintaining a uniform power allocation may not be always possible due to the causal arrival of the harvested energy. Due to this nonuniform availability of harvested energy over blocks, more energy only becomes available for transmission in the latter blocks. Intuitively, we also expect more energy to be allocated for transmission in the latter blocks such that the EH constraints in (2.5) are satisfied. This type of strategy is optimal from Corollary 2.1, which applies since the channel power gain is constant and hence also non-decreasing. Next, consider the case where the channel power gain is non-decreasing over blocks. From the water-filling power allocation algorithm under a single sum power constraint, to achieve the maximum throughput, it is optimal to allocate more power to the latter blocks that have higher channel power gains. This is consistent with the earlier observation that more power should be allocated to the latter blocks such that the constraints in (2.5) are satisfied. Hence, it is also optimal to allocate more power to the latter blocks. In general, however, if the channel power gain is arbitrary, the blocks with higher channel power gains may not correspond to the latter blocks; hence intuitively the converse of Corollary 2.1 may not hold.

2.3.1.3 Efficient Implementation of Staircase Water-Filling Algorithm

Based on the structural properties, we now develop an efficient algorithm to implement the staircase water-filling power allocation to optimally solve problem (2.28). We make the following definition similarly as in the previous section with Gaussian channels. We refer to the ith *block interval*, where $i = 1, \dots, |S|$, as the blocks between the ith and the $(i+1)$th transition block, i.e. $S_i \triangleq \{t_{i-1}+1, \dots, t_i\}$, where we let $t_0 = (0, 1)$ for convenience. Thus, $\bigcup_i S_i = \{(1, 1), \dots, (N, M)\}$ and $S_i \cap S_j = \phi$ for $i \neq j$. The optimal set of transition blocks corresponding to an optimal power allocation is denoted as $S^\star = \{t_1^\star, t_2^\star, \dots, t_{|S^\star|}^\star\}$.

Corollary 2.2 *The optimal power allocation performs staircase water filling as follows: for every ith block interval, where* $i = 1, \dots, |S^\star|$, *conventional water filling is performed subject to the sum power constraint of* $P(i) \triangleq \sum_{(n,m)\in S_i} E_{n,m}$.

Proof: From the second property in Proposition 2.6, all the harvested energy available in the ith block interval, namely, $P(i)$, is used during the ith block interval. This follows by induction for $i = 1, \dots, |S^\star|$. Moreover, the optimal power allocation in (2.40) is equivalent to the conventional water filling. To maximize the throughput, the optimal power allocation must then be to use conventional water filling with sum power constraint of $P(i)$ for every ith block interval. □

From Corollary 2.2, without loss of optimality, the staircase water-filling solution composes of multiple conventional water-filling solutions, one for each block interval. Then, similar as in (2.19) in the Gaussian channel case, the original optimization problem (2.7) can thus be reduced to a search for the optimal transition block set $S^\star$ that has a size from 1 to at most MN:

$$S^\star = \max_{1\le|S|\le MN} \max_{S} \mathcal{T}^{\mathrm{WF}}(\{h_{n,m}\}_{(1,1)}^{t_1}, P(1)) + \mathcal{T}^{\mathrm{WF}}(\{h_{n,m}\}_{t_1+1}^{t_2}, P(2)) + \cdots + \mathcal{T}^{\mathrm{WF}}(\{h_{n,m}\}_{t_{|S|-1}+1}^{t_{|S|}}, P(|S|)) \quad (2.41)$$

subject to the power allocation $P(1), \dots, P(|S|)$, satisfying the constraints in (2.5). It turns out that it is optimal to simply employ a forward-search procedure, starting with the search of the optimal $t_1^\star$, then of the optimal $t_2^\star$, and so on until the last optimal TS $t_{|S^\star|}^\star$ equals (N, M), at which point the optimal size $|S^\star|$ is also obtained.

The first optimal transition block $t_1^\star$ can be found in Lemma 2.2 given below; by induction, the search of the subsequent optimal transition blocks will follow similarly. Lemma 2.2 requires the following *feasible-search procedure* for a given optimization problem (2.28):

(1) Initialize S_1 as an empty set.
(2) For $t_1 = (1,1), \dots, (N, M)$, obtain the optimal power allocation from block $(1,1)$ to block t_1 by using a water-filling algorithm (such as Algorithm 2.2) assuming that all harvested energy is available, i.e. the sum power constraint is $P_{\max} = \sum_{(i,j)=(1,1)}^{t_1} E_{i,j}$.
(3) Admit t_1 in the set S_1 if the corresponding optimal power allocation satisfies the constraint in (2.5) for $(n, m) = (1,1), \dots, t_1$.

Lemma 2.2 *Let S_1 be the feasible set of t_1 obtained by the feasible-search procedure. Then the optimal transition block is given by the largest element in S_1, i.e.*

$$t_1^\star = \max_{t_1 \in S_1} t_1. \quad (2.42)$$

We now propose Algorithm 2.3 to solve problem (2.41) (and thus problem (2.28)), which is optimal according to Proposition 2.7. Briefly, Algorithm 2.3 performs a forward-search procedure (starting from block (1, 1)) in each of the outer iteration for $t_1^\star, t_2^\star, \dots,$ until $t_{|S^\star|}^\star = (N, M)$. Given that $t_1^\star, \dots, t_{i-1}^\star$ is found, to obtain $t_i^\star$, an inner iteration is performed via a *backward-search* procedure, starting from block $t_{i-1}^\star, t_{i-1}^\star - 1, \dots,$ until block (1, 1).

Proposition 2.7 *Algorithm 2.3 obtains the optimal $S^\star$ that solves the optimization problem in (2.41) and thus equivalently problem (2.28).*

Note that the proofs of Lemma 2.2 and Proposition 2.7 are similar to those for Lemma 2.1 and Proposition 2.2 and thus are omitted here for brevity. More details can be found in [4].

2.3.2 The Case with Causal CSIT and ESIT

Next, we consider the other case with only causal CSIT and ESIT, i.e. at each block (n, m), the transmitter only knows the knowledge of current and previous E_j's, $j = 1, \dots, m$, and $h_{i,j}$'s, $(i,j) = (1,1), \dots, (n,m)$, but is not aware of future E_j's, $j = m+1, \dots, M$, and $h_{i,j}$'s, $(i,j) = (n+1, m), \dots, (N, M)$. Under this setup, we consider the online optimization for problem (2.28).

Algorithm 2.3: Finding the optimal transition blocks.

Input: $t_0^\star = (0, 0)$
Output: $t_i^\star$

for $i = (1, 1), \dots, (N, M)$ **do**
  for $k = (N, M), \dots, t_{i-1}^\star + 1$ **do**
    Use Algorithm 2.2 for block $t_{i-1}^\star + 1$ to block k, i.e. $\mathcal{T}^{\text{WF}}(\{h_{n,m}\}_{t_{i-1}^\star+1}^{k}, \sum_{j=t_{i-1}^\star+1}^{k} E_j)$ to obtain $\{P_{t_{i-1}^\star+1}, \dots, P_k\}$
    if $\{P_{1,1}, \dots, P_k\}$ satisfy the constraints in (2.5) **then**
      $t_i^\star \leftarrow k$
    end if
    if $t_i^\star = (N, M)$ **then**
      Exit the algorithm
    end if
  end for
end for

2.3.2.1 Dynamic Programming

First, we consider dynamic programming to solve problem (2.28) online, where the harvested energy E_m's and the channel power gain $h_{n,m}$'s follow stochastic processes with known distributions. In this case, the transmitter aims to maximize the expected throughput over the finite horizon of MN communication blocks, i.e. $\mathbb{E}\left(\sum_{m=1}^{M}\sum_{n=1}^{N}\log(1 + h_{n,m}P_{n,m})\right)$, subject to the EH constraints in (2.5).

At each block (n, m), we denote the state of the system as the channel power gain $h_{n,m}$, the harvested energy $E_{n,m}$, and the energy storage level, denoted by $B_{n,m}$, at that block. Here, $B_{n,m}$'s are given as

$$B_{n,m} = \sum_{j=1}^{m-1}\sum_{i=1}^{N}(E_{i,j} - P_{i,j}) + \sum_{i=1}^{n}(E_{i,m} - P_{i,m}),$$
$$n = 1, \dots, N, m = 1, \dots, M, \tag{2.43}$$

where $B_{1,1} = 0$. Note that with $B_{n,m}$ at hand, the EH constraints in (2.5) can be re-expressed as

$$P_{n,m} \le B_{n,m} + E_{n,m}, \quad n = 1, \dots, N, \ m = 1, \dots, M. \tag{2.44}$$

Then we have the following proposition.

Proposition 2.8 *Given E_1 and $B_{1,1}$, the optimal value achieved by maximizing $\mathbb{E}\left(\sum_{m=1}^{M}\sum_{n=1}^{N}\log(1 + h_{n,m}P_{n,m})\right)$ subject to the EH constraints in (2.5) is given by $J_{1,1}(h_{1,1}, E_1, B_{1,1})$, which can be computed recursively based on the following Bellman equations, starting from $J_{N,M}(h_{N,M}, E_M, B_{N,M})$, $J_{N-1,M}(h_{N-1,M}, E_M, B_{N-1,M})$, and so on until $J_{1,1}(h_{1,1}, E_1, B_{1,1})$:*

$$J_{N,M}(h_{N,M}, E_M, B_{N,M}) = \max_{P_{N,M}\ge 0} \log(1 + h_{N,M}P_{N,M})$$
$$\text{s.t. } P_{N,M} \le B_{N,M} + E_M, \tag{2.45}$$

$$J_{n,m}(h_{n,m}, E_m, B_{n,m}) = \max_{P_{n,m}\geq 0} \log(1 + h_{n,m}P_{n,m}) + \bar{J}_{(n,m)+1}(B_{n,m} - P_{n,m})$$
$$\text{s.t. } P_{n,m} \leq B_{n,m} + E_m, \tag{2.46}$$

for $(n, m) = (1, 1), \ldots, (N, M) - 1$, *where*

$$\bar{J}_{(n,m)+1}(B_{n,m} - P_{n,m}) = \mathbb{E}_{h_{n+1,m},E_{\hat{m}}}(J_{n+1,m}(h_{n+1,m}, E_{\hat{m}}, B_{n+1,m})), \tag{2.47}$$

where $\mathbb{E}_{h_{n+1,m},E_{\hat{m}}}(\cdot)$ *denotes the expectation over* $h_{n+1,m}$ *and* $E_{\hat{m}}$, *with* $\hat{m} = m$ *if* $n < N$ *and* $\hat{m} = m + 1$ *if* $n = N$. *Note that if* $n < N$, *then* $E_{\hat{m}} = E_m$ *is deterministic as it is known. An optimal policy is accordingly given by* $\pi^\star = \{P^{\text{DP}}_{n,m}(h_{n,m}, E_m, B_{n,m})\}$, *where* $P^{\text{DP}}_{n,m}(h_{n,m}, E_m, B_{n,m})$ *is the optimal solution to problem (2.45) for* $(n, m) = (N, M)$ *and (2.46) for* $(n, m) < (N, M)$.

Proof: This proposition can be similarly proved as for Proposition 2.3. The proof follows directly by applying Bellman equations [15] and thus is omitted here for brevity. □

Note that we can interpret the throughput maximization problem in (2.46) as a tradeoff between the present and future rewards. This is because the mutual information $\log_2(1 + h_{n,m}P_{n,m})$ represents the present reward, while $\bar{J}_{(n,m)+1}(B_{n,m} - P_{n,m})$ is the expected future mutual information accumulated from block $(n, m) + 1$ until block (N, M).

Next, we obtain structural properties of the optimal policy $\pi^\star$ in the following two propositions, where the proofs can be referred to in [4].

Proposition 2.9 *Given* $h_{n,m}$ *and* E_m, *at each block* (n, m), *it follows that:*

(1) $J_{n,m}(h_{n,m}, E_m, B_{n,m})$ *in (2.46) is concave in* $B_{n,m}$.
(2) $\bar{J}_{(n,m)+1}(B_{n,m} - P_{n,m})$ *in (2.47) is concave in* $B_{n,m}$.

Thus, $J_{1,1}(h_{1,1}, E_1, B_{1,1})$ *is concave in* $B_{1,1}$.

Proposition 2.10 *Given* $h_{n,m}$ *and* E_m *at each block* (n, m), *then the optimal power allocation* $P^{\text{DP}}_{n,m}(h_{n,m}, E_m, B_{n,m})$ *that solves (2.46) is non-decreasing in* $B_{n,m}$.

The structural properties in Propositions 2.9 and 2.10 simplify the numerical computation of the optimal power allocation solution in Proposition 2.8, as shown next.

From (2.45), we get the optimal power allocation solution for the last block (N, M) as $P^{\text{DP}}_{N,M}(h_{N,M}, E_M, B_{N,M}) = B_{N,M} + E_M$, i.e. the transmitter should use all available energy for transmission in this block. Next, we consider the problem of finding the optimal $P^{\text{DP}}_{n,m}(h_{n,m}, E_m, B_{n,m})$ to obtain $J_{n,m}(h_{n,m}, E_m, B_{n,m}), (1, 1) \leq (n, m) < (N, M)$. Let us fix the channel power gain and harvested energy as $h_{n,m}$ and E_m, respectively, and drop these arguments when possible to simplify notations. Consider the *unconstrained maximization* over all $P_{n,m} \geq 0$, i.e. not subject to any EH constraints:

$$P^{\dagger}_{n,m} = \arg\max_{P_{n,m}\geq 0} g(P_{n,m}), \tag{2.48}$$

where we denote $g(P_{n,m}) = \log(1 + h_{n,m}P_{n,m}) + \bar{J}_{(n,m)+1}(B_{n,m} - P_{n,m})$. Since $\log(1 + h_{n,m}P_{n,m})$ is concave, $\bar{J}_{(n,m)+1}(B_{n,m} - P_{n,m})$ is concave due to Proposition 2.9. Thus, the maximization over all $P_{n,m}$ gives a unique solution $P^{\dagger}_{n,m}$, which can be solved using numerical techniques such as a bisection search [14]. Also, Proposition 2.10 helps to reduce the search space by restricting the search to be in one direction for different $B_{n,m}$. Alternatively, if $g(P_{n,m})$ is differentiable and available in closed form, $P^{\dagger}_{n,m}$ is given by solving $g'(P_{n,m}) = 0$. Finally, we get the optimal solution for (2.46) by restricting the maximization in (2.48) to be over $0 \leq P_{n,m} \leq B_{n,m} + E_m$ to give

$$P^{\mathrm{DP}}_{n,m}(h_{n,m}, E_m, B_{n,m}) = \begin{cases} 0 & P^{\dagger}_{n,m} \leq 0 \\ B_{n,m} + E_m & P^{\dagger}_{n,m} \geq B_{n,m} + E_m \\ P^{\dagger}_{n,m} & 0 < P^{\dagger}_{n,m} < B_{n,m} + E_m. \end{cases} \tag{2.49}$$

This is because if $P^{\dagger}_{n,m} \leq 0$, the (concave) objective function $g(P_{n,m})$ must be decreasing for $P_{n,m} \geq 0$; if $P^{\dagger}_{n,m} \geq B_{n,m} + E_m$, the objective function must be increasing for $P_{n,m} \leq B_{n,m} + E_m$.

Figure 2.5 compares the throughput per EH block with noncausal ESIT versus that with causal ESIT, where the system parameters are set similarly as those in Figure 2.3 and the i.i.d. Rayleigh fading channel is considered. Furthermore, we set $N = 1$, and each EH block corresponds to a communication block. Figure 2.5 can be explained similarly as for Figure 2.3.

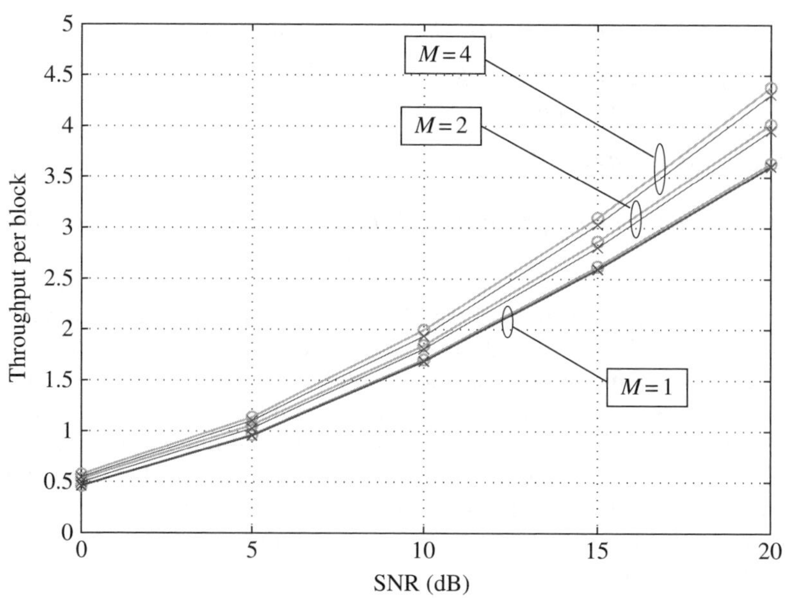

Figure 2.5 Fading channel. Optimal throughput when causal ESIT (dark gray with "×" markers) or noncausal ESIT (light gray with "o" markers) is available for $M = 1, 2, 4$.

2.3.2.2 Heuristic Online Solutions

Furthermore, we consider two heuristic online schemes that use only causal ESIT and CSIT yet can be easily implemented in practice, namely, the naive scheme and the power-halving scheme.

In the *naive scheme*, all the available energy $B_{n,m} + E_m$ is used in every block (n, m), i.e. $P_{n,m} = B_{n,m} + E_m$. This is equivalent to the case of $N = M = 1$ in our optimization problem regardless of whether causal ESIT and CSIT are available (see Proposition 2.8) or noncausal ESIT and CSIT are available (see Algorithm 2.3). In both cases, it is optimal to use all the stored energy. As seen in Figures 2.3 and 2.5, the case of $M = 1$ performs significantly worse than the optimal schemes for $M > 2$ in both cases. To obtain further improvement in the per block throughput, we need to further exploit the causal CSIT and ESIT available.

In the *power-halving scheme*, all the stored power is used for transmission in the last block, while for all other blocks a half of the available energy is used, i.e. $P_{n,m} = w_{n,m}(B_{n,m} + E_m)$, where $w_{n,m} = 1$ if $(n, m) = (N, M)$ and $w_{n,m} = 1/2$ otherwise. This scheme is simple to implement. Intuitively, the present throughput is traded equally with the future throughput by splitting the available energy $B_{n,m} + E_m$ into two halves. We note that this scheme also implicitly exploits causal information of the harvested energy. Moreover, the power-halving scheme satisfies the following characteristics and thus likely improves the throughput when only causal ESIT is available:

(1) $P_{n,m}$ increases with $B_{n,m} + E_m$, in accordance with Proposition 2.10 in the causal ESIT case.
(2) More stored energy is deferred to be used in the latter blocks, thus resembling the optimal policy with staircase WLs in the noncausal ESIT case.

Figure 2.6 shows the throughput per block obtained by averaging the numerical results from 2×10^4 independent runs of Monte Carlo simulations, for both Gaussian channels and Rayleigh fading channels. We set $N = 1$. We fix the signal-to-noise ratio (SNR) at 20 dB. As benchmarks, we only plot the optimal throughput when noncausal ESIT and CSIT are available. This is because the computational complexity in solving Bellman's equations in Proposition 2.8 when causal ESIT is available becomes prohibitive for large M; moreover we observed earlier that the performance with causal ESIT and CSIT is close to the performance with noncausal ESIT and CSIT. The results show that the power-halving scheme is able to improve on the per block throughput as M is increased and is only within about 0.2 bps Hz^{-1} away from the case when noncausal ESIT and CSIT are available. Moreover, we see that at small M, the gap is even closer, as suggested by the second characteristic mentioned above. Similar results are obtained at lower SNR, with an even smaller throughput degradation compared with the case when noncausal ESIT and CSIT are available. Further performance gain may also be obtained by optimizing this tradeoff by considering the channel conditions explicitly.

2.3.3 Other ESIT and CSIT Cases

For Cases 3 and 4, to our best knowledge, the throughput maximization problems have not been studied in the literature yet. Here, we provide brief discussions on

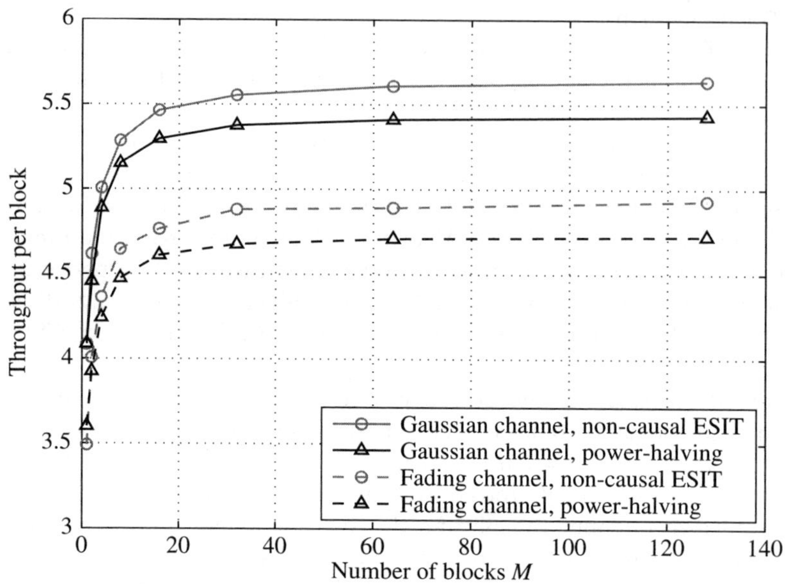

Figure 2.6 Throughput based on the power-halving scheme. The optimal throughput when noncausal ESIT is also plotted for comparison.

these two cases to motivate future investigations. With causal CSIT and noncausal ESIT in Case 3, the transmitter aims to maximize the expected throughput over the randomness of channel realizations, subject to a set of deterministic EH constraints. In general, such a problem can be optimally solved by the dynamic programming technique similarly as in Case 2, based on which the transmitter needs to decide its power allocations with the updated CSI block by block. By exploiting dynamic programming and the deterministic EH constraints, it may be feasible to further obtain insightful and well-structured solutions, at least under some specific channel distributions, e.g. i.i.d. channel gains, which is an interesting problem worth pursuing.

In Case 4, the throughput optimization is generally a very challenging problem that remains open. The reason is that due to the lack of CSIT, the transmitter cannot adapt its transmit power and hence rate based on the instantaneous CSI, and therefore the receiver may fail to decode the information if the channel is too weak at that block. Thus, the total throughput may not be achievable in general. Despite this difficulty, one tractable case of this problem is when the number of communication blocks is sufficiently large within each EH block (say, $N \rightarrow \infty$). In this case, the ergodic capacity is achievable by letting the transmit power remain constant over each EH block and using a sufficiently long capacity-achieving code. With such a transmission scheme, the throughput utility achieved by each EH block can be expressed explicitly (although not in closed form for general fading distributions); hence, problem (2.28) can be formulated as an ergodic throughput maximization problem over M EH blocks with noncausal or causal ESIT, which could be solved in the similar way to that for Cases 1 or 2.

2.4 Outage Probability Minimization with EH Transmitter

Outage occurs when there is a failure in decoding the data packet at the receiver, which is mainly due to the fact that the received signal undergoes a "deep" channel fading such that its power is not sufficient to combat against the receiver noise at the decoder. In the EH wireless channel, besides the channel fading, the uncertainty in the amount of harvested energy could be another source for transmission outage due to insufficient transmit power (or even insufficient circuit power; see [12]) at the EH transmitter.

To better address the outage issue, let $\mathcal{F}_{n,m}(P_{n,m})$ denote the outage probability function of the (n, m)th block, which depends on both the transmit power $P_{n,m}$ and the channel power gain $h_{n,m}$. For example, by using the instantaneous mutual information $I_{n,m}(h_{n,m}, P_{n,m})$ in (2.3) as the achievable rate, and assuming that the required transmission rate at all NM blocks, denoted by R, is constant, the outage probability $\mathcal{F}_{n,m}(P_{n,m})$ at the (n, m)th block can be expressed as the probability that the achievable rate is less than the required transmission rate [11], i.e.

$$\begin{aligned}\mathcal{F}_{n,m}(P_{n,m}) &= \Pr\{I_{n,m}(h_{n,m}, P_{n,m}) < R\} \\ &= \Pr\left\{h_{n,m} < \frac{2^R - 1}{P_{n,m}}\right\}.\end{aligned} \tag{2.50}$$

Following this definition, the utility function $\mathcal{U}_{n,m}(P_{n,m})$ could be modeled by the non-outage probability for the (n, m)th block, i.e. $\mathcal{U}_{n,m}(P_{n,m}) = 1 - F_{n,m}(P_{n,m})$. It is thus straightforward to see that problem (2.4) is equivalent to minimizing the average outage probability over the transmissions. Mathematically, the average outage probability minimization problem over the finite horizon of NM communication blocks is formulated as

$$\min_{\{P_{n,m}\}} \frac{1}{NM} \sum_{n=1}^{N} \sum_{m=1}^{M} \mathcal{F}(P_{n,m}) \tag{2.51}$$

$$\text{s.t. } P_{n,m} \geq 0, \quad 1 \leq n \leq N, 1 \leq m \leq M \tag{2.52}$$

$$(2.5).$$

For the throughput utility maximization problems (2.7) and (2.28) investigated in the previous two sections, their optimal solution structures crucially rely on the concavity of the rate or utility function. Similarly, we expect that the optimal power allocation for the outage probability minimization problem also critically depends on the properties of the outage probability function $\mathcal{F}_{n,m}(P_{n,m})$. Here, $\mathcal{F}_{n,m}(P_{n,m})$ is determined based on the availability of CSIT and the probability distribution of the channel gain $h_{n,m}$, as will be specified for different cases in the following. For simplicity, we assume in this section that the channel gains $\{h_{n,m}\}$ are i.i.d. over different communication blocks.

2.4.1 The Case with No CSIT and Noncausal ESIT

In this section, we consider the case with no CSIT and with noncausal ESIT (Case 4). For the time horizon of our interest, we assume that the number of communication blocks per EH block N and the number of EH blocks M are both finite integers and will address the case of $N \to \infty$ later.

Without CSIT, the outage probability function $\mathcal{F}_{n,m}(P_{n,m})$ generally depends on the probability distribution of the fading channel gain $h_{n,m}$ and the transmission rate R. As the wireless channels are i.i.d. and the transmission rate R is constant over different blocks, the outage probability function $\mathcal{F}_{n,m}(P_{n,m})$ is identical for different blocks, i.e. $\mathcal{F}_{n,m}(\cdot) = \mathcal{F}(\cdot), n = 1, \dots, N,\ m = 1, \dots, M$. We assume that $\mathcal{F}(\cdot)$ is strictly decreasing over its domain of $P_{n,m} \geq 0$. Furthermore, we assume that the EH rate levels of all the M EH blocks are known prior to the transmissions. Since the constraints in (2.52) are all linear, the convexity of problem (2.28) is determined by that of the function $\mathcal{F}(\cdot)$, which will be discussed later in Section 2.4.1.1. Here, we show some properties of the optimal solution with an arbitrary $\mathcal{F}(\cdot)$, which are summarized in the following.

First, note that the optimal solution of problem (2.51) may not be unique. This is due to the fact that if the transmit power values over two arbitrary blocks are decreasing over time, we can switch them without violating the EH constraints and still achieve the same objective value. Thus, similar to the staircase power allocation in throughput maximization over Gaussian channels in Section 2.2, we are only interested in obtaining the optimal power allocation of problem (2.51) that is non-decreasing over time.

Second, it is worth noting that the last EH constraint (i.e. the EH constraint for the last (N, M)th block) in (2.5) is always satisfied with equality by the optimal solution of problem (2.51). This is due to the fact that any residual energy after the last communication block should be used in the last block to yield a lower average outage probability, since the outage probability function $\mathcal{F}(\cdot)$ is assumed to be strictly decreasing.

Before solving problem (2.51) under general values of M and N, we first present some important properties for the outage probability function $\mathcal{F}(\cdot)$, then apply them to derive optimal and suboptimal power allocations for the special case of $M = 1$ with only one EH block, and finally consider the general case with $M > 1$ EH blocks.

2.4.1.1 Properties of Outage Probability Function

We show some interesting properties of the outage probability function $\mathcal{F}(\cdot)$ and accordingly define two types of outage probability functions for different fading channel distributions. Without loss of generality, we adopt Weibull fading [17] as an example, from which we later draw a general result for other fading distributions. For convenience, we omit the subscript (n, m) for $h_{n,m}$ and $P_{n,m}$ in this subsection.

With Weibull fading, the complex channel coefficient $\tilde{h}$ (with $h = |\tilde{h}|^2$) can be written as [17]

$$\tilde{h} = (X + jY)^{2/\beta}, \tag{2.53}$$

where X and Y are i.i.d. Gaussian random variables with zero mean and identical variance satisfying $\mathbb{E}(|\tilde{h}|^2) = 1$ and β is a parameter controlling the severity or the diversity of the channel fading [18], with $\beta > 0$. It is observed that when $\beta = 2$, Weibull fading degrades to the well-known Rayleigh fading. The probability density function (PDF) of $|\tilde{h}|$ is given as $f(r) = \beta r^{\beta-1} \exp(-r^\beta)$, with $r = |\tilde{h}| \geq 0$, and thus the corresponding outage probability function is given as [17]

$$\mathcal{F}(P) = 1 - \exp\left(-\left(\frac{2^R - 1}{P}\right)^{\beta/2}\right). \tag{2.54}$$

In Figure 2.7, we plot the outage probability function versus transmit power for Weibull fading. It is observed and also can be verified (see Proposition 2.11) that for all

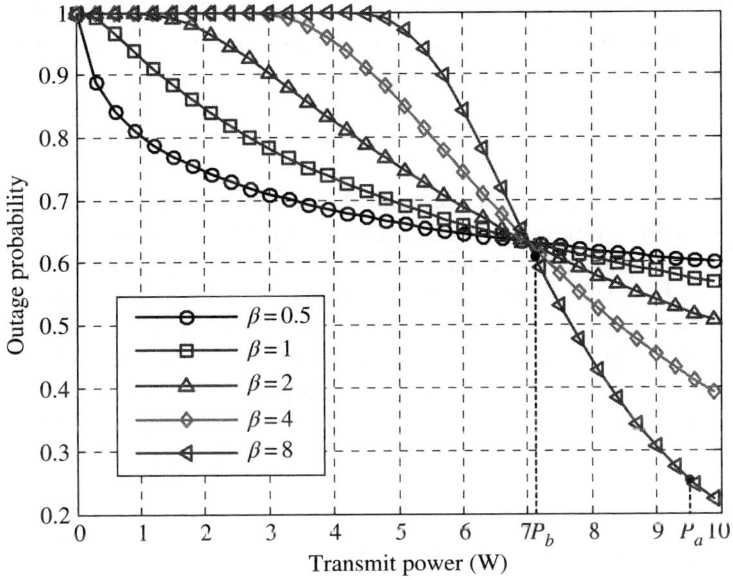

Figure 2.7 Outage probability versus transmit power for Weibull fading with different fading parameters β, with $R = 3$ bps Hz^{-1}.

cases with different values of β, the outage probability function is non-convex. Recall that β is an indicator of the "diversity order" [18] of the Weibull fading channel (larger values of β imply higher diversity orders). Thus, the Weibull fading model is quite general for modeling practical fading channels with different degrees of diversity.

Next, we obtain the following result on the convexity of the outage probability function with Weibull fading.

Proposition 2.11 *The outage probability function given in (2.54) is concave over $P \in [0, P_b]$ and convex over $P \in [P_b, +\infty)$, where*

$$P_b = \left(\frac{\beta/2}{\beta/2+1}\right)^{\beta/2} (2^R - 1). \tag{2.55}$$

Proof: By obtaining the second-order derivative of $\mathcal{F}(P)$ in (2.54), it follows that

$$\mathcal{F}''(P) = -\frac{\beta}{2}(2^R - 1)^{\beta/2} P^{-\beta/2-2} \exp\left(-\left(\frac{2^R-1}{P}\right)^{\beta/2}\right) \cdot \left(\frac{\beta}{2}(2^R - 1)^{\beta/2} P^{-\beta/2} - \left(\frac{\beta}{2}+1\right)\right).$$

We thus have $\mathcal{F}''(P) \leq 0$ over $0 < P \leq P_b$ and $\mathcal{F}''(P) > 0$ over $P > P_b$, which means that $\mathcal{F}(P)$ is concave when $P \geq P_b$ and convex when $0 < P \leq P_b$. Thus, this proposition is proved. □

Besides P_b, we now give another important parameter for the outage probability function as follows, for which the proof can be found in [11].

Proposition 2.12 *There exists a value P_a, with $P_a > P_b$, such that all the points $(P, \mathcal{F}(P))$ on the outage probability curve given in (2.54) are above the straight line passing through the two points $(0, 1)$ and $(P_a, \mathcal{F}(P_a))$, i.e.*

$$\mathcal{F}(P) \geq \frac{\mathcal{F}(P_a) - 1}{P_a} P + 1, \quad \forall P \geq 0. \tag{2.56}$$

It is worth noting that beyond the existence of point P_a, we can further show that the point satisfying this property is unique. This claim can be validated by contradiction. Suppose that this property holds at two different points P_a^1 and P_a^2. It is then easy to check that for an arbitrary $P_0 \in [P_a^1, P_a^2]$, this property is also satisfied, which means that the function $\mathcal{F}(P)$ over the range $[P_a^1, P_a^2]$ is actually a line segment connecting the two points $(P_a^1, \mathcal{F}(P_a^1))$ and $(P_a^2, \mathcal{F}(P_a^2))$. It is thus true that the first-order derivative of function $\mathcal{F}(P)$ is a constant over $[P_a^1, P_a^2]$, while its second-order derivative is zero over this regime. However, the above cannot be true in general for a practical continuous fading distribution (see, e.g. (2.53)). Thus, the presumption cannot be true, and the uniqueness of P_a is verified.

In general, it is difficult to obtain the closed-form expression for P_a. However, we can obtain the value of P_a by a simple bisection search summarized in Algorithm 2.4, for which the main ideas are given as follows: for a given point $\tilde{P}_a$, denote the function of the straight line connecting points $(0, 1)$ and $(\tilde{P}_a, \mathcal{F}(\tilde{P}_a))$ as $g(\tilde{P}_a; x)$, which is defined as

$$g(\tilde{P}_a; x) = \frac{\mathcal{F}(\tilde{P}_a) - 1}{\tilde{P}_a} x + 1. \tag{2.57}$$

If $\tilde{P}_a < P_a$, it is observed that there exists a $\delta > 0$ such that $g(\tilde{P}_a; x) < \mathcal{F}(x)$ for $x \in (\tilde{P}_a - \delta, \tilde{P}_a)$ and $g(\tilde{P}_a; x) > \mathcal{F}(x)$ for $x \in (\tilde{P}_a, \tilde{P}_a + \delta)$; otherwise if $\tilde{P}_a > P_a$, it is observed that there exists a $\delta > 0$ such that $g(\tilde{P}_a; x) > \mathcal{F}(x)$ for $x \in (\tilde{P}_a - \delta, \tilde{P}_a)$ and $g(\tilde{P}_a; x) < \mathcal{F}(x)$ for $x \in (\tilde{P}_a, \tilde{P}_a + \delta)$. According to the above property, we design Algorithm 2.4 to search for P_a, which returns an $x \in (P_a - \epsilon, P_a + \epsilon)$ with a prescribed tolerable error $0 < \epsilon \ll 1$.

Algorithm 2.4: Bisection search for P_a defined in Proposition 2.12.

Input: $a_{\text{low}} = 0$, a_{up} (any sufficiently large positive number), and the error tolerance ϵ
Output: P_a

while $a_{\text{up}} - a_{\text{low}} > \epsilon$ **do**
  $P_a \leftarrow \frac{a_{\text{up}} + a_{\text{low}}}{2}$
  if $g(P_a; P_a - \epsilon) < \mathcal{F}(P_a - \epsilon)$ **and** $g(P_a; P_a + \epsilon) > \mathcal{F}(P_a + \epsilon)$ **then**
    $a_{\text{low}} \leftarrow P_a$
  else
    $a_{\text{up}} \leftarrow P_a$
  end if
end while

In Figure 2.7, for the case of $\beta = 8$, we show the two points P_a and P_b. For other types of fading channel distributions, e.g. Rician fading and Nakagami fading, it can be shown that their outage probability functions have similar properties to the above for Weibull fading; however, unlike Weibull fading, P_b may no longer be expressible in a closed form as that in (2.55), and can only be computed numerically. Moreover, for certain fading distributions, it is possible that the outage probability function is convex over the entire range of $P \geq 0$, which can be regarded as a special case with $P_a = P_b = 0$. With the above observations, we categorize the outage probability functions for fading channel distributions in the following two types.

Definition 2.1 An outage probability function $\mathcal{F}(x)$, $x \geq 0$, is said to be with *Type A*, if $\mathcal{F}(x)$ is convex over $x \geq 0$.

Definition 2.2 An outage probability function $\mathcal{F}(x), x \geq 0$, is said to be with *Type B*, if there exist two unique points $0 < P_b \leq P_a$, such that:

(1) $\mathcal{F}(x)$ is concave over $[0, P_b]$ and convex over (P_b, ∞).
(2) $\mathcal{F}(x)$ is always above the line $g(x)$, passing through the two points $(0, 1)$ and $(P_a, \mathcal{F}(P_a))$, i.e. $\mathcal{F}(x) \geq g(x), \quad x \geq 0$.

Here, a Type B function is with a "concave–convex" shape, while a Type A function is a special case of Type B functions with $P_a = P_b = 0$. It can be verified that Weibull fading, Rician fading, Nakagami fading, and double Rayleigh fading in general lead to Type B outage probability functions. However, we need to point out that an arbitrary outage probability function can be neither Type A nor Type B, which is out of the scope of this book.

Remark 2.1 It is worth noting that if $\mathcal{F}(\cdot)$ is with Type A, problem (2.51) is convex. Thus, it can be shown that the staircase power allocation algorithm (e.g. Algorithm 2.1) can be applied to solve problem (2.51) optimally. As such, we are mainly interested in the case of Type B outage probability functions in this section.

2.4.1.2 Optimal Offline Power Allocation with $M = 1$

In this subsection, we solve problem (2.51) for the special case of $M = 1$, based on the previously derived outage probability function properties. Here, since $M = 1$, we use the notation P_n instead of $P_{n,1}$ for convenience.

Instead of solving problem (2.51) with $M = 1$ directly, we consider the following problem by removing the first $N - 1$ constraints in (2.5), i.e.

$$\min_{\{P_n \geq 0\}} \frac{1}{N} \sum_{n=1}^{N} \mathcal{F}(P_n)$$

$$\text{s.t.} \sum_{n=1}^{N} P_n \leq NE_1. \tag{2.58}$$

Obviously, problem (2.58) is a relaxed version of problem (2.51) with $M = 1$, which provides a lower bound on the optimal value of problem (2.51). We will show next that since there exists a non-decreasing optimal solution of problem (2.58), which is guaranteed to satisfy the omitted $N - 1$ constraints in (2.5), the solution to problem (2.58) is also optimal for problem (2.51) with $M = 1$.

It is easy to see that problem (2.58) is also a power allocation problem to minimize the average outage probability over an N-block block-fading channel with an average power constraint. If $\mathcal{F}(P_n)$ is a non-convex function (e.g. Type B function), problem (2.58) is non-convex [14] and thus difficult to be solved by conventional convex optimization techniques. To provide a solution, we first present some structural properties of the optimal power solution of problem (2.58) as follows.

Proposition 2.13 *For the optimal solution to problem (2.58), there is at most one strictly positive power value $P_n^\star$ that is below P_b defined in Proposition 2.11.*

Proof: This proposition is proved by contradiction. Suppose that for the optimal power solution $P_n^\star$'s of problem (2.58), there exist n_1 and n_2, $1 \le n_1, n_2 \le N$, such that $0 \le P_{n_1}^\star \le P_{n_2}^\star < P_b$. Since $\mathcal{F}(x)$ is assumed to be concave over $[0, P_b]$, we define $0 < \epsilon < \min(P_{n_1}^\star, P_b - P_{n_2}^\star)$, and it is easy to check that $\mathcal{F}(P_{n_1}^\star) + \mathcal{F}(P_{n_2}^\star) \ge \mathcal{F}(P_{n_1}^\star - \epsilon) + \mathcal{F}(P_{n_2}^\star + \epsilon)$, which means that a new power allocation, with $\widehat{P}_{n_1} = P_{n_1}^\star - \epsilon$ and $\widehat{P}_{n_2} = P_{n_2}^\star + \epsilon$, yields a lower outage probability and this contradicts with the optimality of $P_n^\star$'s for problem (2.58). Therefore, this proposition is proved. □

Proposition 2.14 *For problem (2.58), all optimal power values $P_m^\star$'s, which are above P_b defined in Proposition 2.11, are identical.*

Proof: The proof is similar to that of Proposition 2.13, by considering the convexity of the function $\mathcal{F}(x)$ over the region $[P_b, +\infty)$. □

Remark 2.2 Based on Propositions 2.13 and 2.14, it follows that the optimal solution to problem (2.58) should have the following structure: there are at most one block assigned with power $\widehat{P}_0$, $0 < \widehat{P}_0 < P_b$, $k^\star$ blocks, $0 \le k^\star \le M$, assigned with identical power $\frac{ME_1 - \widehat{P}_0}{k^\star} \ge P_b$, and the rest blocks with zero power. Thus, solving problem (2.58) is equivalent to finding the values for $k^\star$ and $\widehat{P}_0$.

Besides P_b, P_a defined in Proposition 2.12 also plays an important role in solving problem (2.58). In this case, it is always desirable to allocate the available power to be close to P_a when $E_1 < P_a$. As such, the desired $k^\star$ in Remark 2.2 should be either $k^\star = \left\lfloor \frac{E_1}{P_a} N \right\rfloor$ or $k^\star = \left\lfloor \frac{E_1}{P_a} N \right\rfloor + 1$, where $\lfloor x \rfloor$ denotes the flooring operation and $\widehat{P}_0$ can be obtained by a one-dimensional search. Therefore, we obtain the following proposition on the optimal solution to problem (2.58), for which the detailed proof is given in [11] and omitted here.

Proposition 2.15 *The optimal solution to problem (2.58) is given as follows:*

(1) If $E_1 \ge P_a$, $P_n^\star = E_1$, $1 \le n \le N$.
(2) If $E_1 < P_a$,

$$P_n^\star = \begin{cases} 0 & 1 \le n \le N - k_0 - 1 \\ \widehat{P}_0 & n = N - k_0 \\ \frac{NE_1 - \widehat{P}_0}{k_0} & M - k_0 + 1 \le n \le M, \end{cases} \tag{2.59}$$

where k_0 and $\widehat{P}_0$ are given as follows:

$$k_0 = \left\lfloor \frac{E_1}{P_a} N \right\rfloor, \tag{2.60}$$

$$\hat{P}_0 = \arg\min_{P \in \mathcal{P}} \mathcal{F}(P) + k_0 \mathcal{F}\left(\frac{NE_1 - P}{k_0}\right) \tag{2.61}$$

with $\mathcal{P} = [0, P_b) \bigcup \frac{NE_1}{k_0+1}$.

It is worth noting that only a one-dimensional search is needed to compute the optimal power allocation for problem (2.58) when $E_1 < P_a$. This is due to the fact that we cannot claim any monotonicity results on the right-hand side (RHS) of (2.61), and therefore, an exhaustive search is necessary.

Remark 2.3 It is worth pointing out that the outage capacity results in the literature for fading channels without CSIT are usually based on uniform power allocation over all the communication blocks. However, Proposition 2.15 reveals that uniform power allocation can be suboptimal for certain fading channel distributions (e.g. Type B fading) in the case of $E_1 < P_a$, which correspond to a low-power regime or equivalently a high-outage regime.

Since the solution given in Proposition 2.15 is non-decreasing and always satisfies the first $N-1$ constraints in (2.5), we obtain the following corollary.

Corollary 2.3 *The optimal solution obtained in Proposition 2.15 for problem (2.58) is also optimal for problem (2.51) with $M = 1$.*

Remark 2.4 Proposition 2.15 shows that in the low-power regime, uniform power allocation may be nonoptimal, while on–off power allocation achieves the minimum outage probability. It is easy to check that as N goes to infinity, the optimal power allocation for the case of $E_1 \leq P_a$ converges to the following binary power allocation: the transmitter transmits with power P_a over $\frac{E_1}{P_a}$ fraction of the total blocks and keeps silent in the rest of blocks. This is due to the following facts: (i) $\lim_{N\to\infty} \left| \lfloor \frac{E_1}{P_a} N \rfloor - \frac{E_1}{P_a} N \right| = 0$, which means that $k_0 \to \frac{E_1}{P_a} N$ as $N \to \infty$, and (ii) $\frac{NQ_1}{k_0+1} \to P_a$ and thus $\frac{NQ_1 - \hat{P}_0}{k_0} \to P_a$ as $N \to \infty$ for any finite $\hat{P}_0$, which implies that to satisfy the average power constraint, we should have $\hat{P}_0 \to 0$.

Remark 2.5 Compared with the conventional uniform power allocation [19], the performance gain in the low-power regime with the optimal power allocation is expected to be more substantial as P_a increases. Taking Weibull fading as an example, it is easy to see that a larger β corresponds to a larger P_a. Therefore, we conclude that for the fading channels with higher diversity orders, the performance gain will become more notable, for any given $E_1 < P_a$.

2.4.1.3 Suboptimal Power Allocation with $M = 1$

From Proposition 2.15, it is observed that the threshold P_a plays an important role in the optimal power allocation: if $E_1 < P_a$, all the nonzero power values, except at most one that is below P_b, are identical and as close to P_a as possible. This implies that an on–off two-level power allocation strategy may perform close to the optimal power allocation. To avoid the exhaustive search in obtaining the optimal solution, we can simply allocate the power uniformly in the "on" state, such that we only need to determine how many blocks should be in the "on" state. As such, we propose the following on–off power allocation scheme for the case of $E_1 < P_a$ as follows:

$$P_n = \begin{cases} 0 & 1 \le n \le N - k_0 \\ \tilde{P} & N - k_0 + 1 \le n \le N, \end{cases} \tag{2.62}$$

where

$$k_0 = \left\lfloor \frac{E_1}{P_a} N \right\rfloor, \quad \tilde{P} = \frac{NE_1}{k_0}, \tag{2.63}$$

while for the case of $E_1 \ge P_a$, the transmitter transmits with power E_1 for all N blocks, the same as Proposition 2.15. Obviously, when N goes to infinity, (2.62) is asymptotically optimal according to Remark 2.4, i.e. $\tilde{P} \to P_a$. We thus have the following proposition.

Proposition 2.16 *The power allocation in (2.62) is asymptotically optimal for problem (2.51) with $M = 1$ as N goes to infinity.*

2.4.1.4 Optimal Power Allocation for the General Case of $M > 1$

In this subsection, we derive the optimal solution to problem (2.51) for the general case of $M > 1$ with noncausal ESI. First, we give the following results on the structures of the optimal solution to problem (2.51) with $M > 1$.

Proposition 2.17 *For the optimal solution of problem (2.51), there is at most one strictly positive power $P_{n,m}$ that is below P_b defined in Proposition 2.11.*

Proof: The proof is similar to that of Proposition 2.13 and thus omitted for brevity. □

Proposition 2.17 implies that if the available transmit power for any communication block is below the value P_b, the transmitter should not transmit with at most one exceptional case until more energy is harvested to make the available power above P_b. In addition, for problem (2.51), if the optimal power values $P^\star_{n,m}$ and $P^\star_{n+1,m}$ (with $P_{n+1,m}$ denoting $P_{1,m+1}$ if $n = N$) for two consecutive communication blocks are both larger than P_b, we obtain the following properties for a non-decreasing optimal power allocation.

Proposition 2.18 *For the non-decreasing optimal solution of problem (2.51), any two consecutive transmit power values that are both larger than P_b must satisfy the following two conditions:*

(1) If the EH constraint at the (n, m)th block is not achieved with equality, we have $P^\star_{n,m} = P^\star_{n+1,m}$.
(2) From (1), we infer that if $P^\star_{n,m} < P^\star_{n+1,m}$, the EH constraint at the (n, m)th block is achieved with equality.

Proof: To show the first condition, we first assume that $P^\star_{n,m} < P^\star_{n+1,m}$. It is easy to check that a new power solution, defined as $\tilde{P}_{n,m} = P^\star_{n,m} + \epsilon$ and $\tilde{P}_{n+1,m} = P^*_{n+1,m} - \epsilon$, with $0 < \epsilon < \frac{P^*_{n+1,m} - P^*_{n,m}}{2}$, leads to a lower average outage probability, which contradicts with the optimality presumption. Therefore, the first condition is proved. Then, the second condition could be easily proved by using the first condition and the non-decreasing property. □

Remark 2.6 From Propositions 2.17 and 2.18, we conclude that the optimal solution of problem (2.51) for the case of $M > 1$ with the non-decreasing power solution must possess the following "save-then-transmit" structure: initially, the transmitter keeps silent for a certain number of communication blocks; then, it (possibly) transmits with a power smaller than P_b over one communication block; after that, it keeps transmitting with power larger than P_b and increases power levels right after the EH block where the harvested energy is exhausted due to (2) in Proposition 2.18.

Based on the above structure, we propose an algorithm, namely, Algorithm 2.5, to compute the globally optimal solution of problem (2.51) with $M > 1$, which is summarized in Algorithm 2.5. The proposed algorithm mainly addresses the following two questions: (i) when the transmission should start and (ii) how to determine the following parameters –the first positive transmission power $\tilde{P}_0$, which might be smaller than P_b, and k_0, which is the number of communication blocks with identical power value $\tilde{P}_1$ immediately after the first communication block with positive power value $\tilde{P}_0$. If these two issues are solved, the remaining power allocation can be computed efficiently by the staircase water-filling algorithm, since the remaining part will be operated in the convex region of the function $\mathcal{F}(\cdot)$. It will be shown next that the above two problems are equivalent to finding an EH block with index $i_{\tilde{t}}$ such that the result in the second condition of Proposition 2.15 could be applied from the first to the $i_{\tilde{t}}$th EH blocks to obtain the corresponding parameters k_0, $\tilde{P}_0$, and $\tilde{P}_1$.

Before presenting the general algorithm, we show an example in Figure 2.8 to illustrate the main ideas involved. The proposed algorithm implements a forward search from the first to the Mth EH block, in order to determine a particular EH block (indexed by $i_{\tilde{t}}$) within which there is one (if any) communication block with a positive power value below P_b. As shown in Figure 2.8, the following cases may occur in this search:

(1) The EH rate values E_1 and E_2 over the first and second EH blocks are below P_a. It is then claimed that $i_{\tilde{t}} > 2$ and we should continue the searching.
(2) At the third EH block, it is observed that $E_3 < P_a$ and $E_4, E_5, E_6 \geq P_a$. Therefore, we claim that the optimal power allocation may be no smaller than P_a after the third EH block, i.e. $i_{\tilde{t}} = 3$, and thus the power allocation from the first to the third EH blocks can be computed similarly as the second condition of Proposition 2.15 with

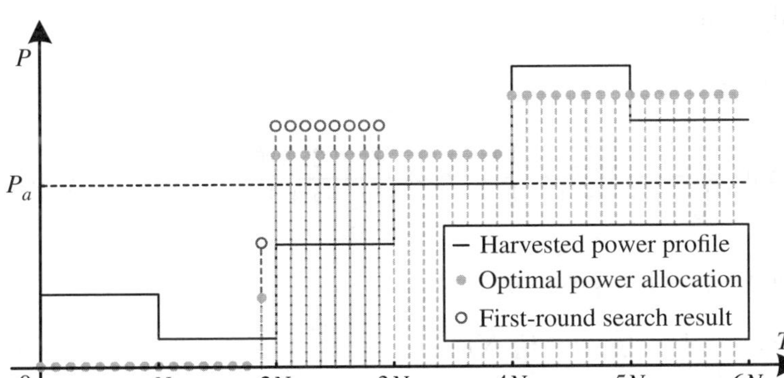

Figure 2.8 An example of the optimal power allocation solution for problem (2.51) with $N = 6$.

equivalent average power $\frac{1}{3}(E_1 + E_2 + E_3)$ and the number of blocks $3N$, by which we obtain k_0, $\tilde{P}_0$, and $\tilde{P}_1$. Note that the range of $\tilde{P}_0$ is determined by P_b.
Next, we allocate the obtained power values to the blocks starting from the $(N, 3)$th block in a backward manner to obtain a non-decreasing (in terms of a forward direction) power profile, as shown by the circles in Figure 2.8. It is easy to see that this power profile satisfies all the EH constraints up to the end of the third EH block. However, since the obtained power value in the third EH block is larger than the EH rate in the fourth EH block (which violates the non-decreasing power allocation for the optimal solution), we conclude that it cannot be optimal. Thus, we claim $i_{\tilde{t}} = 4$ and use Proposition 2.15 to update k_0, $\tilde{P}_0$, and $\tilde{P}_1$ with the new average power $\frac{1}{4}(E_1 + E_2 + E_3 + E_4)$ and the total number of blocks $4N$. Similarly, we obtain the new power profile from the first to the fourth EH blocks, as shown by the dots in Figure 2.8. Then, it is observed that the obtained power values satisfy the following two conditions: (i) the power values in the fourth EH block are smaller than the average EH rate in the fifth and sixth EH blocks (to comply with the optimal condition given in Proposition 2.18), and (ii) even if we raise the power value $\tilde{P}_1$ by $\frac{\tilde{P}_0}{k_0}$ and correspondingly set $\tilde{P}_0 = 0$, the new $\tilde{P}_1$ is still no larger than the average EH rate in the fifth and sixth EH blocks (this condition guarantees that no larger-scale search is needed). As such, we conclude that $i_{\tilde{t}} = 4$ is optimal for problem (2.51).

(3) In the fifth and sixth EH blocks, the harvested average power is larger than both P_a and the allocated power in the fourth EH block, and thus the transmitter should transmit at this average power. Note that in the fifth EH block, the transmitter needs to save certain power for the sixth EH block.

By generalizing the above procedure, we summarize the optimal solution for problem (2.51) with $M > 1$ in Algorithm 2.5.

Algorithm 2.5: Compute the optimal non-decreasing power allocation for problem (2.51) with $M > 1$.

Input: $i = 1$ and temp $= 0$

Output: The optimal power allocation $P^*_{p,q}$

Compute P_a by Algorithm 2.4

repeat

Compute i_s and $\widehat{P}_i$ by using

$$i_s = \arg\min_{i \le m \le M} \left\{ \frac{Q_0 + \sum_{k=i}^{m} E_k}{m - i + 1} \right\}, \quad \widehat{P}_i = \frac{Q_0 + \sum_{k=i}^{i_s} E_k}{i_s - i + 1} \tag{2.64}$$

if $\widehat{P}_i \ge P_a$ **then**

The optimal power profile is given by

$$P^*_{p,q} = \widehat{P}_i, \quad (i, 1) \le (p, q) \le (i_s, N) \tag{2.65}$$

$$i \leftarrow i_s + 1 \tag{2.66}$$

else

Recursively compute i_s and $\widehat{P}_i$ by using

$$i_{s+t} = \arg \min_{i_{s+t-1}+1 \le m \le M} \left\{ \frac{\sum_{k=i_{s+t-1}+1}^{m} E_k}{m - i_{s+t-1}} \right\} \tag{2.67}$$

$$\widehat{P}_{i+t} = \frac{\sum_{k=i_{s+t-1}+1}^{i_{s+t}} E_k}{i_{s+t} - i_{s+t-1}} \tag{2.68}$$

end if
if $t_0 \ge 1$ **and** $\widehat{P}_{i+t_0} < P_a$ **and** $\widehat{P}_{i+t_0+1} \ge P_a$ **then**
$\tilde{t} = t_0$, and compute $\widetilde{P}_0$ and $\widetilde{P}_1$ by (2.69) and (2.70)
while $i_{s+\tilde{t}} < M$ **and** $\left((i_0, j_0) = (N, i_{s+\tilde{t}}) \text{ or } \widetilde{P}_1 > \widehat{P}_{i+\tilde{t}+1}\right)$ **do**
$\tilde{t} \leftarrow \tilde{t} + 1$
Update $\widetilde{P}_0$ and $\widetilde{P}_1$ by (2.69) and (2.70)
end while
temp $= \tilde{t}$
while $i_{s+\tilde{t}} < M$ **and** $\widetilde{P}_0 > k_0\left(\widehat{P}_{i+\tilde{t}+1} - \widetilde{P}_1\right)$ **do**
$\tilde{t} \leftarrow \tilde{t} + 1$
update $\widetilde{P}_0$ and $\widetilde{P}_1$ by (2.69) and (2.70)
if $\widetilde{P}_0 + \left((i_{s+\tilde{t}-1} - i_0 + 1)N - j_0\right)\widetilde{P}_1 \le N\sum_{i=1}^{i_{s+\tilde{t}-1}} E_i$ **and** $p_{\text{temp}} + \sum_{r=\text{temp}+1}^{\tilde{t}} \left(i_{s+r} - i_{s+r-1}\right) N\mathcal{F}(\widehat{P}_{i+r}) > p_{\tilde{t}}$ **then**
temp $= \tilde{t}$
end if
end while
Update $\tilde{t} =$ temp, and compute $\widetilde{P}_0$ and $\widetilde{P}_1$ by (2.69) and (2.70), respectively. The optimal power profile is given as (2.71), and let $i = i_{s+\tilde{t}} + 1$
else
$i_{s+\tilde{t}} \leftarrow M$
Compute $\widetilde{P}_0$ and $\widetilde{P}_1$ by

$$p_{\tilde{t}} = \min_{P \in \mathcal{P}} \mathcal{F}(P) + k_0 \mathcal{F}\left(P_0 + \frac{P_0 - P}{k_0}\right) \tag{2.69}$$

$$\widetilde{P}_1 = P_0 + \frac{P_0 - \widetilde{P}_0}{k_0} \tag{2.70}$$

The optimal power allocation is given by

$$P^*_{p,q} = \begin{cases} 0 & (1, i) \le (p, q) \le (j_0, i_0) - 1 \\ \widetilde{P}_0 & (p, q) = (j_0, i_0) \\ \widetilde{P}_1 & (j_0, i_0) + 1 \le (p, q) \le (N, i_{s+\tilde{t}}), \end{cases} \tag{2.71}$$

where $(n, m) - 1$ is defined as the index of the previous communication block before the (n, m)th one, and let $i = M + 1$
end if
until $i > M$

Proposition 2.19 *The solution obtained by Algorithm 2.5 is optimal for problem (2.51).*

Proof: This proposition is proved by contradiction, i.e. for both the cases of $\widehat{P}_i \geq P_a$ and $\widehat{P}_i < P_a$, we can show that the optimal power allocation with any non-decreasing profile cannot be larger nor smaller than that obtained by Algorithm 2.5. Please refer to the details in [11]. □

Remark 2.7 In Section 2.2, we have investigated the power allocation problem to maximize the throughput over Gaussian channels with noncausal ESIT, where the throughput function is assumed to be concave. The optimal power profile for the throughput maximization problem is shown to have a continuous, non-decreasing, and piecewise-constant or staircase-like structure. In this section, we consider a different problem by minimizing the outage probability over fading channels with noncausal ESIT, which is non-convex due to the concave–convex shape of the outage probability function (Type B). Our results show that an on–off transmission strategy is optimal, for which the transmitter should only transmit when the available power is sufficiently large. It is worth noting that our problem will degrade to the one considered in Section 2.2 if the convex Type A outage probability function is considered.

Remark 2.8 In Algorithm 2.5, it is observed that exhaustive searches are needed to solve the optimization problem defined in (2.69). Since we need to repeat the one-dimensional search in steps (26–28) of Algorithm 2.5, these search operations are executed at most M times. Note that except these one-dimensional searches, the computation complexity of Algorithm 2.5 is on the order of $\mathcal{O}(M^2)$ (see (2.64) and (2.67)). Thus, the main computation burden of the proposed algorithm is due to the one-dimensional searches in (2.69). However, compared with searching exhaustively the optimal power allocation of problem (2.51) over the total number of NM communication blocks, the computation complexity is greatly reduced with the proposed algorithm.

Remark 2.9 By a similar argument to Remark 2.4, the power allocation profile given in Algorithm 2.5 converges to the following threshold-based transmission scheme as $N \to \infty$: at the ith EH block, first use (2.64) to compute the possible transmission power $\widehat{P}_i$. If $\widehat{P}_i$ is no smaller than P_a, we transmit with power $\widehat{P}_i$ over the ith to the i_sth EH blocks; otherwise, keep the first $\tilde{k} = \left\lfloor (1 - \frac{\widehat{P}_i}{P_a})(i_s - i + 1)N \right\rfloor$ communication blocks silent, and then transmit with power $\frac{(i_s - i+1)N}{(i_s-i+1)N-\tilde{k}}\tilde{P}_i$ over the rest communication blocks from the ith to the i_sth EH blocks.

2.4.1.5 Suboptimal Offline Power Allocation with $M > 1$

In this subsection, we propose a suboptimal algorithm for problem (2.51) with lower complexity than that of Algorithm 2.5 and show that it is asymptotically optimal as N goes to infinity.

Similar to the case of $M = 1$ in Section 2.4.1.3, we propose a suboptimal power allocation algorithm, namely, Algorithm 2.6, for problem (2.51) with $M > 1$, which is shown in Algorithm 2.6. The main idea of this algorithm is described as follows. From the first EH block, we search the index of the next possible power exhausting EH block by (2.64). If

$\widehat{P}_i \geq P_a$, we claim that the transmitter should transmit with its best effort, i.e. the power allocation is given as (2.65); otherwise, an on–off transmission is adopted to guarantee that the allocated power is equal to or larger than P_a. The power allocation is thus given as

$$P_{p,q} = \begin{cases} 0 & (1,i) \leq (p,q) \leq (j_0, i_0) \\ \frac{(i_s-i+1)N}{k_0}\widehat{P}_i & (j_0, i_0) < (p,q) \leq (N, i_s), \end{cases} \tag{2.72}$$

where i_0, j_0, and k_0 are computed by

$$i_0 = i_{s+\tilde{t}} - \left\lfloor \frac{k_0}{N} \right\rfloor, \tag{2.73}$$

$$j_0 = N - \left(k_0 - \left\lfloor \frac{k_0}{N} \right\rfloor N \right), \tag{2.74}$$

$$k_0 = \left\lfloor \frac{N \sum_{t=0}^{\tilde{t}} (i_{s+t} - i_{s+t-1})\widehat{P}_{i+t}}{P_a} \right\rfloor. \tag{2.75}$$

Note that from the ith to the i_sth EH blocks, the power profile is obtained by the suboptimal solution proposed in Section 2.4.1.3 with $M = 1$, while in Algorithm 2.6, we still use (2.64) to determine i_s as in the optimal solution given in Algorithm 2.5.

Algorithm 2.6: Suboptimal power allocation for problem (2.51) with $M > 1$.

Input: $i = 1$
Output: Suboptimal power allocation for problem (2.51)

```
while i < M do
    Compute P̂_i and i_s by using (2.64)
    if P̂_i ≥ P_a then
        The power allocation is given by (2.65)
    else
        The power allocation is given by (2.72)
    end if
    i ← i_s + 1
end while
```

Proposition 2.20 *The solution obtained by Algorithm 2.6 is asymptotically optimal for problem (2.51) with $M > 1$ as N goes to infinity.*

Proof: The proof is similar to that of Proposition 2.16 for $M = 1$ and thus omitted for brevity. □

2.4.2 The Case with No CSIT and Causal ESIT

Now, we consider the other case with no CSIT and with causal ESIT, where we assume that only the past and current EH rate levels are known at the transmitter. Furthermore, we assume that the EH process $\{E_m\}$ is a first-order stationary Markov process over m, and its distribution $\mathcal{P}(E_{m+1}|E_m)$ is known at the transmitter. We define the storage state $B_{n,m}$ as the energy level at the beginning of the (n,m)th communication block (prior to any EH and transmission in the block). Then, the storage state $B_{n,m} \geq 0$ at the beginning of the (n,m)th communication block is given by

$$B_{n+1,m} = B_{n,m} + E_m - P_{n,m}, \tag{2.76}$$

where $B_{n+1,m}$ denotes the energy storage level at the beginning of the $((n,m)+1)$th communication block, with $(n,m)+1$ being defined as the index of the next communication block after the (n,m)th one. Thus, it is easy to check that $\{B_{n,m}\}$ also follows a first-order Markov model. For convenience, we assume zero initial energy storage, i.e. $B_{1,1} = 0$. From (2.76), it is easy to see that $P_{n,m} \leq B_{n,m} + E_m$, since $B_{n+1,m} \geq 0$.

Therefore, for the mth EH block, the minimum average outage probability with given E_m is obtained by solving the following problem:

$$\mathcal{T}_m^* = \min_{\{P_{1,m},\ldots,P_{N,m}\}} \sum_{i=1}^{N} \mathcal{F}(P_{i,m}) + \mathbb{E}\left[\sum_{j=m+1}^{M}\sum_{i=1}^{N} \mathcal{F}(P_{j,i}) \middle| E_n\right] \tag{2.77}$$

$$\text{s.t. } 0 \leq P_{n,j} \leq B_{n,j} + E_j, \quad 1 \leq n \leq N, \quad m \leq j \leq M. \tag{2.78}$$

It is worth noting that only E_m is known to the transmitter and $E_{m+1}, \ldots, E_M$ are random variables. The expectation in problem (2.77) is taken over $\{E_{m+1}, \ldots, E_M\}$; and $P_{i,j}$, $1 \leq i \leq N$ and $m+1 \leq j \leq M$, is a function of $\{E_{m+1}, \ldots, E_j\}$ and $\{P_{1,m}, \ldots, P_{N,m}\}$. It is easy to verify that the group of problems in (2.77), $m = 1, \ldots, M$, corresponds to a Markov decision process (MDP) problem, for which the optimal solution will be studied next.

In the following, we consider the group of problems in (2.77), $m = 1, \ldots, M$, for the case with only causal ESIT at the transmitter and $M > 1$. First, we show that the optimal solution of these problems can be obtained by applying dynamic programming and the results in Section 2.4.1.3 for the case of $M = 1$. Then, we propose a suboptimal online power allocation algorithm.

2.4.2.1 Optimal Online Power Allocation

In general, the optimization of $P_{i,j}$ for problem (2.77) with $N > 1$ cannot be performed independently over each EH block, since the storage states defined in (2.76) are coupled over time. Thus, we adopt a dynamic programming method to solve this problem, as stated in the following proposition.

Proposition 2.21 *For problem (2.77), $1 \leq m \leq M$, and the given initial states E_m and $B_m = B_{1,m}$, the minimum average outage probability $\mathcal{T}_m^*$ is given by $J_m(E_m, B_m)$, which can be computed recursively from $J_M(E_M, B_M)$ to $J_{M-1}(E_{M-1}, B_{M-1})$, until $J_m(E_m, B_m)$. The sequence of optimization problems are constructed as*

$$J_M(E_M, B_M) = \min_{\{P_{i,M}\}} \sum_{i=1}^{N} \mathcal{F}(P_{i,M}) \tag{2.79}$$

$$\text{s.t.} \sum_{i=1}^{N} P_{i,M} \leq B_M + NE_M, \tag{2.80}$$

and for $1 \leq j < M$,

$$J_j(E_j, B_j) = \min_{0 \leq B_{j+1} \leq B_j + NE_j} \min_{\{P_{i,j}\}} \sum_{i=1}^{N} \mathcal{F}(P_{i,j}) + \bar{J}_{j+1}(E_j, B_{j+1}) \tag{2.81}$$

$$\text{s.t.} \sum_{i=1}^{N} P_{i,j} \leq B_j - B_{j+1} + NE_i, \tag{2.82}$$

where

$$\bar{J}_{i+1}(E_i, B_{i+1}) = \mathbb{E}_{E_{i+1}}[J_{i+1}(E_{i+1}, B_{i+1}) | E_i]. \tag{2.83}$$

Proof: The proof directly follows by applying Bellman's equation [15] and using (2.76). Note that in each problem in (2.81), there should be N EH constraints, i.e. $\sum_{i=1}^{n} P_{i,j} \leq B_i - B_{i+1} + nE_i,\ 1 \leq n \leq N$; however, by a similar argument as in Section 2.4.1.3, we could eliminate the first $N-1$ EH constraints by finding an optimal solution $P_{i,j}$ in a non-decreasing manner. □

Note that problem (2.81), $1 \leq j \leq M-1$, can be solved by first applying Proposition 2.15 with fixed B_{j+1} and then searching over all possible B_{j+1}'s, with $0 \leq B_{j+1} \leq B_j + NE_i$. Therefore, the above MDP problems can be solved by dynamic programming and applying our previous results for the case of $M = 1$.

2.4.2.2 Suboptimal Online Power Allocation

In this subsection, we propose a suboptimal online power allocation algorithm called "q-block look-ahead," which is based on the current storage state and the predicted power to be harvested for the next $q-1$ EH blocks, with $q \geq 2$.

We assume that the EH process $\{E_j\}$ is a discrete-time first-order Markov process, and the future ESI can be predicted as

$$\hat{E}_j = \mathbb{E}(E_j | E_i), \quad i+1 \leq j \leq \min\{i+q-1, M\}, \tag{2.84}$$

where the integer q indicates the prediction window length from the current EH block. Here we only assume that the mean values of the future harvested energy are known, while the exact distribution may not be known to the transmitter, which greatly relaxes the requirements for computing the online power allocation.

Then, with the current energy profile $\hat{E}_i = E_i + \frac{B_i}{N}$ and the predicted ones, $\hat{E}_j$, $i+1 \leq j \leq \min\{i+q-1, N\}$, we compute the power allocation by using either the optimal or the suboptimal algorithms, i.e. Algorithm 2.5 or 2.6, and thereby adopt the power allocation for the current ith EH block, i.e. from the $(1, i)$th to the (N, i)th communication blocks. Similarly, at the next EH block, we repeat the above procedure to obtain its corresponding power allocation, until the Mth EH block is reached. Evidently, if $q = 1$, the proposed scheme becomes a "greedy" power allocation, i.e. at the end of each EH block, all the stored harvested energy is used up. In the next subsection, we will show the performance of this suboptimal online algorithm.

2.4.3 Numerical Results

This subsection presents numerical results to compare the outage probabilities under different number of EH blocks M and different ESIT cases. For the purpose of exposition, we consider Weibull fading with $\beta = 8$.

2.4.3.1 The Case of $M = 1$

First, we show some numerical results to validate our analysis in Section 2.4.1.3 for the case of $M = 1$. In Figure 2.9, we plot the outage probability versus average transmit power with different numbers of communication blocks, N. It is observed that the outage probability with the optimal power allocation is no larger than that with uniform power allocation over the region $E_1 < P_a$. In particular, the minimum outage probability with the optimal allocation for the case of $N \to \infty$ over this region is a straight line connecting the points $[0, 1]$ and $[P_a, \mathcal{F}(P_a)]$. Moreover, as N increases, the minimum outage probability converges to that with $M \to \infty$.

In Figure 2.10, we plot the average outage probability versus N with both the proposed optimal and suboptimal power allocation schemes. It is observed that the outage probabilities of both schemes converge to the same value as N goes to infinity, which is in accordance with Proposition 2.16.

2.4.3.2 The Case of $M > 1$

In this subsection, we compare the performance of the proposed offline algorithm against that of the online power allocation algorithm, for the case of $M > 1$ with noncausal versus causal ESI, respectively. We consider a first-order Markov EH process, where the EH rate at each communication block has three possible states: 0, P, and $2P$. Denote the probability that the EH rate at the ith block changes from one state to any

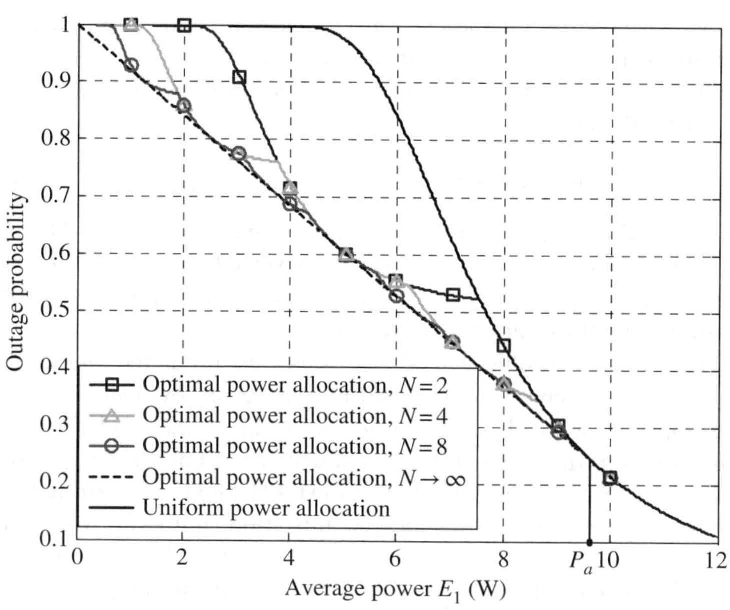

Figure 2.9 Outage probability over the Weibull fading, with $\beta = 8$ and $R = 3$ bps Hz^{-1}.

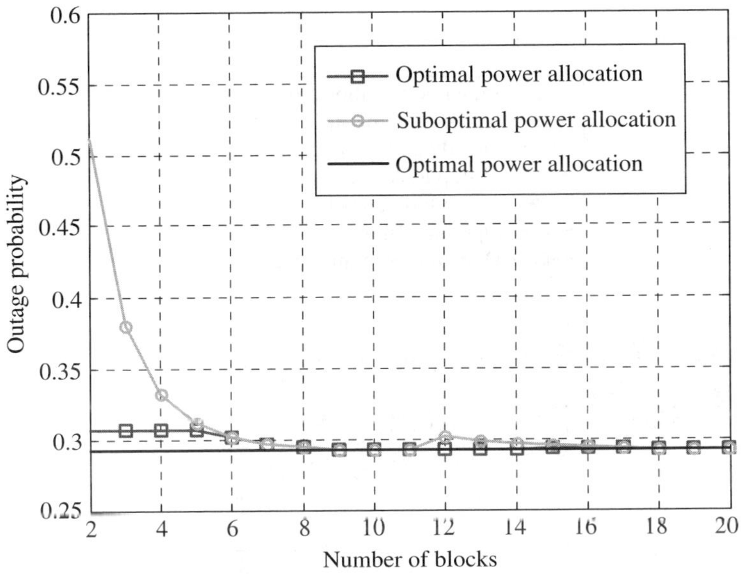

Figure 2.10 Outage probability comparison for the optimal and suboptimal power allocation schemes, with $\beta = 8$, $E_1 = 9$, and $R = 3$ bps Hz^{-1}, where the lowest curve corresponds to the case with $N \to \infty$.

of the other two as δ, with $0 \leq \delta \leq 1/2$, and thus the probability that the EH rate keeps unchanged from the previous EH block is given as $1 - 2\delta$. At the first communication block, we assume that E_1 takes one of the three power values with an equal probability of $1/3$. For other parameters, we adopt $N = 1$, $M = 20$, and $R = 0.5$ bps Hz^{-1}. Besides the proposed online algorithm in Section 2.4.2, we consider the following three alternative online power allocation schemes for comparison:

(1) Receding horizon scheme [20]: Instead of applying dynamic programming over the current and all the rest of communication blocks as in our proposed optimal online algorithm, we apply it for only the current block and its immediate next block.
(2) Equal power allocation: Uniformly allocate all the available energy to the rest of the blocks.
(3) Greedy power allocation: Allocate all the available power to the current block.

As shown in Figures 2.11–2.13, we compare the outage probability of different power allocation algorithms with the average EH rate P, for $\delta = 0.1$, $1/3$, and 0.4, respectively. Note that when $\delta = 1/3$, this case in fact corresponds to the i.i.d. energy arrival process over time. As a good approximation for the continuous-value MDP problem in (2.77), we model the storage states by a finite set, where the difference between any two adjacent discrete states is set as 0.01. It is observed that in the low EH rate regime, the M-block look-ahead scheme in general performs very close to the optimal online algorithm by dynamic programming; in the high EH rate regime, the equal power allocation scheme performs the best among all the suboptimal online algorithms for the case of $\delta = 0.1$; and the receding horizon scheme is asymptotically optimal in the sense of achieving the dynamic programming bound for larger δ's as the average EH rate becomes large.

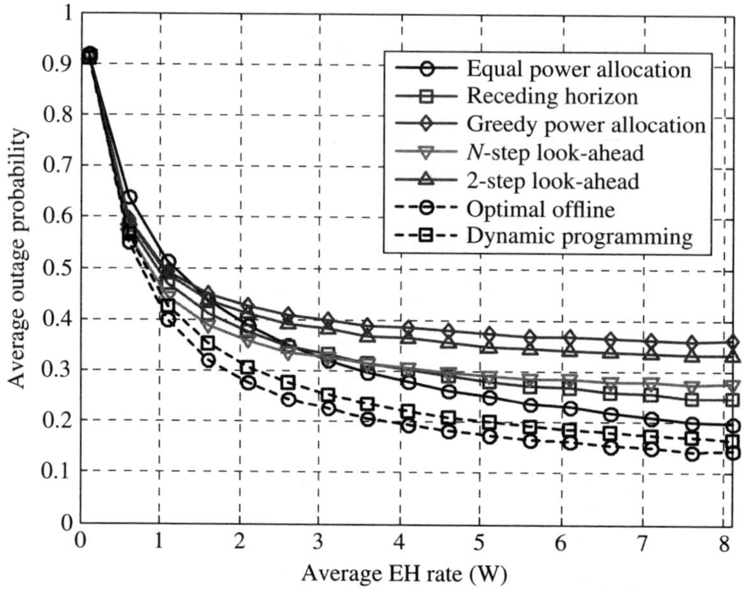

Figure 2.11 Comparison of various power allocation algorithms, with $\delta = 0.1$, $R = 0.5$ bps Hz^{-1}, $M = 20$, and $N = 1$.

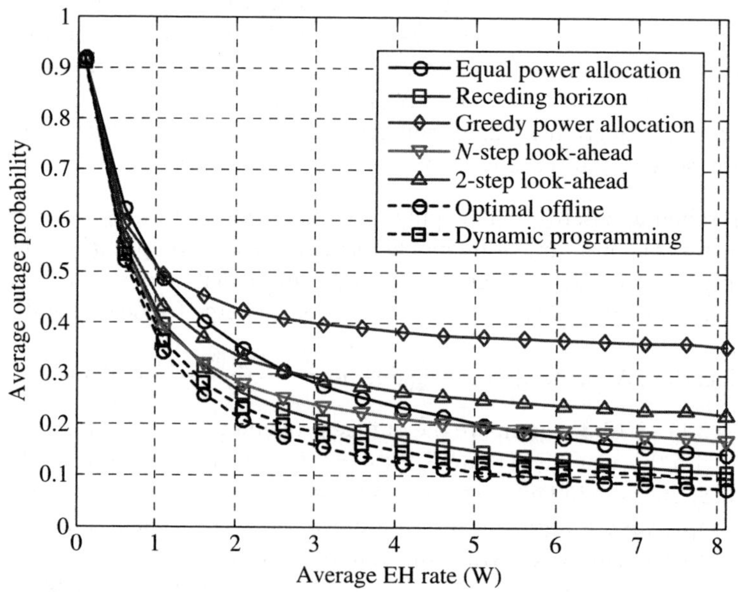

Figure 2.12 Comparison of various power allocation algorithms, with $\delta = 1/3$, $R = 0.5$ bps Hz^{-1}, $M = 20$, and $N = 1$.

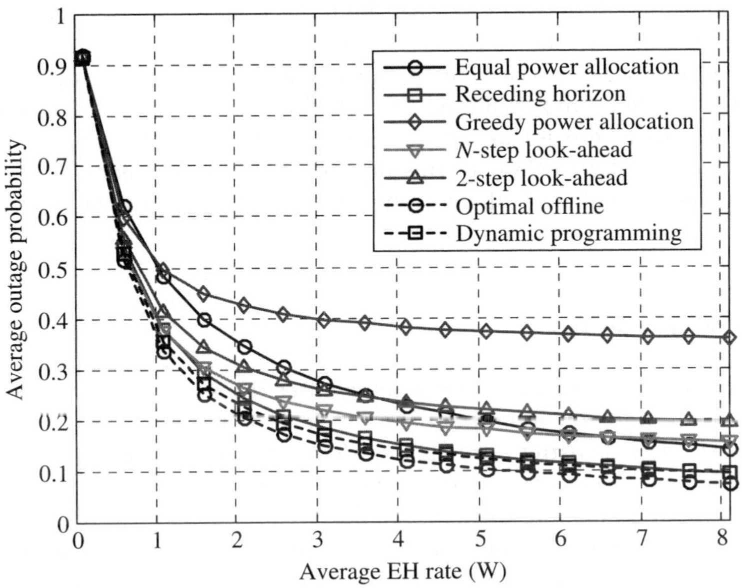

Figure 2.13 Comparison of various power allocation algorithms, with $\delta = 0.4$, $R = 0.5$ bps Hz^{-1}, $M = 20$, and $N = 1$.

2.4.4 Other CSIT and ESIT Cases

For the other three cases (i.e. Cases 1–3), to our best knowledge, how to find the optimal power allocation for the outage probability minimization has not yet been investigated. In the following, we provide some intuitions that may be helpful to solve this problem under different cases.

Considering Case 1 with noncausal CSIT and ESIT, the outage probability function $\mathcal{F}_{n,m}(P_{n,m})$ becomes an indicator function of the transmit power $P_{n,m}$ for any given (n, m)th block, i.e. $\mathcal{F}_{n,m}(P_{n,m}) = 1$ if the achievable rate with power $P_{n,m}$ is strictly less than the required transmission rate; otherwise, $\mathcal{F}_{n,m}(P_{n,m}) = 0$. With this simplification, the average outage probability minimization is equivalent to minimizing the sum value of the indicator functions over $1 \leq n \leq N$ and $1 \leq m \leq M$. Although such a problem is non-convex in general, it has the following structures. First, it is evident that the allocated transmit power at any communication block should be either zero for an outage event or the minimum required power for a non-outage transmission (with the resulting achievable rate equal to the required transmission rate), since the energy would be wasted otherwise. Another fact is that the objective value of each communication block is either zero or one. With these two observations, we conjecture that the optimal power allocation in this case could be found by first ordering the channel gains from the best to the worst and then allocating the nonzero transmit power iteratively based on the order of channel gains subject to the EH constraints.

Finally, for Cases 2 and 3 with causal CSIT, the outage probability minimization corresponds to minimizing the expected outage probability over either the randomness of EH rates (Case 2) or deterministic EH constraints (Case 3). In general, the optimal power allocations for these two cases can be derived via standard dynamic programming techniques, and in each communication block, the transmitter should decide its current power allocation based on the binary objective value for outage. Interestingly, since the current CSIT is always known, the transmit power at each block (n, m) should be either zero for an outage event or the minimum one for a non-outage transmission, similar to Case 1. This decision is made by balancing the tradeoff between avoiding the outage in the current block and minimizing the future outage probabilities.

2.5 Limited Battery Storage

In EH wireless communications, the EH wireless transmitter relies on a rechargeable battery storage device to store the renewable energy for communications. While previous sections assume that the battery storage is perfect with infinite storage capacity, in practice it is subject to various constraints such as finite capacity, battery leakage, and imperfect charging and discharging efficiency. In this section, we discuss the effect of battery capacity on the power allocation of the EH transmitter. In particular, let $E_{\max}$ denote the rechargeable battery capacity, and accordingly, at any time instant the EH transmitter is subject to an energy overflow constraint, that is, the remaining energy at the battery cannot exceed the battery capacity $E_{\max}$, since otherwise it will be wasted. Explicitly, in the (n, m)th block, the following battery capacity constraint is imposed:

$$N \sum_{j=1}^{m-1} E_j + nE_m - \sum_{j=1}^{m-1} \sum_{i=1}^{N} P_{i,j} - \sum_{i=1}^{n} P_{i,m} \leq E_{\max},$$
$$n = 1, \dots, N, \quad m = 1, \dots, M. \tag{2.85}$$

This new battery capacity constraint fundamentally changes the power allocation principle at the EH transmitter, since in this case, the EH transmitter needs to ensure that the battery has sufficient room to store the incoming energy in the future to avoid the energy being wasted. To reveal such an effect, we particularly consider the power allocation at the EH transmitter for maximizing the throughput utility over Gaussian channels and fading channels, respectively, under the case with noncausal CSIT and ESIT (i.e. Case 1). We will show how the staircase power allocation and the staircase water filling will be modified to deal with the battery capacity constraints.

2.5.1 Throughput Maximization over Gaussian Channel with Noncausal ESIT

First, we consider the case with Gaussian channels, where we have $h_{n,m} = h, n = 1, \dots, N, m = 1, \dots, M$ and the number of communication blocks is $N = 1$ per EH block. In this case, the throughput maximization problem is expressed as follows by incorporating the battery capacity constraint (2.85) into problem (2.7) and setting $N = 1$:

$$\max_{\{P_m \geq 0\}} \sum_{m=1}^{M} \log_2(1 + hP_m) \tag{2.86}$$

$$\text{s.t.} \sum_{j=1}^{m} E_j - \sum_{j=1}^{m} P_j \leq E_{\max}, \quad m = 1, \dots, M \tag{2.87}$$

(2.8).

Problem (2.86) is a convex optimization problem, and we use the KKT condition to reveal the structural property of the optimal solution. Let $\lambda_m \geq 0$ denote the Lagrange multiplier associated with the mth EH constraint in (2.8), $\theta_m \geq 0$ denote the Lagrange multiplier associated with the mth battery capacity constraint in (2.87), and $\mu_m \geq 0$ denote the Lagrange multiplier associated with the power constraint $P_m \geq 0$, respectively, $m = 1, \dots, M$. Then the Lagrangian associated with problem (2.86) is

$$\mathcal{L}_{2.5}(\{P_m\}, \{\lambda_m\}, \{\theta_m\}, \{\mu_m\}) = \sum_{m=1}^{M} \log_2(1 + hP_m)$$
$$- \sum_{m=1}^{M} \lambda_m \left(\sum_{j=1}^{m} P_j - \sum_{j=1}^{m} E_j \right) - \sum_{m=1}^{M} \theta_m \left(\sum_{j=1}^{m} E_j - \sum_{j=1}^{m} P_j - E_{\max} \right)$$
$$+ \sum_{m=1}^{M} \mu_m P_m. \tag{2.88}$$

The necessary and sufficient conditions for $\{P_m^\star\}$, $\{\lambda_m^\star\}$, $\{\theta_m^\star\}$, and $\{\mu_m^\star\}$ to be the primal and dual optimal solutions to (2.86) are given by the following KKT conditions:

$$0 \leq P_m^\star, \tag{2.89}$$

$$0 \geq \sum_{j=1}^{m-1} P_j^\star - \sum_{j=1}^{m-1} E_j, \tag{2.90}$$

$$0 \geq \sum_{j=1}^{m} E_j - \sum_{j=1}^{m} P_j - E_{\max}, \tag{2.91}$$

$$0 \leq \lambda_m^\star, \tag{2.92}$$

$$0 \leq \mu_m^\star, \tag{2.93}$$

$$0 \leq \theta_m^\star, \tag{2.94}$$

$$0 = \lambda_m^\star \left(\sum_{j=1}^{m-1} P_j^\star - \sum_{j=1}^{m-1} E_j \right), \tag{2.95}$$

$$0 = \theta_m^\star \left(\sum_{j=1}^{m} E_j - \sum_{j=1}^{m} P_j^\star - E_{\max} \right), \tag{2.96}$$

$$0 = \mu_m^\star P_m^\star, \tag{2.97}$$

$$0 = \frac{\partial \mathcal{L}_{2.5}(\{P_m^\star\}, \{\lambda_m^\star\}, \{\theta_m^\star\}, \{\mu_m^\star\})}{\partial P_m^\star}, \tag{2.98}$$

for all $m = 1, \dots, M$. From (2.89), (2.93), (2.97), and (2.98), we obtain the optimal power allocation as

$$P_m^\star = [\nu_m - 1]^+, \quad m = 1, \dots, M, \tag{2.99}$$

where $\nu_m \triangleq \left(\ln 2 \sum_{j=m}^{M} (\lambda_j^\star - \theta_j^\star)\right)^{-1} \geq 0$ and the $\lambda_j^\star$'s and $\theta_j^\star$'s satisfy the above KKT conditions. Note that based on (2.95) and (2.96), it follows that for any $m = 1, \dots, M$, the $\lambda_j^\star$ and $\theta_j^\star$ cannot be nonzero at the same time.

It is easy to observe from (2.99) that under the battery capacity constraint, the monotonic non-decreasing property does not hold for the optimal power allocation $P_m^\star$. Indeed, when $\lambda_m^\star - \theta_m^\star < 0$ (i.e. $\theta_m^\star > 0$ and $\lambda_m^\star = 0$ as they cannot be positive at the same time) at a certain EH block m, we have $\nu_m > \nu_{m+1}$ and $P_m^\star < P_{m+1}^\star$, i.e. the transmit power decreases after the mth EH block. The optimal power allocation solution in this case is referred to as *modified staircase power allocation*.

Furthermore, there generally exist two types of *transition blocks* in this case. An EH block $t \in \{1, \dots, M\}$ is called a *Type I* transition block if the transmit power increases after EH block t, i.e. $P_t^\star < P_{t+1}^\star$, while an EH block t is named a *Type II* transition block if the transmit power decreases after EH block t, i.e. $P_t^\star > P_{t+1}^\star$. We define the last EH block M also as a Type I transition block by defining $P_{M+1}^\star$ to be infinity; hence there is at least one Type I transition block. We collect all transition blocks as the set $\mathcal{S} = \{t_1, \dots, t_{|\mathcal{S}|}\}$, where $t_i < t_j$ for $i < j$ and $t_{|\mathcal{S}|} = M$, and denote the corresponding type of the transition blocks as $\Delta = \{\delta_{t_1}, \dots, \delta_{t_{|\mathcal{S}|}}\}$, where $\delta_{t_i} \in \{\mathrm{I}, \mathrm{II}\}$ and $\delta_{t_{|\mathcal{S}|}} = \mathrm{I}$. Then, we have the following structural properties for the optimal power allocation $\{P_m^\star\}$ to problem (2.86).

Proposition 2.22 *The optimal power allocation solution to problem (2.86) has the following properties:*

- *If EH block t is a Type I transition block (i.e. $P_t^\star < P_{t+1}^\star$), then the battery storage is empty after the tth block, i.e. the accumulative energy consumed up to EH block t equals that harvested up to then, i.e. the EH constraint in (2.90) holds with equality for t.*
- *If EH block t is a Type II transition block (i.e. $P_t^\star > P_{t+1}^\star$), then the battery storage is full after the tth block, i.e. the battery capacity constraint in (2.91) holds with equality for t.*

Proof: Before proving this proposition, note that at each EH block m, the EH constraint in (2.90) and the battery capacity constraint in (2.91) cannot be met with equality at the same time. By using this fact together with the complementary slackness conditions in (2.95) and (2.96), it follows that $\lambda_j^\star$ and $\theta_j^\star$ cannot be strictly positive at the same time, i.e. $\lambda_j^\star \cdot \theta_j^\star = 0$.

Next, consider that the EH block t is a Type I transition block, where it follows that $\nu_m < \nu_{m+1}$ and accordingly $\lambda_j^\star - \theta_j^\star > 0$. By using that $\lambda_j^\star \geq 0$, $\theta_j^\star > 0$, and $\lambda_j^\star \cdot \theta_j^\star = 0$, we have $\lambda_t^\star > 0$ and $\theta_t^\star = 0$. Based on the complementary slackness condition in (2.95), it is evident that (2.90) holds with equality for t (i.e. the battery storage is empty after the tth block).

Finally, consider that the EH block t is a Type II transition block, where it follows that $v_m > v_{m+1}$ and accordingly $\lambda_j^\star - \theta_j^\star < 0$. By using $\lambda_j^\star \geq 0$, $\theta_j^\star > 0$, and $\lambda_j^\star \cdot \theta_j^\star = 0$, we have $\lambda_t^\star = 0$ and $\theta_t^\star > 0$. Based on the complementary slackness condition in (2.96), it is evident that (2.91) holds with equality for t (i.e. the battery storage is full after the tth block). □

Based on the structural property of the optimal solution in Proposition 2.22, we design an efficient algorithm to implement the modified staircase power allocation in the following.

Note that between any two transition blocks t_{i-1} and t_i, the transmit power at the transmitter remains constant. Suppose that the initial battery storage $\hat{E}_{t_{i-1}}$ at block t_{i-1} is given, where $\hat{E}_{t_{i-1}} = 0$ if block t_{i-1} is of Type I, while $\hat{E}_{t_{i-1}} = E_{\max}$ if block t_{i-1} is of Type II. In this case, if the subsequent transition block t_i is of Type I, then the transmit power between the two transition blocks is given as $P_{t_i}^{\mathrm{I}} = \dfrac{\hat{E}_{t_i} + \sum_{m=t_{i-1}+1}^{t_i} E_m}{t_i - t_{i-1}}$; otherwise, if the transition block t_i is of Type II, then the transmit power is given as $P_{t_i}^{\mathrm{II}} = \dfrac{\hat{E}_{t_i} + \sum_{m=t_{i-1}+1}^{t_i} E_m - E_{\max}}{t_i - t_{i-1}}$. Due to the one-to-one mapping between the transmit power allocation and the transition blocks, similar to that in Section 2.2, the derivation of the optimal power allocation solution to problem (2.86) is equivalent to obtaining the optimal set of the transition blocks $\mathcal{S}$, together with the types of each transition block Δ. Let $\mathcal{S}^\star = \{t_1^\star, \dots, t_{|\mathcal{S}^\star|}^\star\}$ denote the optimal set of transition blocks and $\Delta^\star = \{\delta_1^\star, \dots, \delta_{|\mathcal{S}^\star|}^\star\}$ denote the corresponding set collecting the optimal type of each transition block.

The algorithm to find the optimal transition blocks for the modified staircase power allocation to optimize problem (2.86) is given in Algorithm 2.7, which is briefly explained as follows. In this algorithm, the iteration procedure is similar to that in the staircase power allocation in Algorithm 2.1, while their main difference is that in addition to finding the transition blocks $t_i^\star$'s, the modified staircase power allocation also needs to determine the type of each transition block $\delta_{t_i^\star}^\star \in \{\mathrm{I}, \mathrm{II}\}$. Toward this end, in each iteration i, we need to compute two constant transmit powers $P_{t_i}^{\mathrm{I}}$ and $P_{t_i}^{\mathrm{II}}$ by assuming the subsequent transition block t_i is of Type I and Type II, respectively, (see lines 2 and 3). In lines 4 and 5, t_i^{I} corresponds to the maximum index of Type I transition blocks such that the transmit power can maintain constant between transition blocks t_i^{I} and $t_{i-1}^\star$ without violating the EH constraints, while t_i^{II} corresponds to the maximum index of Type II transition blocks such that the transmit power can maintain constant between the two transition blocks without violating the battery capacity constraints. In lines 6–12, if $P_{t_i^{\mathrm{I}}}^{\mathrm{I}} \geq P_{t_i^{\mathrm{II}}}^{\mathrm{II}}$, then the battery capacity constraint is loose, and the optimal power allocation over the blocks $t_{i-1}^\star, \dots, t_i^{\mathrm{I}}$ is same as that without such a constraint. Therefore, we have $t_i^\star = t_i^{\mathrm{I}}$ and $\delta_{t_i^\star}^\star = \mathrm{I}$ in this case. If $P_{t_i}^{\mathrm{I}} < P_{t_i}^{\mathrm{II}}$, then we always have $t_i^{\mathrm{I}} \neq t_i^{\mathrm{II}}$. In this case, we should choose the constant transmit power corresponding to the earlier transition block (i.e. the block $\min(t_i^{\mathrm{I}}, t_i^{\mathrm{II}})$), so as to avoid violating the EH constraints or battery capacity constraints. Accordingly, the type of this transition block is determined as $\delta_{t_i^\star}^\star = \arg\min_{j \in \{\mathrm{I},\mathrm{II}\}} t_i^j$. The optimality of such an algorithm can be verified based on the structural properties in Proposition 2.22, for which the details are omitted here. Readers can refer to [21] for the detail of a similar proof.

Algorithm 2.7: Finding the optimal transition blocks for the modified staircase power allocation.

Input: $t_0^\star = 0$ and $\hat{E}_{t_0^\star} = 0$
Output: $t_i^\star$

for $i = 1, 2, \dots, M$ **do**
  $P^{\mathrm{I}}_{t_i} \leftarrow \frac{\hat{E}_{t^\star_{i-1}} + \sum_{j=t^\star_{i-1}+1}^{m} E_j}{m - t^\star_{i-1}}$
  $P^{\mathrm{II}}_{t_i} \leftarrow \frac{\hat{E}_{t^\star_{i-1}} + \sum_{j=t^\star_{i-1}+1}^{m} E_j - E_{\max}}{m - t^\star_{i-1}}, \quad t_i = t^\star_{i-1} + 1, \dots, M$
  $t_i^{\mathrm{I}} \leftarrow \arg\min_{t^\star_{i-1}+1 \le t_i \le M} P^{\mathrm{I}}_{t_i}$
  $t_i^{\mathrm{II}} \leftarrow \arg\max_{t^\star_{i-1}+1 \le t_i \le M} P^{\mathrm{II}}_{t_i}$
  if $P^{\mathrm{I}}_{t_i^{\mathrm{I}}} \ge P^{\mathrm{II}}_{t_i^{\mathrm{II}}}$ **then**
    $t_i^\star \leftarrow t_i^{\mathrm{I}}$
    $\delta^\star_{t_i^\star} \leftarrow \mathrm{I}$
  else
    $t_i^\star \leftarrow \min(t_i^{\mathrm{I}}, t_i^{\mathrm{II}})$
    $\delta^\star_{t_i^\star} \leftarrow \arg\min_{j \in \{\mathrm{I},\mathrm{II}\}} t_i^j$
  end if
  if $\delta^\star_{t_i^\star} = \mathrm{I}$ **then**
    the remaining stored energy at the end of transition block $t_i^\star$ as $\hat{E}_{t_i^\star} \leftarrow 0$
  else
    if $\delta^\star_{t_i^\star} = \mathrm{II}$ **then**
      $\hat{E}_{t_i^\star} \leftarrow E_{\max}$
    end if
  end if
  if $t_i^\star = M$ **then**
    Exit the algorithm
  end if
end for

2.5.2 Throughput Maximization over Fading Channels with Noncausal CSIT and ESIT

For fading channels, the throughput maximization problem is given as

$$\max_{\{P_{n,m} \ge 0\}} \sum_{m=1}^{M} \sum_{n=1}^{N} \log_2(1 + h_{n,m} P_{n,m}) \tag{2.100}$$
$$\text{s.t. (2.5) and (2.85).}$$

With noncausal ESIT and CSIT, the optimization problem (2.100) is convex and hence can be solved by the KKT condition [14]. Let $\lambda_{n,m}$ denote the Lagrange multiplier associated with the (n, m)th EH constraint in (2.5), $\theta_{n,m}$ denote the Lagrange multiplier associated with the (n, m)th battery capacity constraint in (2.85), and $\mu_{n,m}$ denote the Lagrange multiplier associated with the power constraint $P_{n,m} \ge 0$. The Lagrangian associated with problem (2.100) is

$$\begin{aligned}
&\mathcal{L}_{2.6}(\{P_{n,m}\}, \{\lambda_{n,m}\}, \{\theta_m\}, \{\mu_{n,m}\}) \\
&= \sum_{m=1}^{M}\sum_{n=1}^{N} \log_2(1 + h_{n,m}P_{n,m}) \\
&- \sum_{m=1}^{M}\sum_{n=1}^{N} \lambda_{n,m}\left(\sum_{j=1}^{m-1}\sum_{i=1}^{N} P_{i,j} + \sum_{i=1}^{n} P_{i,m} - N\sum_{j=1}^{m-1} E_j - nE_m\right) \\
&- \sum_{m=1}^{M}\sum_{n=1}^{N} \theta_{n,m}\left(N\sum_{j=1}^{m-1} E_j + nE_m - \sum_{j=1}^{m-1}\sum_{i=1}^{N} P_{i,j} - \sum_{i=1}^{n} P_{i,m} - E_{\max}\right) \\
&+ \sum_{m=1}^{M}\sum_{n=1}^{N} \mu_{n,m}P_{n,m}.
\end{aligned} \tag{2.101}$$

Then the necessary and sufficient conditions for $\{P^\star_{n,m}\}$, $\{\lambda^\star_{n,m}\}$, $\{\theta^\star_{n,m}\}$, and $\{\mu^\star_{n,m}\}$ to be the primal and dual optimal solutions to problem (2.100) are given by the following KKT conditions:

$$\begin{aligned}
&0 \le P^\star_{n,m}, \\
&0 \ge \sum_{j=1}^{m-1}\sum_{i=1}^{N} P^\star_{i,j} + \sum_{i=1}^{n} P^\star_{i,m} - N\sum_{j=1}^{m-1} E_j - nE_m, \\
&0 \ge N\sum_{j=1}^{m-1} E_j + nE_m - \sum_{j=1}^{m-1}\sum_{i=1}^{N} P^\star_{i,j} - \sum_{i=1}^{n} P^\star_{i,m} - E_{\max}, \\
&0 \le \lambda^\star_{n,m}, \\
&0 \le \theta^\star_{n,m}, \\
&0 \le \mu^\star_{n,m}, \\
&0 = \lambda^\star_{n,m}\left(\sum_{j=1}^{m-1}\sum_{i=1}^{N} P^\star_{i,j} + \sum_{i=1}^{n} P^\star_{i,m} - N\sum_{j=1}^{m-1} E_j - nE_m\right), \\
&0 = \theta^\star_{n,m}\left(N\sum_{j=1}^{m-1} E_j + nE_m - \sum_{j=1}^{m-1}\sum_{i=1}^{N} P^\star_{i,j} - \sum_{i=1}^{n} P^\star_{i,m} - E_{\max}\right), \\
&0 = \mu^\star_{n,m}P^\star_{n,m}, \\
&0 = \frac{\partial \mathcal{L}_{2.6}(\{P^\star_{n,m}\}, \{\lambda^\star_{n,m}\}, \{\theta^\star_{n,m}\}, \{\mu^\star_{n,m}\})}{\partial P^\star_{n,m}},
\end{aligned}$$

for all $m = 1, \ldots, M$ and $n = 1, \ldots, N$. Therefore, we obtain the optimal power allocation as

$$P^\star_{n,m} = \left[\nu^\star_{n,m} - \frac{1}{h_{n,m}}\right]^+, \quad m = 1, \ldots, M, \tag{2.102}$$

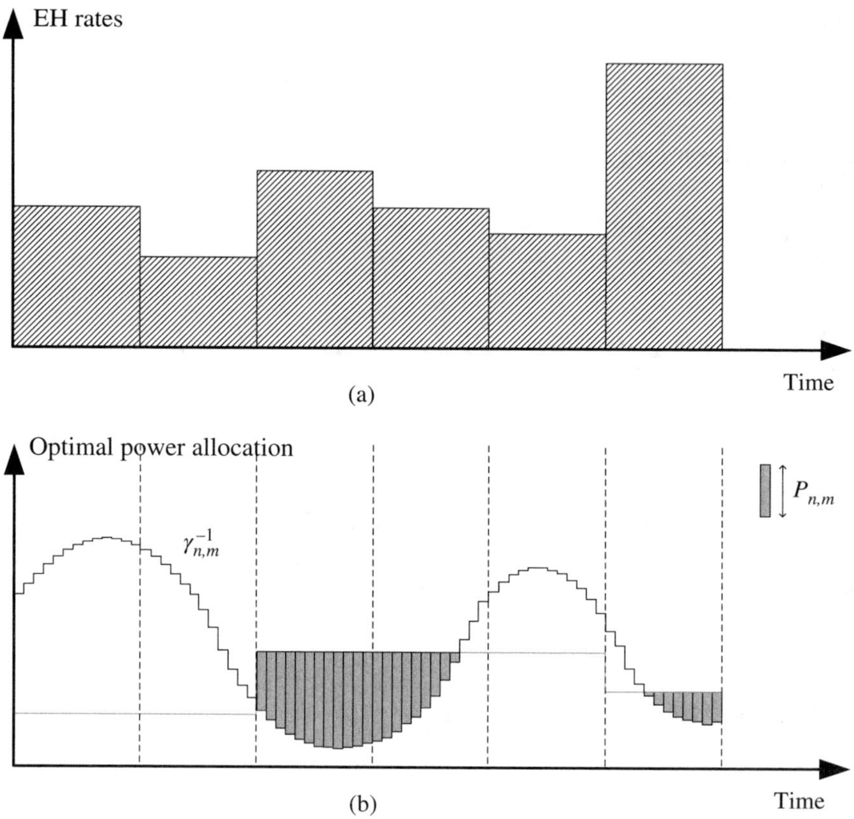

Figure 2.14 Modified staircase water-filling power allocation. (a) The EH rates over time. (b) The optimal modified staircase water filling over time, where there are two transition blocks and hence three different WLs.

where $v_{n,m}^{\star} \triangleq \left(\ln 2\left(\sum_{j=m+1}^{M}\sum_{i=1}^{N}(\lambda_{i,j}^{\star}-\theta_{i,j}^{\star})+\sum_{i=n}^{N}(\lambda_{i,m}^{\star}-\theta_{i,m}^{\star})\right)\right)^{-1} \geq 0$ is the WL for the (n,m)th block and the $\lambda_{i,j}^{\star}$'s and $\theta_{i,j}^{\star}$ satisfy the above KKT conditions. It is worth noting that different from the case with infinite batter capacity that the WL is non-decreasing and piecewise constant (staircase) over time, the WL in (2.102) may decrease over certain blocks as shown in Figure 2.14. We refer to the optimal solution as a *modified staircase water filling*.

Similar to the case with Gaussian channel case, we define two types of *transition blocks* here. An EH block $t \in \{(1,1),\dots,(N,M)\}$ is a *Type I* transition block if the WL increases after EH block t, i.e. $v_t < v_{t+1}$, while an EH block t is a *Type II* transition block if the WL decreases after EH block t, i.e. $v_t > v_{t+1}$. We define the last EH block (N,M) also as a Type I transition block by defining $v_{N+1,M}$ to be infinity; hence there is at least one Type I transition block. We collect all transition blocks as the set $S = \{t_1,\dots,t_{|S|}\}$, where $t_i < t_j$ for $i<j$ and $t_{|S|} = NM$, and denote the corresponding type of the transition blocks as $\Delta = \{\delta_{t_1},\dots,\delta_{t_{|S|}}\}$, where $\delta_{t_i} \in \{\mathrm{I},\mathrm{II}\}$ and $\delta_{t_{|S|}} = \mathrm{I}$.

Then, we have the following structural properties for the optimal WL $\{v_m^{\star}\}$.

Proposition 2.23 *The optimal WL to problem (2.100) has the following properties:*

- *If EH block t is a Type I transition block (i.e. $v_t^\star < v_{t+1}^\star$), then the battery storage is empty after the tth block, i.e. the accumulative energy consumed up to communication block t equals that harvested up to then, i.e. (2.90) holds with equality for t.*
- *If EH block t is a Type II transition block (i.e. $v_t^\star > v_{t+1}^\star$), then the battery storage is full after the tth block, i.e. the accumulative energy consumed up to communication block t minus that harvested up to then equals the battery capacity $E_{\max}$, i.e. (2.91) holds with equality for t.*

Proof: This proposition can be similarly proved as Proposition 2.22, for which the details are omitted for brevity. □

Based on Proposition 2.23, it is evident that finding the optimal WL $\{v_m^\star\}$ to solve problem (2.100) is equivalent to searching for the optimal set of transition blocks $S^\star = \{t_1^\star, \dots, t_{|S^\star|}^\star\}$, as well as the optimal type of each transition block $\Delta^\star = \{\delta_1^\star, \dots, \delta_{|S^\star|}^\star\}$. Indeed, by combining the ideas of Algorithm 2.3 for the staircase water filling and Algorithm 2.7 for the modified staircase power allocation, we can obtain a new modified staircase water-filling algorithm to solve problem (2.100) optimally. While the essential ideas are similar, we refer interested readers to [4, 5] for details.

2.5.3 Other Cases

In the previous two subsections, we have studied the throughput utility maximization for the point-to-point EH channel, by taking into account the practical battery capacity constraints and considering the case with noncausal ESIT and CSIT (Case 1). As for other CSIT and ESIT cases (e.g. Cases 2–4), the effect of the battery capacity constraints on the throughput maximization still remains as open problems. Intuitively, for the causal ESIT and CSIT case (Case 2), the dynamic programming technique (see Sections 2.2.2 and 2.3.2) is applicable to find the optimal online power allocation for throughput maximization. Nevertheless, for Cases 3 and 4, even the throughput maximization without battery capacity constraints has not been solved in the literature yet (see Section 2.3.3). For the two cases, the incorporation of these new constraints makes the problem more challenging and difficult to solve.

On the other hand, it is interesting to investigate the effect of battery capacity constraints on the outage probability minimization problem under various CSIT and ESIT cases. To draw some insights, we take the case with no CSIT and with noncausal ESIT as an example, in which the outage probability function can either be convex (with Type A) or follow a concave–convex shape (with Type B), as shown in Section 2.4.1.1. If the outage probability function is convex with Type A, then the outage probability minimization problem corresponds to the minimization of the sum of a set of convex functions, which has similar properties as the throughput maximization problem in Section 2.5.1. In this case, the optimal modified staircase power allocation for throughput maximization could be also applicable for outage probability minimization. By contrast, if outage probability function follows a concave–convex shape with Type B, then the outage probability minimization has different properties from the throughput maximization. In this case, we intuitively discuss the optimal power allocation when the EH rates are generally low (such that $\tilde{P}_i < P_a$ in (2.64)) and the EH rates are generally high (such that $\tilde{P}_i \geq P_a$ in (2.64)), respectively. When $\tilde{P}_i < P_a$, the transmitter may prefer sending information once it accumulates enough energy (see the power allocation given in (2.71)). Accordingly, energy overflow of the battery may rarely occur, as the capacity of the battery is normally larger than the value P_a in practice. In this case, the battery capacity constraints are generally inactive, and therefore, the algorithm without battery capacity constraints (i.e. Algorithm 2.5) may be still applicable here. By contrast, in the high EH rate scenario when $\tilde{P}_i \geq P_a$, the transmitter is likely to adopt relatively high transmit power under which the outage probability function is convex. In this case, the outage probability minimization problem can be approximated as the minimization of a set of convex functions, for which the power allocation algorithm in Section 2.5.1 could be applicable. For general scenarios with moderate EH rates, how to optimally allocate transmit power for outage probability minimization still requires further investigation.

2.6 Imperfect Circuits

Previous sections consider the transmit power consumption as the sole energy budget at the EH wireless transmitter. In practice, however, in addition to the transmit power consumption, a practical wireless transmitter also consumes nonideal circuit power, which can be comparable with the transmit power, especially for short-distance communications. In this section, we will show that the nonideal circuit power consumption has a significant impact on the optimal power allocation design at the EH transmitter.

2.6.1 Practical Power Consumption for Wireless Transmitters

In this subsection, we present a practical power consumption model for a wireless transmitter and discuss its impact on the design of wireless communication systems powered by fixed energy supplies to provide insights on the design of EH wireless communication systems.

For a practical wireless transmitter, besides the transmit power P, it also consumes *nonideal* circuit power when $P > 0$, which accounts for the power consumptions at, e.g. the AC/DC converter and the analog radio frequency (RF) amplifier, and thus amounts to a significant part of the total power consumption at the transmitter. By contrast, when there is no data transmission, i.e. $P = 0$, the transmitter can turn into a *micro-sleep* mode [22] by switching off the power amplifier to reduce the circuit power consumption. For the ease of description, the transmitter status with $P > 0$ and $P = 0$ is referred to as the *on* and *off* modes, respectively. Denote the nonideal circuit power during an "on" mode as $\alpha \geq 0$ in Watt, the efficiency of the RF chain as $0 < \eta \leq 1$, and the power consumed during an "off" mode as $\zeta \geq 0$ in Watt. A practical power consumption model for the transmitter is given by [23]

$$P_{\text{total}} = \begin{cases} \frac{P}{\eta} + \alpha & P > 0 \\ \zeta & P = 0, \end{cases} \tag{2.103}$$

where P_{total} denotes the total power consumed at the transmitter. In practice, ζ is generally much smaller as compared with α and thus can be ignored for simplicity [24–26]. In this case, we assume $\zeta = 0$. Therefore, from (2.103) it follows that without loss of generality, we can further assume $\eta = 1$ since η is only a scaling constant.

It is worth noting that the nonideal circuit power at the transmitter has significant impact on the design principle of the wireless communication. We take the green communications with fixed power supply as an example, in which it is fundamental to characterize the tradeoff between the *spectrum efficiency* (SE) (i.e. the number of transmitted bps Hz^{-1}) and the *energy efficiency* (EE) (i.e. the number of transmitted bits delivered per joule). We show the effect of the nonideal circuit power on the important SE–EE tradeoff with fixed power supply in the following.

Considering a Gaussian channel with unit bandwidth and noise power spectral density, employing Shannon's capacity formula, the SE is expressed as $\xi_{SE} = \log_2(1 + P)$, with P denoting the transmit power. When the transmit power P is the sole power budget, the EE is given as $\xi_{EE} = \log_2(1 + P)/P$. It thus follows that the optimal EE–SE tradeoff is characterized by $\xi_{EE} = \frac{\xi_{SE}}{(2^{\xi_{SE}}-1)N_0}$, where ξ_{EE} is a monotonically decreasing function of ξ_{SE}, as shown in the left sub-figure of Figure 2.15. In this case, any SE increment will inevitably result in a decrement in EE. Next, consider the case with nonideal circuit power consumption $\alpha > 0$. In this case, the EE can be re-expressed as $\xi_{EE} = \log_2(1 + P)/(P + \alpha)$ for $P > 0$, and the EE–SE tradeoff is shown in the right sub-figure of Figure 2.15 for a given $\alpha > 0$, from which it is seen that the nonideal circuit power completely changes the behavior of the EE–SE tradeoff as compared with the case of $\alpha = 0$.

Motivated by the result that the nonideal circuit power modifies the EE–SE tradeoff considerably in the conventional case with fixed power supply as shown in Figure 2.15, we expect that it also plays an important role in the EE–SE tradeoff characterization for the new EH wireless communications with random power supply at the EH transmitter. In this section, we particularly focus on the throughput maximization problem for a point-to-point Gaussian channel with an EH transmitter. We investigate the effects of the nonideal circuit power with $\alpha > 0$ on the throughput optimal power allocation as well as the resulting EE–SE tradeoff.

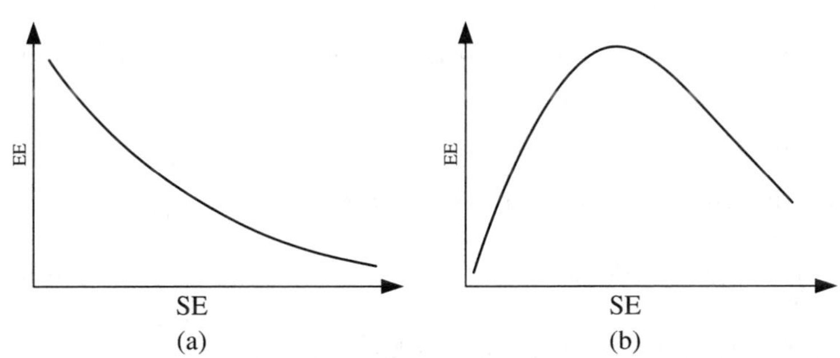

Figure 2.15 Tradeoff between EE and SE for the ideal circuit power case of $\alpha = 0$ (a) and the nonideal circuit power case of $\alpha > 0$ (b).

In the following, we will first consider the *offline* optimization under the assumption with noncausal ESIT. We show that the optimal power allocation to maximize the average throughput under this setup is a *non-convex* optimization problem, due to the nonideal circuit power. However, we derive an efficiently implementable optimal solution for this problem, which is shown to correspond to a *two-phase* transmission in general: the first phase with an *EE-maximizing* on–off power allocation and the second phase with an *SE-maximizing* staircase power allocation that is non-decreasing over time. Thus, we reveal an interesting result that both the EE and SE optimizations are unified in an EH-powered wireless transmission.

Next, inspired by the offline optimization, we will propose a heuristic *online* algorithm under the other case with causal ESIT. It is shown by simulation that the proposed online algorithm has a small performance gap from the upper bound achieved by the optimal offline solution and also outperforms other heuristically designed online algorithms.

2.6.2 The Case with Noncausal ESIT

In the Gaussian channel case, we have the number of communication blocks as $N = 1$, and the wireless channels remain constant over the whole time horizon consisting of M EH blocks, with the channel power gain being expressed as $h_m = h, m = 1, \dots, M$.

With the practical on–off nonideal circuit power taken into account, it is not efficient for the wireless transmitter to work in the on mode with a constant transmit power over each EH block (as assumed in the case without circuit power before). For example, if the EH rate is lower than the circuit power, the harvested energy is not able to support any positive transmit power, thus making the wireless communication infeasible. To tackle this issue, the EH transmitter is allowed to adaptively adjust the transmit power within each block, such that it can stay in the off mode for some time to save energy for communication in the remaining time. Without loss of generality, over the time horizon $(0, T]$ with duration T, we denote the transmit power at any time instant $t \in (0, T]$ as $P(t)$. Accordingly, by considering the on–off transmitter power model in (2.103) with $\zeta = 0$ and $\eta = 1$, we have the total power consumption of the transmitter at time instant t as

$$P_{\text{total}}(t) = \begin{cases} P(t) + \alpha & P(t) > 0 \\ 0 & P(t) = 0. \end{cases} \tag{2.104}$$

Similarly, we denote the achievable rate as the instantaneous mutual information in (2.3) (in bps Hz^{-1}) at time instant $t \in (0, T]$, denoted by

$$R(P(t)) = \log_2(1 + hP(t)), \tag{2.105}$$

and the throughput over the whole time horizon is expressed as

$$\int_0^T R(P(t))\text{d}t = \int_0^T \log_2(1 + hP(t))\text{d}t. \tag{2.106}$$

Furthermore, we define the EH rate at time instant t as $\hat{E}(t)$ with

$$\hat{E}(t) = E_m, t \in \left(\frac{(m-1)T}{M}, mT\right], \quad m = 1, \dots, M. \tag{2.107}$$

Since the accumulatively consumed energy up to any time at the EH transmitter cannot exceed the energy accumulatively harvested, the EH constraint on the total consumed power is given by

$$\int_0^{\hat{t}} P_{\text{total}}(t)\text{d}t \leq \int_0^{\hat{t}} \hat{E}(t)\text{d}t, \quad \forall \hat{t} \in (0, T]. \tag{2.108}$$

Thus, the throughput maximization problem over the time horizon with duration T can be formulated as

$$\max_{\{P(t)\geq 0\}} \int_0^{T} R(P(t))\text{d}t \tag{2.109}$$

$$\text{s.t.} \int_0^{\hat{t}} P_{\text{total}}(t)\text{d}t \leq \int_0^{\hat{t}} \hat{E}(t)\text{d}t, \quad \forall \hat{t} \in (0, T]. \tag{2.110}$$

For the offline optimization of problem (2.109) with noncausal ESIT, it is easy to see that the objective function is concave; however, the constraint is non-convex in general since $P_{\text{total}}(t)$ in (2.104) is a concave function of $P(t)$ if $\alpha > 0$. As a result, the problem is in general non-convex and thus cannot be solved by standard convex optimization techniques. Therefore, new approaches are needed to solve this problem by exploiting the specific problem structure.

It is worth noting that for the offline optimization, problem (2.109) can be shown to be convex if $\alpha = 0$. In this case, problem (2.109) can be reduced to problem (2.7), in which the throughput optimal power allocation is shown to follow a non-decreasing piecewise-constant (staircase) function over time. As will be shown later, the nonideal circuit power with $\alpha > 0$ will change the optimal power allocation for this problem considerably.

2.6.2.1 Problem Reformulation

To facilitate solving problem (2.109), we reformulate problem (2.109) based on the following lemma.

Lemma 2.3 *During any mth EH block* $\left(\frac{(m-1)T}{M}, mT\right]$, $m = 1, \ldots, M$, *the optimal solution to problem (2.109) is given as follows by separating this block into two portions:*

- *In the first portion of the mth EH block with duration* l_m^{off}, *the transmitter remains silent with* $P(t) = 0, \forall t \in \left(\frac{(m-1)T}{M}, \frac{(m-1)T}{M} + l_m^{\text{off}}\right]$. *This portion is termed as an "off" period, where* $0 \leq l_m^{\text{off}} \leq T/M$.
- *For the second portion of the mth EH block with duration* $l_i^{\text{on}} = T/M - l_m^{\text{off}}$, *the transmitter adopts a constant transmit power with* $P(t) = P_m > 0, \forall t \in \left(\frac{(m-1)T}{M} + l_m^{\text{off}}, mT\right]$. *This portion is termed as an "off" period, and the variable* P_m *is a parameter to be optimized.*

Note that Lemma 2.3 can be verified based on the concavity of the throughput function $R(P(t))$ together with the "on–off" property of the total power consumption model, and the details can be found in [9]. According to Lemma 2.3 and by denoting the constant transmit power $P_m > 0$ for the "on" period with length $0 \leq l_m^{\text{on}} \leq T/M$ in the mth EH block, problem (2.109) is reformulated as

$$\max_{\{P_m, l_m^{\text{on}}\}} \sum_{m=1}^{N} l_m^{\text{on}} R(P_m)$$
$$\text{s.t. } P_m > 0, \quad m = 1, \dots, M$$
$$0 \leq l_m^{\text{on}} \leq T/M, \quad m = 1, \dots, M$$
$$\sum_{j=1}^{m} (P_j + \alpha) l_j^{\text{on}} / T \leq \sum_{j=1}^{m} E_j, \quad m = 1, \dots, M. \tag{2.111}$$

However, the above problem is still non-convex due to the coupling between P_m's and l_m^{on}'s. In the following, we first solve this problem for the special case of $M = 1$ and then generalize the solution to the case with $M \geq 1$.

2.6.2.2 Single-Block Case with $M = 1$

In the single-block case with $M = 1$, problem (2.109) is reduced to

$$\max_{l_1^{\text{on}}, P_1} l_1^{\text{on}} R(P_1)$$
$$\text{s.t. } P_1 > 0$$
$$0 \leq l_1^{\text{on}} \leq T$$
$$l_1^{\text{on}} (P_1 + \alpha) \leq E_1 T. \tag{2.112}$$

The solution of the above problem is given in the following proposition.

Proposition 2.24 *The optimal solution of $P_1^\star$ and $l_1^{\text{on}\star}$ to problem (2.112) is expressed as*

$$P_1^\star = \max(P_{\text{ee}}, E_1 - \alpha), \tag{2.113}$$

$$l_1^{\text{on}\star} = \frac{E_1 T}{P_1^\star + \alpha}, \tag{2.114}$$

where P_{ee} is given by

$$P_{\text{ee}} = \arg\max_{P_1 > 0} \frac{R(P_1)}{P_1 + \alpha}. \tag{2.115}$$

Proof: First, note that the third inequality constraint of problem (2.112) must be met with equality by the optimal solution, since otherwise the throughput can be further improved by increasing P_1. Thus, by substituting $l_1^{\text{on}} = \frac{E_1 T}{P_1 + \alpha}$ into the objective function as well as the constraint $l_1^{\text{on}} \leq T$, the problem becomes equivalent to finding

$$P_1^\star = \arg \max_{P_1 > 0, P_1 \geq E_1 - \alpha} \frac{E_1 T}{P_1 + \alpha} R(P_1) = \arg \max_{P_1 > 0, P_1 \geq E_1 - \alpha} \frac{R(P_1)}{P_1 + \alpha}. \tag{2.116}$$

Consider first the following problem with the relaxed power constraint:

$$\max_{P_1 > 0} \frac{R(P_1)}{P_1 + \alpha}. \tag{2.117}$$

This problem has been studied in [26], where the globally optimal solution is known as the EE-maximizing power allocation, denoted by P_{ee}. It was also shown in [26] that given $\alpha > 0$, $\frac{R(P_1)}{P_1+\alpha}$ is monotonically increasing with P_1 if $0 \leq P_1 < P_{ee}$ and monotonically decreasing with P_1 if $P_1 > P_{ee}$. Thus, the optimal solution to problem (2.116) is obtained as $P_1^\star$ as in (2.113). Accordingly, the optimal on-period is given as $l_1^{on\star}$ in (2.114). Thus, this proposition is proved. □

It is worth noting that P_{ee} given in (2.115) is the optimal power allocation that maximizes the EE of the Gaussian channel under the nonideal circuit power model as shown in [26]. From Proposition 2.24, it follows that if $P_{ee} > E_1 - \alpha$, we have $P_1^\star = P_{ee}$, $l_1^{on\star} < T$, and $l_1^{off\star} = T - l_1^{on\star} > 0$, which corresponds to an on–off transmission. However, if $P_{ee} \leq \frac{E_1}{T} - \alpha$, we have $P_1^\star = \frac{E_1}{T} - \alpha$, $l_1^{on\star} = T$, and $l_1^{off\star} = 0$, which corresponds to a continuous transmission. We will see in the next subsection that the EE-maximizing power allocation P_{ee} plays an important role in the general case with $M \geq 1$. Also note that the RHS of (2.115) is a quasi-concave function of P_1 since it is concave over linear [14]; thus, P_{ee} can be efficiently obtained by a simple bisection search [14].

2.6.2.3 General Multi-Block Case with $M \geq 1$

Inspired by the solution for the single-block case, we derive the optimal solution to problem (2.109) in the general case with $M \geq 1$, as given by the following proposition.

Proposition 2.25 *The optimal solution to problem (2.109), denoted by* $[P_1^\star, \dots, P_M^\star]$ *and* $[l_1^{on\star}, \dots, l_M^{on\star}]$*, is obtained as follows. Denoting*

$$\begin{aligned} m_{ee,0} &= 0, \\ m_{ee,j} &= \min\left\{ m \middle| \frac{\sum_{k=m_{ee,j-1}+1}^{m} E_k}{m - m_{ee,j-1}} - \alpha \leq P_{ee}, m = m_{ee,j-1}+1, \dots, M \right\}, \quad j \geq 1, \\ J &= \arg\max_{j\geq 0} m_{ee,j}, \\ m_{ee} &= m_{ee,J}, \end{aligned} \tag{2.118}$$

then the optimal power allocation for EH blocks $1, \dots, m_{ee}$ *is given by*

$$P_m^\star = P_{ee}, \quad m = 1, \dots, m_{ee}, \tag{2.119}$$

and the optimal on-period of the EH block $l_m^{on\star}$, $m = 1, \dots, m_{ee}$ *is given as*

$$(P_{ee} + \alpha) \sum_{m=1}^{m_{ee}} l_m^{on\star} = \sum_{m=1}^{m_{ee}} E_m T/M. \tag{2.120}$$

$$(P_{ee} + \alpha) \sum_{j=1}^{m} l_j^{on\star} \leq \sum_{j=1}^{m} E_j T/M, \quad m = 1, \dots, m_{ee}. \tag{2.121}$$

Moreover, for EH blocks $m_{ee}+1, \dots, M$*, the optimal solution is given by*

$$l_m^{on\star} = T/M, \quad m = m_{ee}+1, \dots, M, \tag{2.122}$$

$$P_m^\star = \frac{\sum_{j=n_{m-1}+1}^{n_m} E_j}{n_m - n_{m-1}} - \alpha, \quad m = m_{ee}+1, \dots, M, \tag{2.123}$$

where

$$n_{m_{ee}} = m_{ee},$$

$$n_m = \arg \min_{n_{m-1}+1 \leq j \leq M} \left\{ \frac{\sum_{k=n_{m-1}+1}^{j} E_k}{j-n_{m-1}} - \alpha \right\}, \quad m = m_{ee} + 1, \dots, M. \tag{2.124}$$

Note that Proposition 2.25 can be proved by using Proposition 2.24 together with the throughput maximization with zero circuit power in Section 2.2, and the details are shown in [9]. It is interesting to take note that the optimal transmission policy given in Proposition 2.25 has a *two-phase* structure, which is explained as follows in more detail:

- *Blocks* $0 < m \leq m_{ee}$: In the first phase, the optimal transmission policy is an on–off power allocation with a constant power P_{ee} for all their on-blocks. Note that P_{ee} is the EE-maximizing power allocation given in (2.115). Also note that the optimal on-blocks $l_m^{on\star}, m = 1, \dots, m_{ee}$, may not be unique provided that they satisfy the conditions given in (2.120) and (2.121). Without loss of generality, we assume that in each EH block, the EH transmitter chooses to be on at the on-period with power P_{ee} provided that its stored energy is not used up, i.e.

  $$l_1^{on\star} = \min(E_1/(P_{ee} + \alpha), T/M),$$
  $$l_2^{on\star} = \min((E_2 + E_1 - (P_{ee} + \alpha)l_1^{on\star})/(P_{ee} + \alpha), T/M),$$

  and so on.
- *Blocks* $m_{ee} < m \leq M$: In the second phase, a continuous transmission is optimal, i.e. $l_m^{on\star} = T/M$, $m = m_{ee} + 1, \dots, M$. Therefore, problem in (2.111) for $m = m_{ee} + 1, \dots, M$, is reduced to problem (2.7), for which the optimal solution has been shown in Section 2.2 to follow a non-decreasing piecewise-constant (staircase) function. In this case, the optimal solution to problem (2.109) in (2.123) and (2.124) can be understood based on the staircase power allocation in Algorithm 2.1 for the case without circuit power. It is worth noting that the staircase power allocation achieves the maximum SE for an equivalent Gaussian channel subject to a sequence of EH constraints (modified to take into account the circuit power α) for $m = m_{ee} + 1, \dots, M$.

From the above discussion, it is revealed that for the throughput maximization in an EH transmission system subject to the nonideal circuit power, the optimal transmission unifies both the EE and SE maximization policies independently developed in [26] and [4, 6], respectively.

To illustrate the optimal two-phase transmission given in Proposition 2.25, we consider an example Gaussian channel. It is assumed that the time horizon of our interest has a length of $T = 20$ seconds and consists of a total of $M = 20$ EH blocks, with each block of 1 second. Here, the EH rates are shown in Figure 2.16a, where the EH rate E_m's are given as 1/8 W from blocks 1 to 4, 1/4 W for blocks 5 and 6, 1/10 W from blocks 7 to 11, 1/3 W from blocks 12 to 14, 1/4 W from blocks 15 and 16, 3/8 W for blocks 17 and 18, and 1/4 W for blocks 19 and 20. We set the channel power gain (normalized by the noise power) as 100 (20 dB). Then, we compare the optimal allocation of the totally consumed power by Proposition 2.25 versus that obtained by Algorithm 2.1 in the case with ideal circuit power $\alpha = 0$. Here, we apply this algorithm to obtain a

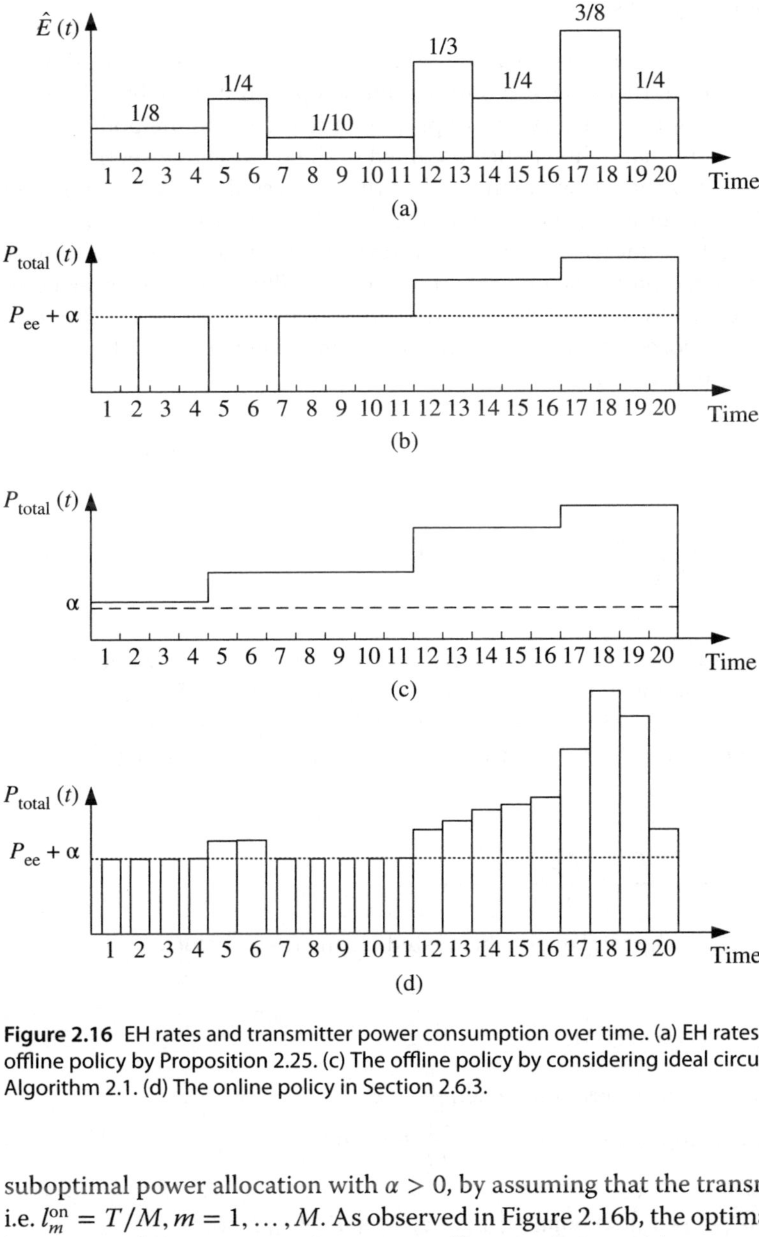

Figure 2.16 EH rates and transmitter power consumption over time. (a) EH rates. (b) The optimal offline policy by Proposition 2.25. (c) The offline policy by considering ideal circuit power $\alpha = 0$ via Algorithm 2.1. (d) The online policy in Section 2.6.3.

suboptimal power allocation with $\alpha > 0$, by assuming that the transmitter is always on, i.e. $l_m^{on} = T/M, m = 1, \dots, M$. As observed in Figure 2.16b, the optimal power allocation has a two-phase structure, i.e. an on–off transmission with transmit power P_{ee} when $m = 1, \dots, 11$ followed by a continuous transmission with non-decreasing staircase power allocation when $m = 12, \dots, 20$, which is in accordance with Proposition 2.25. By contrast, as observed in Figure 2.16c, the suboptimal power allocation by Algorithm 2.1 with the transmitter always on results in a continuous transmission with non-decreasing staircase power allocation during the entire block, i.e. $m = 1, \dots, 20$. In addition, it can be shown that the proposed optimal solution achieves the total throughput 63.14 bits Hz^{-1}, while the suboptimal solution achieves only 55.80 bits Hz^{-1}, over $T = 20$ seconds.

2.6.3 The Case with Causal ESIT

Next, we consider the other case with causal ESIT and present the online optimization for problem (2.109). In this case, the optimal online optimization is based on the dynamic programming. In this subsection, we propose a heuristic online policy based on the structure of the optimal offline policy obtained in the previous subsection.

For the purpose of exposition, we assume that the harvested energy is modeled by a compound Poisson process, where EH rates are i.i.d. distributed with mean $\bar{E}$. It is assumed that $\bar{E}$ is known at the transmitter. We propose an online power allocation algorithm based on the structure of the optimal offline solution revealed in Proposition 2.25. Specifically, considering the start time of each block, from Proposition 2.25, we obtain the closed-form solution for the optimal offline power allocation at $m = 1$ in the following proposition.

Proposition 2.26 *The optimal offline power allocation solution for (2.111) at $m = 1$ is given by*

$$P_1^\star = \max\left(\min_{m=1,\dots,M}\left(\frac{\sum_{k=1}^{m} E_k}{m} - \alpha\right), P_{\text{ee}}\right). \tag{2.125}$$

Note that in (2.125), E_1 is available at the transmitter at $m = 1$, while M and $E_i, i = 2, \dots, M$ are all unknown at the transmitter due to the causal ESI. As a result, we cannot compute $P_1^\star$ in (2.125) at $m = 1$ for the online policy. Nevertheless, we can approximate the expression of $P_1^\star$ based on the statistical knowledge of the energy arrival process, i.e. $\bar{E}$, as follows.

Denote

$$\frac{\sum_{k=1}^{m} E_k}{m} = \frac{E_1 + \sum_{k=2}^{m} E_k}{m} = \frac{E_1}{m} + \frac{\sum_{k=2}^{m} E_k}{m}. \tag{2.126}$$

For any $m \leq M$, $\sum_{k=2}^{m} E_k$ is the total energy harvested during the EH blocks $2, \dots, m$. We thus have

$$\frac{\sum_{k=2}^{m} E_k}{m} \approx \bar{E}, \quad \forall 1 < m \leq M, \tag{2.127}$$

where the approximation becomes exact when $m \to \infty$.

Using (2.127), we can approximate $\frac{\sum_{k=1}^{m} E_k}{m}$ as $\frac{E_1}{m} + \bar{E}$, and

$$\min_{m=1,\dots,M}\left(\frac{\sum_{k=1}^{m} E_k}{m} - \alpha\right) \approx \min_{m=1,\dots,M}\left(E_1 - \alpha, \frac{E_1}{m} + \bar{E} - \alpha\right)$$
$$= \min\left(E_1 - \alpha, \frac{E_1}{M} + \bar{E} - \alpha\right),$$

where the term $E_1 - \alpha$ is exactly known by the transmitter at the beginning of the first block. Then, we obtain

$$P_1^\star \approx \max\left(\min\left(\frac{E_1}{M} + \bar{E} - \alpha, E_1 - \alpha\right), P_{\text{ee}}\right). \tag{2.128}$$

Since E_1 and $\bar{E}$ are both known at the transmitter at $m = 1$, (2.128) can be computed online. In this case, the corresponding on-period is expressed as

$$l_1^{\text{on}\star} = \frac{T \min(E_1/P_1^\star, 1)}{M}, \tag{2.129}$$

and the remaining energy level at the storage device at the end of block $m = 1$ is given as $\tilde{E}_2 = E_1 T/M - (P_1^\star + \alpha)l_1^{\text{on}\star}$. We define $\tilde{E}_1 = 0$ for notational convenience.

For any $m = 2, \ldots, M$, we can view the online optimization at the EH block m as a throughput maximization problem over the remaining $M - m + 1$ EH blocks, with the initial energy level in the storage device being $\tilde{E}_m$. In this case, we can distribute the initial energy amount $\tilde{E}_m$ in the first block m and consider the EH rate at this block m as $\frac{M\tilde{E}_m}{T(M-m+1)} + E_m$. Therefore, by replacing M and E_1 in (2.128) with $M - m + 1$ and $\frac{M\tilde{E}_m}{T(M-m+1)} + E_m$, respectively, we obtain the following online transmit power allocation policy:

$$\begin{aligned} P_m^{\text{online}} &= \max\left(\min\left(\frac{M\tilde{E}_{m-1}}{T(M-m+1)} + \frac{E_m}{M-m+1} + \bar{E} - \alpha, \right.\right. \\ &\qquad \left.\left. \frac{M\tilde{E}_{m-1}}{T} + E_m - \alpha\right), P_{\text{ee}}\right), \\ l_m^{\text{online}} &= \min\left(\frac{\frac{M\tilde{E}_m}{M-m+1} + TE_m}{P_m^{\text{online}} + \alpha}, \frac{T}{M}\right), \end{aligned} \tag{2.130}$$

for any $m = 1, \ldots, M$.

The online policy in (2.130) provides some useful insights. Note that $\frac{M\tilde{E}_m}{T(M-m+1)} + \frac{E_m}{M-m+1} + \bar{E} - \alpha$ can be viewed as the "expected" available transmit power for the remaining EH blocks in the whole horizon, while $\frac{M\tilde{E}_{m-1}}{T} + E_m - \alpha$ is the maximum constant power that can be supported by the transmitter over block m. Thus, if the minimum of the two values is less than the EE-maximizing power allocation P_{ee}, the transmitter should transmit with P_{ee} to save energy; however, if the inequality is reversed, the transmitter should transmit more power to maximize the SE. Moreover, as compared with the optimal offline power allocation for the single-block case given in Proposition 2.24, we see that the online policy (2.130) bears a similar structure, by noting that $E_1 - \alpha$ in (2.113) for the single-block case is also the available transmit power for the remaining time in each block.

To illustrate the proposed online power allocation policy in (2.130), we consider the same channel setup and harvested energy process for the offline case in the example for Figure 2.16. In Figure 2.16d, we show the total transmitter power consumption by the proposed online policy assuming that the exact average harvested power $\bar{E} = (\sum_{i=2}^{M} E_i)/T = 217.10$ mW is known at the transmitter. It is observed that the online power allocation is no more piecewise constant like the optimal offline power allocation in Figure 2.16b. Nevertheless, it is also observed that these two policies result in very similar power allocation patterns, i.e. starting with an on–off power allocation followed by a non-decreasing (in the sense of average power profile for the online policy case) power allocation. This suggests that the proposed online policy captures the essential features of the optimal offline policy. Moreover, it can be shown that the proposed online policy achieves the total throughput 62.57 bits Hz^{-1} over $T = 20$ seconds, which is only 1.53 bits Hz^{-1} from 63.14 bits Hz^{-1} of the optimal offline policy. In addition, it can be verified that the total throughput obtained by the proposed online policy is very robust to the assumed average harvested power value $\bar{E}$. For example, by setting $\bar{E}$ to be 150 or 200 mW, the proposed online policy obtains the throughput 61.78 and 62.54 bits Hz^{-1}, respectively, which result in very small loss in each case.

2.7 Power Allocation with EH Receiver

So far, we have focused on the case with an EH transmitter by assuming the receiver is supplied by a fixed energy source such as conventional battery. In practice, wireless transmitters and receivers are equally likely to be powered by EH, but only limited works (see, e.g. [27–29]) have investigated the case with EH receiver. This may be due to the fact that there do not exist generic power consumption models for wireless receivers, as the power consumption at each receiver depends on various issues such as the circuit technologies employed [29]. In certain wireless communication systems (e.g. short-range communications), the power consumed at the transmitter side may be negligible as compared with that at the receiver side for decoding. In this case, it becomes important to study the power allocation policies for the EH receiver, as will be discussed in this section.

2.7.1 Power Consumption Model for a Wireless Receiver

To analyze the transmission strategies at an EH receiver, we first need to understand the power consumption model. For a typical wireless receiver, the information reception generally consists of two stages including the analog processing for sampling and digital processing for decoding [29, 30]. In general, the analog sampling stage corresponds to the process of transforming the received RF signal into the digital baseband signal and includes all the RF front-end processing. On the other hand, the digital decoding stage corresponds to the process of converting the baseband digital signal into information that includes demodulation, channel decoding, etc. In practice, the sampling power consumption can dominate the decoding power consumption in the case without sophisticated digital signal processing implemented (e.g. uncoded communications) [30], while the sampling power consumption can also be comparable as the decoding power consumption in the case with complicated digital signal processing such as iterative decoding performed [29]. In the following, we provide detailed discussions about the analog sampling and digital decoding, respectively, and then present an abstract receiver power consumption model.

First, we consider the analog sampling stage of a wireless receiver as shown in Figure 2.17. At the receiver side, the RF signal is first filtered and amplified by the low noise amplifier (LNA), then cleaned by the antialiasing filter and downconverted by the mixer, then filtered again before going through the intermediate frequency amplifier (IFA) whose gain is adjustable, and finally converted back to a digital signal via the analog-to-digital converter (ADC). Note that although this model is based on a generic low-IF transceiver structure, the framework can be easily modified to analyze

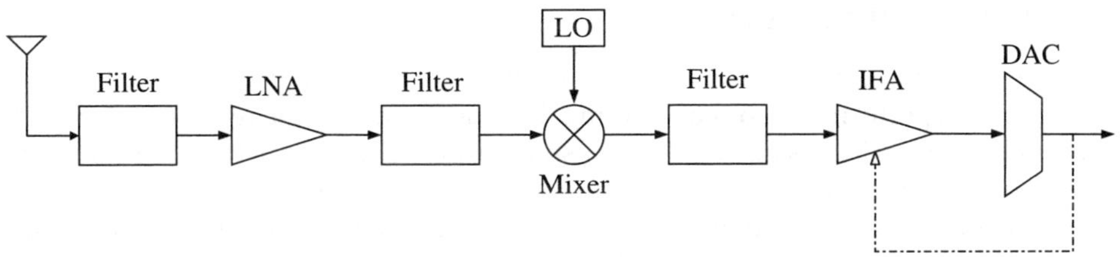

Figure 2.17 Circuit blocks (for analog sampling) of a wireless receiver. Source: Cui et al. 2005 [30]. Reproduced with permission of IEEE.

other architectures as well. In this case, the receiver power consumption in analog sampling consists of the mixer power consumption P_{mix}, the frequency synthesizer power consumption P_{syn}, the LNA power consumption P_{LNA}, the active filter power consumption P_{fil}, the IFA power consumption P_{IFA}, and the ADC power consumption P_{ADC}. In particular, based on an estimation model in [31], the power consumption of a Nyquist-rate ADC is given as

$$P_{\text{ADC}} = \frac{3V_{\text{dd}}^2 L_{\text{min}}(2B + f_{\text{cor}})}{10^{-0.1525\text{ENOB}+4.838}}, \tag{2.131}$$

where B is the system bandwidth, V_{dd} is the power supply, L_{min} is the minimum channel length for the given complementary metal-oxide-semiconductor (CMOS) technology, f_{cor} is the corner frequency of the noise, and ENOB is the effective number of bits of the ADC. Note that when the communication is employed under constant bandwidth B and the receiver uses constant parameters $L_{\text{min}}, f_{\text{cor}}$, and ENOB, the analog sampling power consumption can be viewed as a constant value. In addition, the power consumption values of the mixers P_{mix}, the frequency synthesizers P_{syn}, the filters P_{fil}, and the LNA P_{LNA} are approximated as constants [30]. By combining the power consumption of these different components, we denote the analog sampling power consumption as $P_{\text{sam}} = P_{\text{ADC}} + P_{\text{mix}} + P_{\text{syn}} + P_{\text{fil}} + P_{\text{LNA}}$.

Next, as for the decoding stage, the corresponding power consumption, denoted by P_{dec}, is related to the code rate employed at the transmitter and the associated channel capacity, as well as the specific channel coding methods employed. Specifically, consider that the transmitter sends at a code rate R and the channel capacity from the transmitter to the receiver is given as C. Then the decoding power consumption can be described through the capacity gap, given by $\delta = 1 - R/C$. It is shown in [27, 28] that the decoding power consumption grows exponentially with $1/\delta$ for Forney's concatenated codes, scales like $(1/\delta)\log(1/\delta)$ for low-density parity-check (LDPC) and turbo codes, and grows polynomially in $1/\delta$ for polar codes. Moreover, the decoding power consumption can be viewed as an increasing function of the code rate R that diverges as R approaches the capacity C, and it can also be viewed as a decreasing function of the capacity C for a fixed code rate R.

By combining the analog sampling power consumption and the digital decoding power consumption, an abstract model for the wireless receiver is obtained.

2.7.2 The Case with Only EH Receiver

In this subsection, we discuss the receiver power allocation to improve the end-to-end communication performance by considering the case when only the receiver is powered by EH but the transmitter is powered by constant power supply. To better reveal insights on the receiver power allocation subject to the EH constraint, we focus on a simplified scenario with Gaussian channels and consider that the transmitter uses a constant transmit power that leads to a constant channel capacity C over different EH blocks. Furthermore, we consider that the sampling power consumption P_{sam} is a constant, while the decoding power consumption P_{dec} is a variable that depends on the code rate R, denoted as a function $P_{\text{dec}}(R)$. In this case, in order for the EH receiver to adaptively adjust the decoding power consumption, it needs the coordination of the transmitter to change the corresponding code rate R at that block.

Similar as in previous sections, we consider a finite time horizon with M blocks, in which the code rate employed at the transmitter is denoted as R_m for the mth block and the correspondingly required decoding power is given by $P_{\text{dec}}(R_m)$. Our objective is to maximize the end-to-end throughput $\sum_{m=1}^{M} R_i$, and the EH receiver is subject to the EH constraint given as

$$\sum_{j=1}^{m}(P_{\text{dec}}(R_i) + P_{\text{sam}}) \leq \sum_{j=1}^{m} E_j, \quad m = 1, \dots, M. \tag{2.132}$$

In this case, the throughput maximization problem for an EH receiver is expressed as

$$\max_{\{R_m \geq 0\}} \sum_{m=1}^{M} R_m \tag{2.133}$$

$$\text{s.t.} \sum_{j=1}^{m}(P_{\text{dec}}(R_j) + P_{\text{sam}}) \leq \sum_{j=1}^{m} E_j, \quad m = 1, \dots, M. \tag{2.134}$$

Note that the optimization variable in problem (2.133) is the code rate R_m's over different blocks, while R_m at each block correspondingly determines the decoding power $P_{\text{dec}}(R_m)$ and thus the total receiver power consumption $P_{\text{dec}}(R_m) + P_{\text{sam}}$. Then, we address problem (2.133) by considering the case when the transmitter has the noncausal ESI of the EH receiver, and thus it can use an offline optimization approach to determine the code rate R_m's over time. More specifically, we consider that the decoding power $P_{\text{dec}}(R_m)$ is a monotonically increasing and convex function of the code rate R_m for simplicity, while the throughput maximization under other types of the function $P_{\text{dec}}(R_m)$ (by considering specific coding methods) will be interesting topics to be investigated in the future.

Indeed, the throughput maximization problem via optimizing the code rate allocation (i.e. optimizing R_m's) can be equivalently reformulated as a receive power allocation problem (i.e. optimizing the decoding power) by considering the one-to-one mapping between the decoding power and the code rate (under given channel capacity). To implement this, we denote the decoding power (to be optimized) at the mth block as $P_{\text{dec},m}$ and accordingly denote the corresponding code rate as $R(P_{\text{dec},m})$. Note that the code rate function $R(\cdot)$ is the inverse function of $P_{\text{dec}}(\cdot)$. As $P_{\text{dec}}(\cdot)$ is a monotonically increasing and convex function, the inverse function $R(\cdot)$ is a monotonically increasing and concave function. In this case, the throughput maximization problem (2.133) can be recast as a decoding power allocation problem:

$$\max_{\{P_{\text{dec},m}\geq 0\}} \sum_{m=1}^{M} R(P_{\text{dec},m}) \tag{2.135}$$

$$\text{s.t.} \sum_{j=1}^{m} (P_{\text{dec},j} + P_{\text{sam}}) \leq \sum_{j=1}^{m} E_j, \quad m = 1, \ldots, M. \tag{2.136}$$

Before proceeding to solve problem (2.135), we need to first check its feasibility by examining whether the harvested energy can always supply the analog sampling, i.e. whether the following set of inequalities is satisfied or not. If they cannot be satisfied, then the EH receiver has to be turned off at certain time instants:

$$mP_{\text{sam}} \leq \sum_{j=1}^{m} E_j, \quad m = 1, \ldots, M. \tag{2.137}$$

We focus on the scenario when the EH constraint is feasible. In this case, the EH constraint in (2.134) can be recast as

$$\sum_{j=1}^{m} P_{\text{dec},j} \leq \sum_{j=1}^{m} \tilde{E}_j, \quad m = 1, \ldots, M, \tag{2.138}$$

where $\tilde{E}_m$ denotes a modified EH rate at block m that corresponds to the original EH rate E_m offset by the sampling power P_{sam}. Then problem (2.135) corresponds to maximizing the sum utility (with the utility function being concave), subject to a set of EH constraint in (2.138). Evidently, this problem has a similar form as the throughput maximization problem in (2.7) for Gaussian channels, and as a result, the staircase power allocation algorithm in Algorithm 2.1 can be applied here to solve this problem efficiently. In this case, the receive power (decoding power plus a constant sampling power) follows a monotonically increasing and piecewise-constant (i.e. a staircase-like) fashion, similarly as that in Figure 2.2.

Furthermore, we can extend problem (2.133) with constant sampling power to a new scenario when the EH receiver can adaptively switch itself into the sleep mode to save power. In particular, at each block m, if the code rate R_m in problem (2.133) (or equivalently the decoding power consumption $P_{\text{dec},m}$ in problem (2.135)) is zero (i.e. without any data to be decoded), the EH receiver can turn into a sleep mode to switch off the analog circuit for sampling (including the filter, LNA, mixer, IFA, and ADC) in Figure 2.17 for saving power. Similarly, the authors in [28] assume that the EH receiver selectively samples over time. By neglecting the power consumed during such a sleep mode, we have the following on–off power consumption model for the EH receiver:

$$P_{\text{total,r}}(P_{\text{dec}}) = \begin{cases} P_{\text{dec}} + P_{\text{sam}} & \text{if } P_{\text{dec}} > 0 \\ 0 & \text{otherwise.} \end{cases} \tag{2.139}$$

In this case, it is not necessary or optimal for the EH receiver to use a constant decoding power over each block. Instead, at the mth block, the receiver can choose to operate in an on–off fashion, where over the first portion of the block (with length $l_m^{\text{off}} \geq 0$), the receiver switches off the analog circuits to save power, while over the remaining portion of the block (with length $l_m^{\text{on}} \geq 0$), the receiver uses a positive decoding power $P_{\text{dec},m} > 0$ to receive information. Here, we have $l_m^{\text{off}} = T/M - l_m^{\text{on}}$, where T is the length of the whole time horizon. In this case with on–off sampling (or receive circuit) power model, the throughput maximization problem via optimizing the receive power allocation becomes

$$\max_{\{P_{\text{dec},m}\},\ \{l_m^{\text{on}}\}} \sum_{m=1}^{M} l_m^{\text{on}} R(P_{\text{dec},m})$$

$$\text{s.t.} \quad \sum_{j=1}^{m} (P_{\text{dec},j} + P_{\text{sam}}) l_j^{\text{on}} \le \sum_{j=1}^{m} E_j, \quad m = 1, \dots, M,$$

$$P_{\text{dec},m} \ge 0, \quad m = 1, \dots, M,$$

$$0 \le l_j^{\text{on}} \le T/M, \quad m = 1, \dots, M,$$

In the case when $R(\cdot)$ is a concave function, problem (2.140) is similar to the throughput maximization problem with nonideal circuit power. As a result, we can use the similar approach for it to solve this problem. Similarly as that in Figure 2.16, the optimal receive power allocation should follow an on–off principle.

2.7.3 The Case with Both EH Transmitter and EH Receiver

Furthermore, it is interesting to consider the case with both EH transmitter and EH receiver, where the EH constraints are imposed at both the transmitter and the receiver sides, i.e. the accumulatively consumed transmit and receive energy consumption cannot exceed that harvested at the EH transmitter and EH receiver, respectively [32]. In this case, both the EH transmitter and the EH receiver need to adaptively adjust their power allocations so as to maximize the end-to-end throughput performance. In particular, note that the adaptation of the transmit power accordingly changes the channel capacity over each block, and therefore, at the mth block, the decoding power consumption at the EH receiver varies as a function of both the transmit power P_m and the employed code rate R_m. Suppose that the channel capacity over the mth block is $C_m = \log(1 + hP_m)$, then the decoding power $P_{\text{dec},m}$ in the mth EH block is a function of the capacity gap, given by

$$\delta = 1 - \frac{R_m}{C_m} = 1 - \frac{R_m}{\log_2(1 + hP_m)}. \tag{2.140}$$

In this case, we can jointly optimize the transmit power P_m's and the code rate R_m's at the EH transmitter to maximize the end-to-end throughput from the EH transmitter to the EH receiver. When the EH transmitter adjusts the code rate together with the transmit power (hence the channel capacity), the decoding and receive power consumption can be adjusted accordingly. By doing so, an efficient EH communication with both EH transmitter and EH receiver can be achieved. How to solve this problem is an open problem in the literature.

2.8 Summary

This chapter considers a point-to-point EH wireless channel when the transmitter and the receiver are powered by EH. Under this setup, we first maximize various system utilities (such as the end-to-end throughput and the non-outage probability) via adaptive power allocation over time subject to the EH constraints at the transmitter, where the accumulatively consumed energy should not exceed that accumulatively harvested at any time. Under various CSIT and ESIT assumptions, we discuss the respective optimal designs. For example, in the noncausal CSIT and ESIT case, the optimal offline power allocation is shown to follow a non-decreasing and piecewise-constant (or staircase-like) pattern for the Gaussian channel case, and the optimal transmit power corresponds to a staircase water-filling power allocation for the fading channel case. In the causal ESIT and CSIT case, applicable online optimization strategies include dynamic programming-based and heuristically designed power allocation strategies. We also point out several open problems under different ESIT and CSIT considerations. For example, the throughput maximization problems under the case with causal CSIT and noncausal ESIT and the case with no CSIT and noncausal/causal ESIT remain unsolved in general, while the outage probability minimization problems under causal CSIT and noncausal/causal ESIT are also open problems at large.

Furthermore, we extend the power allocation policies to other scenarios by considering other practical constraints such as limited battery capacity and imperfect circuits. We show that in the noncausal CSIT and ESIT case, the staircase power allocation with Gaussian channel and the staircase water-filling power allocation with fading channels can be modified to take these new practical constraints into account. With imperfect circuits, the EH transmitter should employ an on–off power allocation, which turns the transmitter into a sleep mode to save power in order to transmit more energy efficiently in the wake-up mode to tradeoff between the transmit energy consumption versus the circuit energy consumption. These new offline power allocation strategies motivate new online power allocation approaches with only causal CSIT and ESIT.

Finally, we provide discussions on the transmit and receive power allocations in the case with EH transmitter and EH receiver, where in addition to the transmitter, the receiver's power consumption is also subject to an EH constraint. Under such a scenario, it is essential for the transmitter and the receiver to jointly optimize the transmit power and the code rate (hence, control the receive power consumption) to improve the end-to-end utility. Note that the case of EH receiver has not been widely investigated in the literature and there are rich open problems (e.g. under causal CSIT and ESIT) worth being pursued in future work.

References

1 Li, H., Xu, J., Zhang, R., and Cui, S. (2015). A general utility optimization framework for energy harvesting based wireless communications. *IEEE Communications Magazine* 53 (4): 79–85. https://doi.org/10.1109/MCOM.2015.7081079.

2 Gunduz, D., Stamatiou, K., Michelusi, N., and Zorzi, M. (2014). Designing intelligent energy harvesting communication systems. *IEEE Communications Magazine* 52 (1): 210–216. https://doi.org/10.1109/MCOM.2014.6710085.

3 Ulukus, S., Yener, A., Erkip, E. et al. (2015). Energy harvesting wireless communications: a review of recent advances. *IEEE Journal on Selected Areas in Communications* 33 (3): 360–381. https://doi.org/10.1109/JSAC.2015.2391531.

4 Ho, C.K. and Zhang, R. (2012). Optimal energy allocation for wireless communications with energy harvesting constraints. *IEEE Transactions on Signal Processing* 60 (9): 4808–4818. https://doi.org/10.1109/TSP.2012.2199984.
5 Ozel, O., Tutuncuoglu, K., Yang, J. et al. (2011). Transmission with energy harvesting nodes in fading wireless channels: optimal policies. *IEEE Journal on Selected Areas in Communications* 29 (8): 1732–1743. https://doi.org/10.1109/JSAC.2011.110921.
6 Yang, J. and Ulukus, S. (2012). Optimal packet scheduling in an energy harvesting communication system. *IEEE Transactions on Communications* 60 (1): 220–230. https://doi.org/10.1109/TCOMM.2011.112811.100349.
7 Tutuncuoglu, K. and Yener, A. (2012). Optimum transmission policies for battery limited energy harvesting nodes. *IEEE Transactions on Wireless Communications* 11 (3): 1180–1189. https://doi.org/10.1109/TWC.2012.012412.110805.
8 Devillers, B. and Gunduz, D. (2012). A general framework for the optimization of energy harvesting communication systems with battery imperfections. *Journal of Communications and Networks* 14 (2): 130–139. https://doi.org/10.1109/JCN.2012.6253061.
9 Xu, J. and Zhang, R. (2014). Throughput optimal policies for energy harvesting wireless transmitters with non-ideal circuit power. *IEEE Journal on Selected Areas in Communications* 32 (2): 322–332. https://doi.org/10.1109/JSAC.2014.141212.
10 Ozel, O., Shahzad, K., and Ulukus, S. (2014). Optimal energy allocation for energy harvesting transmitters with hybrid energy storage and processing cost. *IEEE Transactions on Signal Processing* 62 (12): 3232–3245. https://doi.org/10.1109/TSP.2014.2321733.
11 Huang, C., Zhang, R., and Cui, S. (2014). Optimal power allocation for outage probability minimization in fading channels with energy harvesting constraints. *IEEE Transactions on Wireless Communications* 13 (2): 1074–1087. https://doi.org/10.1109/TW.2013.121813.130953.
12 Luo, S., Zhang, R., and Lim, T.J. (2013). Optimal save-then-transmit protocol for energy harvesting wireless transmitters. *IEEE Transactions on Wireless Communications* 12 (3): 1196–1207. https://doi.org/10.1109/TWC.2013.012413.120488.
13 Zhao, Y., Chen, B., and Zhang, R. (2015). Optimal power management for remote estimation with an energy harvesting sensor. *IEEE Transactions on Wireless Communications* 14 (11): 6471–6480. https://doi.org/10.1109/TWC.2015.2455501.
14 Boyd, S. and Vandenberghe, L. (2004). *Convex Optimization*. Cambridge: Cambridge University Press.
15 Bertsekas, D.P. (2007). *Dynamic Programming and Optimal Control*. Belmont, MA: Athena Scientific.
16 Cover, T.M. and Thomas, J.A. (2006). *Elements of Information Theory*, 2e. Wiley.
17 Sagias, N.C. and Karagiannidis, G.K. (2005). Gaussian class multivariate Weibull distributions: theory and applications in fading channels. *IEEE Transactions on Information Theory* 51 (10): 3608–3619. https://doi.org/10.1109/TIT.2005.855598.
18 Tse, D. and Viswanath, P. (2005). *Fundamentals of Wireless Communication*. Cambridge: Cambridge University Press.
19 Ozarow, L.H., Shamai, S., and Wyner, A.D. (1994). Information theoretic considerations for cellular mobile radio. *IEEE Transactions on Vehicular Technology* 43 (2): 359–378. https://doi.org/10.1109/25.293655.

20 Mayne, D.Q. and Michalska, H. (1990). Receding horizon control of nonlinear systems. *IEEE Transactions on Automatic Control* 35 (7): 814–824. https://doi.org/10.1109/9.57020.

21 Tutuncuoglu, K. and Yener, A. (2012). Communicating with energy harvesting transmitters and receivers. In: *Proceedings of ITA Workshop*, 240–245, IEEE. https://doi.org/10.1109/ITA.2012.6181795.

22 Blume, O., Eckhardt, H., Klein, S. et al. (2010). Energy savings in mobile networks based on adaptation to traffic statistics. *Bell Labs Technical Journal* 15 (2): 77–94. https://doi.org/10.1002/bltj.20442.

23 Kim, H. and De Veciana, G. (2010). Leveraging dynamic spare capacity in wireless systems to conserve mobile terminals' energy. *IEEE/ACM Transactions on Networking* 18 (3): 802–815. https://doi.org/10.1109/TNET.2009.2032238.

24 Chen, Y., Zhang, S., Xu, S., and Li, G.Y. (2011). Fundamental trade-offs on green wireless networks. *IEEE Communications Magazine* 49 (6): 30–37. https://doi.org/10.1109/MCOM.2011.5783982.

25 Isheden, C., Chong, Z., Jorswieck, E., and Fettweis, G. (2012). Framework for link-level energy efficiency optimization with informed transmitter. *IEEE Transactions on Wireless Communications* 11 (8): 2946–2957. https://doi.org/10.1109/TWC.2012.060412.111829.

26 Miao, G., Himayat, N., and Li, G.Y. (2010). Energy-efficient link adaptation in frequency-selective channels. *IEEE Transactions onCommunications* 58 (2): 545–554. https://doi.org/10.1109/TCOMM.2010.02.080587.

27 Mahdavi-Doost, H. and Yates, R.D. (2013). Energy harvesting receivers: finite battery capacity. In: *Proceedings IEEE ISIT, Information Theory Proceedings (ISIT)*, 1799–1803. IEEE. https://doi.org/10.1109/ISIT.2013.6620537.

28 Mahdavi-Doost, H. and Yates, R.D. (2014). Fading channels in energy-harvesting receivers. In: *Proceedings of the 48th Annual Conference on Information Sciences and Systems (CISS)*, 1–6. Princeton, NJ: IEEE. https://doi.org/10.1109/CISS.2014.6814126.

29 Yates, R.D. and Mahdavi-Doost, H. (2015). Energy harvesting receivers: packet sampling and decoding policies. *IEEE Journal on Selected Areas in Communications* 33 (3): 558–570. https://doi.org/10.1109/JSAC.2015.2391834.

30 Cui, S., Goldsmith, A.J., and Bahai, A. (2005). Energy-constrained modulation optimization. *IEEE Transactions on Wireless Communications* 4 (5): 2349–2360. https://doi.org/10.1109/TWC.2005.853882.

31 Lauwers, E. and Gielen, G. (2002). Power estimation methods for analog circuits for architectural exploration of integrated systems. *IEEE Transactions on Very Large Scale Integration (VLSI) Systems* 10 (2): 155–162. https://doi.org/10.1109/92.994993.

32 Zhou, S., Chen, T., Chen, W., and Niu, Z. (2015). Outage minimization for a fading wireless link with energy harvesting transmitter and receiver. *IEEE Journal on Selected Areas in Communications* 33 (3): 496–511. https://doi.org/10.1109/JSAC.2015.2391951.

3

Power Allocation for Multi-node Energy Harvesting Channels

In this chapter, we consider the design of wireless communication systems with multiple transmission nodes powered by energy harvesters. Different from the point-to-point channel case, it is not optimal anymore for the multi-node scenarios to separately optimize the transmission of each energy harvesting (EH) communication link with only local energy state information (ESI), while joint power allocation across all the transmission nodes with global ESI is necessary to achieve the best performance. This phenomenon leads to the so-called energy cooperation among all the EH transmission nodes, where power allocation should be carefully designed to either balance the transmissions among multiple EH transmitters, e.g. in the multiple-access channels (MACs) [1], or to cooperatively transmit the source information, e.g. in relay channels [2] and large relay networks [3].

This chapter first investigates the MAC with conferencing links and shared energy harvesters and discusses the transmission cooperation among the transmitters. Then, we consider a three-node relay channel with EH sources, where it is shown that the proposed cooperative transmission scheme can exploit a new form of diversity arising from the independent source and relay energy availability over time, termed "energy diversity," even over time-invariant channels. Finally, we turn to a large relay network, where multiple relays are powered by energy harvesters, and show that the power allocation for each EH relay should be determined by the overall energy distribution among these relays.

3.1 Multiple-Access Channels

3.1.1 System Model

Consider a two-transmitter Gaussian MAC where the transmitters are connected by wired rate-limited two-way conferencing links, as shown in Figure 3.1. Via the conferencing links, transmitter 1 can talk to transmitter 2 with a rate up to C_{12}, similarly for the opposite direction with a rate up to C_{21}. Compared to the transmissions over wireless links that suffer from heavy path losses, the communications over wired links consume much less energy, which is neglected here for the convenience of analysis. Similar assumptions were adopted in [4–8]. Moreover, the two transmitters share one common EH source with a battery of infinite capacity[1] (e.g. the energy sharing and conferencing

1 As will be shown later, our results can be easily extended to the case with finite battery capacity.

Energy Harvesting Wireless Communications, First Edition.
Chuan Huang, Sheng Zhou, Jie Xu, Zhisheng Niu, Rui Zhang and Shuguang Cui.

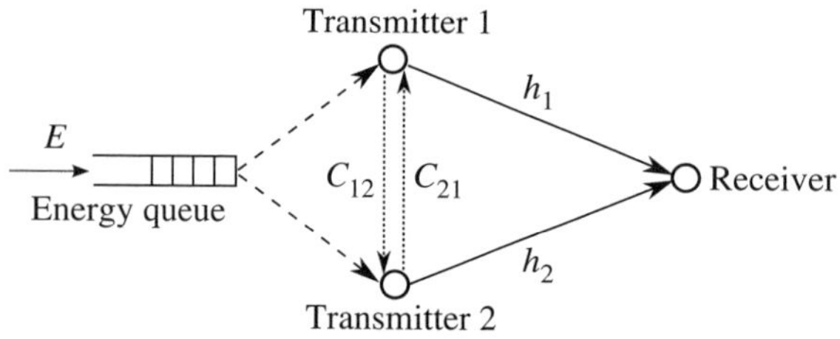

Figure 3.1 MAC with conferencing links and a shared renewable energy source.

could be done over the same wires). The constant channel gains from transmitter 1 and transmitter 2 to the receiver are denoted by h_1 and h_2, respectively. Without loss of generality, we only consider the case of $h_1 \geq h_2$, which indicates that the link from transmitter 1 to the receiver is stronger than or equal to the other one. Moreover, the noise at the receiver is circularly symmetric complex Gaussian (CSCG), with zero mean and unit variance.

In this subsection, we consider a finite time horizon of N slots, each with a duration of T. At the beginning of the nth slot, $n = 1, \ldots, N$, the EH source receives harvested energy (accumulated during the previous time slot) with an amount of $E_n \in [0, E_{\max}]$, where $E_{\max}$ denotes the maximum amount from one energy arrival and is assumed to be known. Denote the sequence of N energy arrival amounts as $\boldsymbol{\omega} = (E_1, \ldots, E_N)$ (which is also called the input sequence) and the set of all possible input sequences as

$$\Omega = \{\boldsymbol{\omega} = (E_1, \ldots, E_N) | 0 \leq E_n \leq E_{\max}, \quad n = 1, \ldots, N\}. \tag{3.1}$$

For the noncausal case, the entire input sequence $\boldsymbol{\omega}$ is known before transmissions; for the causal EH model, at the nth slot, only $E_1, \ldots, E_n$ are known. Denote the data rate and the corresponding transmission power for transmitter i, $i = 1, 2$, at the nth slot, $n = 1, \ldots, N$, as $r_{i,n}$ and $P_{i,n}$, respectively.

The EH constraints for the system model described above require that the total amount of energy consumed by the two transmitters during the first j slots must be smaller than or equal to the total amount of energy harvested in the first j slots, $\forall j \in [1, \ldots, N]$, and could be mathematically modeled as

$$T \sum_{n=1}^{j} (P_{1,n} + P_{2,n}) \leq \sum_{n=1}^{j} E_n, \quad j = 1, \ldots, N. \tag{3.2}$$

3.1.2 Problem Formulation

In this subsection, we first consider a sum power minimization problem and derive the function $g(r_{1,n}, r_{2,n})$, which is the minimum sum power to achieve a given rate pair $(r_{1,n}, r_{2,n})$ across the two users in the nth slot. Then, we rigorously define the maximum departure region and formulate a resource allocation problem to characterize its boundary. In the sequel, we omit the index n whenever it causes no confusion.

For the Gaussian MAC with conferencing links, the coding scheme is described as follows. Transmitter i's message w_i, $i = 1, 2$, can be regarded as comprising two parts: transmitter i's private message w_i^{p} and its common message w_i^{c}. The transmitters exchange their common messages, i.e. w_1^{c} and w_2^{c}, by using the conferencing links as in Willem's scheme [4]. As a result of conferencing, transmitter i, $i = 1, 2$, has the common message $w^c = (w_1^c, w_2^c)$ and its private message w_i^p, which are allocated with power P_i^c and $P_i^p = P_i - P_i^c$, respectively, and sent over the MAC to the receiver.

With the above coding scheme, the capacity region of the Gaussian MAC with conferencing links is given as [5]

$$\begin{cases} r_1 \le C(h_1 P_1^p) + C_{12}, & (3.3) \\ r_2 \le C(h_2 P_2^p) + C_{21}, & (3.4) \\ r_1 + r_2 \le C(h_1 P_1^p + h_2 P_2^p) + C_{12} + C_{21}, & (3.5) \\ r_1 + r_2 \le C\left(h_1 P_1 + h_2 P_2 + 2\sqrt{h_1 P_1^c h_2 P_2^c}\right). & (3.6) \end{cases}$$

It is worth noting that the capacity region defined by (3.3)–(3.6) is convex [9].

Then, the sum power function $g(r_1, r_2)$ can be obtained by solving the following sum power minimization problem:

$$g(r_1, r_2) = \min_{\{P_1^c, P_1^p, P_2^c, P_2^p\}} P_1^c + P_1^p + P_2^c + P_2^p \tag{3.7}$$

$$\text{s.t. } (3.3)\text{–}(3.6),\ \ P_1^c \ge 0,\ \ P_1^p \ge 0,\ \ P_2^c \ge 0,\ \ P_2^p \ge 0. \tag{3.8}$$

It can be proved that problem (3.7)–(3.8) is convex [9] by showing that its feasible region is convex over $(P_1^c, P_1^p, P_2^c, P_2^p)$. By solving problem (3.7)–(3.8), we have the following proposition.

Proposition 3.1 *For a given rate pair (r_1, r_2), the minimum sum power $g(r_1, r_2)$ is given as*

$$g(r_1, r_2) \triangleq \frac{h_2}{h_1(h_1 + h_2)} 2^{(r_1 - C_{12})^+ + (r_2 - C_{21})^+} + \frac{1}{h_1 + h_2} 2^{r_1 + r_2} + \left(\frac{1}{h_2} - \frac{1}{h_1}\right) 2^{(r_2 - C_{21})^+} - \frac{1}{h_2}, \tag{3.9}$$

and the corresponding optimal power allocation for the messages is given as

$$\begin{cases} P_1^{c*} = \dfrac{h_1}{(h_1 + h_2)^2}\left(2^{r_1 + r_2} - 2^{(r_1 - C_{12})^+ + (r_2 - C_{21})^+}\right), & (3.10) \\ P_1^{p*} = \dfrac{1}{h_1} 2^{(r_2 - C_{21})^+}\left(2^{(r_1 - C_{12})^+} - 1\right), & (3.11) \\ P_2^{c*} = \dfrac{h_2}{(h_1 + h_2)^2}\left(2^{r_1 + r_2} - 2^{(r_1 - C_{12})^+ + (r_2 - C_{21})^+}\right), & (3.12) \\ P_2^{p*} = \dfrac{1}{h_2}\left(2^{(r_2 - C_{21})^+} - 1\right). & (3.13) \end{cases}$$

Proof: See [1]. □

Note that $g(r_1, r_2)$ given in (3.9) is jointly convex over r_1 and r_2. By (3.9), the causal EH constraints defined in (3.2) could be rewritten as

$$T\sum_{n=1}^{j} g(r_{1,n}, r_{2,n}) \leq \sum_{n=1}^{j} E_n, \quad j = 1, \dots, N. \tag{3.14}$$

Based on the above analysis, we define the maximum departure region $\mathcal{D}(N)$ as follows.

Definition 3.1 Over the finite time horizon of N slots, the **maximum departure region** $\mathcal{D}(N)$ of the MAC with conferencing links is defined as the union of all achievable bits pair (B_1, B_2) under the EH constraints 3.14, .e.,

$$\mathcal{D}(N) = \left\{ (B_1, B_2) \middle| B_i = T\sum_{n=1}^{N} r_{i,n}, \quad i = 1, 2, \ (3.14) \right\}, \tag{3.15}$$

where B_i is the total amount of transmitted data from transmitter i.

Proposition 3.2 *The maximum departure region $\mathcal{D}(N)$ defined in (3.14) for the MAC with conferencing links and EH constraints is convex.*

Similarly to Lemma 2 in [10], Proposition 3.2 can be proved from the convexity of $g(r_1, r_2)$ and is thus omitted here for brevity.

For the noncausal case, since $\mathcal{D}(N)$ is a convex region defined in the positive orthant, we characterize the boundary of $\mathcal{D}(N)$ by maximizing the weighted sum of B_1 and B_2 as follows:

$$\max_{\{r_{1,n}, r_{2,n}\}} \quad \mu_1 B_1 + \mu_2 B_2 \tag{3.16}$$

$$\text{s.t.} \qquad (3.13), \tag{3.17}$$

where $\mu_1 + \mu_2 = 1$, $\mu_1 \geq 0$, $\mu_2 \geq 0$, are the adjustable weighting factors. It is easy to observe that problem (3.16)–(3.17) is still convex over r_1 and r_2 [9]. The optimal solution of problem (3.16)–(3.17), denoted by $\mathcal{O}$, which achieves the boundary of the maximum departure region $\mathcal{D}(N)$, will be investigated in Section 3.1.3.

3.1.3 The Optimal Offline Scheme

In this section, we first describe the structure of the optimal sum power allocation for problem (3.16)–(3.17), based on which the optimal value for $g(r_{1,n}, r_{2,n})$, $n = 1, \dots, N$, can be obtained. Then, to completely characterize the offline transmission scheme $\mathcal{O}$, we obtain the optimal rate scheduling of the two transmitters.

3.1.4 Optimal Sum Power Allocation

First, we present the structural properties of the optimal sum power allocation for problem (3.16)–(3.17).

Proposition 3.3 *The optimal solution for problem (3.16)–(3.17) satisfies*

$$g(r_{1,n}, r_{2,n}) \leq g(r_{1,n+1}, r_{2,n+1}), \quad \forall n \in \{1, \dots, N-1\}, \tag{3.18}$$

i.e. the optimal sum power is non-decreasing over time.

Proof: The proof is similar to that of Lemma 1 in [10]. If the optimal sum power is decreasing over some adjacent slots, we can always equalize the rate pairs in these slots such that the total amount of energy consumed in the two slots is reduced due to the convexity of $g(r_1, r_2)$, with the EH constraints still satisfied.

Note that the proof is based on the convexity of $g(r_1, r_2)$ and no other claim is made here about the relationship between the rate and the sum power (the optimal rate scheduling given a sum power value will be investigated in the next subsection). □

Proposition 3.4 *The optimal solution for problem (3.16)–(3.17) satisfies that if* $g(r_{1,n}, r_{2,n}) < g(r_{1,n+1}, r_{2,n+1})$, $n \in \{1, \dots, N-1\}$, *there is no residual energy at the end of the nth slot, i.e. when the optimal sum power level changes, all harvested energy must be depleted.*

Proof: Note that the proof for a similar result in [10], which is based on analyzing the Karush–Kuhn-Tucker (KKT) conditions, cannot be directly applied to our case since $g(r_1, r_2)$ defined in (3.9) is not differentiable at some points. Here, we prove this proposition by utilizing the convexity of $g(r_1, r_2)$.

Suppose that under the optimal offline solution, $g(r_{1,\dot{n}}, r_{2,\dot{n}}) < g(r_{1,\dot{n}+1}, r_{2,\dot{n}+1})$ for some $\dot{n} \in \{1, \dots, N-1\}$, and there is some residual energy of amount E_{res} from the $\dot{n}$th slot transmission. The amounts of data sent for the two users over the $\dot{n}$th and $(\dot{n}+1)$th slots under such rate profile are $(r_{1,\dot{n}} + r_{1,\dot{n}+1}, r_{2,\dot{n}} + r_{2,\dot{n}+1})T$.

Define a new rate profile as $(\tilde{r}_{1,\dot{n}}, \tilde{r}_{2,\dot{n}}) = \theta(r_{1,\dot{n}}, r_{2,\dot{n}}) + (1-\theta)(r_{1,\dot{n}+1}, r_{2,\dot{n}+1})$ and $(\tilde{r}_{1,\dot{n}+1}, \tilde{r}_{2,\dot{n}+1}) = (1-\theta)(r_{1,\dot{n}}, r_{2,\dot{n}}) + \theta(r_{1,\dot{n}+1}, r_{2,\dot{n}+1})$, where $\theta \in (0, 1)$.

From the convexity of $g(r_1, r_2)$ over r_1 and r_2, we have

$$g(\tilde{r}_{1,\dot{n}}, \tilde{r}_{2,\dot{n}}) < \theta g(r_{1,\dot{n}}, r_{2,\dot{n}}) + (1-\theta) g(r_{1,\dot{n}+1}, r_{2,\dot{n}+1}),$$
$$g(\tilde{r}_{1,\dot{n}+1}, \tilde{r}_{2,\dot{n}+1}) < (1-\theta) g(r_{1,\dot{n}}, r_{2,\dot{n}}) + \theta g(r_{1,\dot{n}+1}, r_{2,\dot{n}+1}),$$

and therefore

$$(g(\tilde{r}_{1,\dot{n}}, \tilde{r}_{2,\dot{n}}) + g(\tilde{r}_{1,\dot{n}+1}, \tilde{r}_{2,\dot{n}+1}))T < (g(r_{1,\dot{n}}, r_{2,\dot{n}}) + g(r_{1,\dot{n}+1}, r_{2,\dot{n}+1}))T. \tag{3.19}$$

The sum power $g(\tilde{r}_{1,\dot{n}}, \tilde{r}_{2,\dot{n}})$ may be larger than $g(r_{1,\dot{n}}, r_{2,\dot{n}})$. However, since the function $g(r_1, r_2)$ is continuous over (r_1, r_2), there must exist some $\theta \in (0, 1)$ such that $(g(\tilde{r}_{1,\dot{n}}, \tilde{r}_{2,\dot{n}}) - g(r_{1,\dot{n}}, r_{2,\dot{n}}))T \leq E_{\text{res}}$. With such θ, the new rate profile still satisfies the causal EH constraints while the amounts of bits sent for the two users over these two slots are still $(r_{1,\dot{n}} + r_{1,\dot{n}+1}, r_{2,\dot{n}} + r_{2,\dot{n}+1})T$.

Furthermore, (3.19) indicates that the total energy consumed over the $\dot{n}$th and $(\dot{n}+1)$th slots by the new rate profile is less than the original solution. We can increase the rate of one transmitter while keeping the rate of the other one unchanged if we spend the leftover energy uniformly in the $(\dot{n}+1)$th slot, which leads to a larger value of the objective function (3.16). This contradicts the fact that the original rate profile is the optimal solution of problem (3.16)–(3.17). Thus, this proposition is proved. □

As indicated by Propositions 3.3 and 3.4, the optimal sum power sequence for problem (3.16)–(3.17) has the same structure as that for the single-user channel case discussed in [11]. Given an input sequence $\boldsymbol{\omega}$, the optimal sum power sequence $\mathbf{P}^{\mathcal{O}} = [P_1^{\mathcal{O}}, \dots, P_N^{\mathcal{O}}]^T$, where $P_n^{\mathcal{O}}, n = 1, \dots, N$, denotes the optimal sum power in the nth slot, can be obtained recursively as [11]

$$n_k = \arg \min_{n_{k-1} < n \le N} \left\{ \frac{\sum_{j=n_{k-1}+1}^{n} E_j}{(n - n_{k-1})T} \right\}, \tag{3.20}$$

$$g(r_{1,n}, r_{2,n}) = \frac{\sum_{j=n_{k-1}+1}^{n_k} E_j}{(n_k - n_{k-1})T} \doteq P_n^{\mathcal{O}}, \quad \text{for} \quad n_{k-1} < n \le n_k, \tag{3.21}$$

where $n_0 = 0$.

Remark 3.1 When the battery has a finite capacity of $\mathcal{E}$,[2] it is easy to check from the convexity of $g(r_1, r_2)$ that the optimal sum power profile is given the same as that in [12]. As such, the rate scheduling for this case is the same as that of infinite battery.

3.1.4.1 Optimal Rate Scheduling

It is worth noting that the optimal sum power profile obtained by (3.20) and (3.21) is only determined by the harvested energy amounts and is independent of μ_1 and μ_2. Therefore, solving problem (3.16)–(3.17) is equivalent to solving the following problem (3.22)–(3.23) for each $n \in \{1, \dots, N\}$ individually [10]:

$$\max_{r_{1,n}, r_{2,n}} \quad \mu_1 r_{1,n} + \mu_2 r_{2,n} \tag{3.22}$$

$$\text{s.t.} \quad g(r_{1,n}, r_{2,n}) = P_n^{\mathcal{O}}. \tag{3.23}$$

Before presenting the optimal solution of problem (3.22)–(3.23), we show some properties of the curve defined by $g(r_{1,n}, r_{2,n}) = P_n^{\mathcal{O}}$ (denoted as $\mathcal{G}$). We define four possible regions as follows:

$$\mathcal{R}_1 = \{(r_{1,n}, r_{2,n}) \in \mathcal{G} | 0 < r_{1,n} < C_{12}, 0 < r_{2,n} < C_{21}\},$$
$$\mathcal{R}_2 = \{(r_{1,n}, r_{2,n}) \in \mathcal{G} | r_{1,n} > C_{12}, 0 < r_{2,n} < C_{21}\},$$
$$\mathcal{R}_3 = \{(r_{1,n}, r_{2,n}) \in \mathcal{G} | 0 < r_{1,n} < C_{12}, r_{2,n} > C_{21}\},$$
$$\mathcal{R}_4 = \{(r_{1,n}, r_{2,n}) \in \mathcal{G} | r_{1,n} > C_{12}, r_{2,n} > C_{21}\}.$$

Notice that, for any given $P_n^{\mathcal{O}}$, at most three out of the four regions defined above can be non-empty. For the case of $C_{12} > C_{21}$, Figure 3.2 illustrates the curve $\mathcal{G}$ with different $P_n^{\mathcal{O}}$ and $h_1 > h_2$. For the case of $C_{12} < C_{21}$, similar results could be observed and thus is omitted in the following. As will be confirmed later by the tangent line property of $\mathcal{G}$, $\mathcal{R}_1$ is a straight line, whereas $\mathcal{R}_2, \mathcal{R}_3$, and $\mathcal{R}_4$ are curves when $h_1 > h_2$ (when $h_1 = h_2$, $\mathcal{R}_1$ and $\mathcal{R}_4$ are straight lines).

As indicated by the shape of $\mathcal{G}$ in Figure 3.2, for given μ_1 and μ_2, finding the rate pair $(r_{1,n}, r_{2,n})$ that maximizes $\mu_1 r_{1,n} + \mu_2 r_{2,n}$ is equivalent to finding the point $(r_{1,n}, r_{2,n})$ on $\mathcal{G}$, at which the slope of (one of) the tangent line(s) (as will be explained later, the tangent

2 In the finite battery case, E_n, $n = 1, \dots, N$, and thus $E_{\max}$ are truncated at $\mathcal{E}$ since any energy exceeding $\mathcal{E}$ cannot be stored in the battery [12].

line is unique at most points on $\mathcal{G}$ but not at the others) equals $-\frac{\mu_1}{\mu_2}$. Denote the slope of (one of) the tangent line(s) at point $(r_{1,n}, r_{2,n}) \in \mathcal{G}$ by ϕ. By calculating the first-order derivative of (3.9), we conclude the properties of ϕ as follows.

Proposition 3.5 *At the point $(r_{1,n}, r_{2,n})$ in any of the four regions, there exists a unique tangent line, and ϕ satisfies*

$$\phi = \begin{cases} -1 & \text{if } (r_{1,n}, r_{2,n}) \in \mathcal{R}_1 \\ -1 - \dfrac{h_2}{h_1 2^{r_{2,n}+C_{12}}} < -1 & \text{if } (r_{1,n}, r_{2,n}) \in \mathcal{R}_2 \\ -1 + \dfrac{h_1}{h_1 + h_2 2^{r_{1,n}+C_{21}}} > -1 & \text{if } (r_{1,n}, r_{2,n}) \in \mathcal{R}_3 \\ -\left(1 + \dfrac{(h_1^2 - h_2^2) 2^{C_{12}-r_{1,n}}}{h_1 h_2 2^{C_{12}+C_{21}} + h_2^2}\right)^{-1} > -1 & \text{if } (r_{1,n}, r_{2,n}) \in \mathcal{R}_4. \end{cases} \quad (3.24)$$

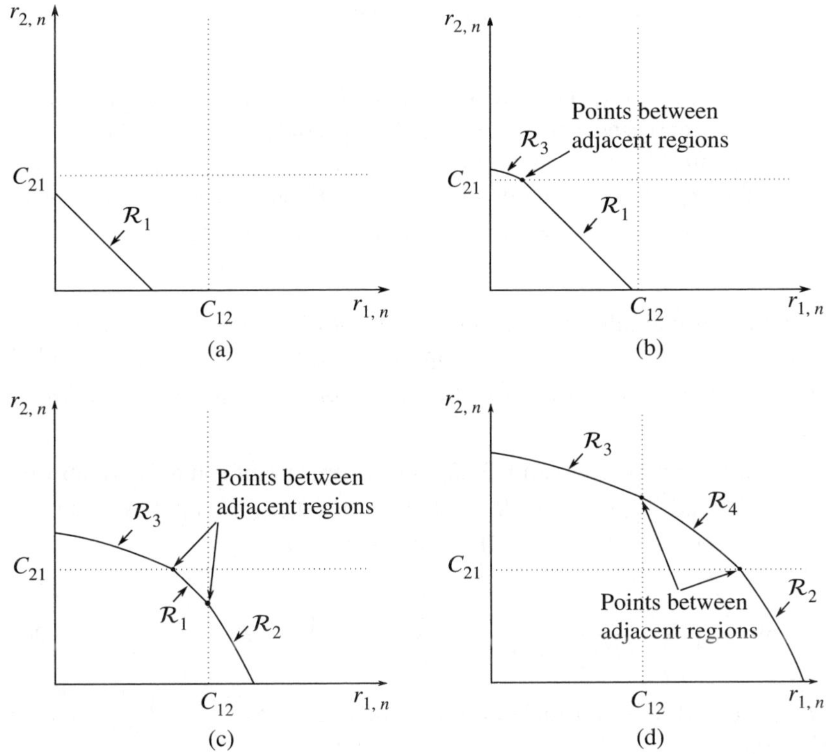

Figure 3.2 Illustrations of curve $\mathcal{G}$ under $C_{12} > C_{21}$ and $h_1 > h_2$. (Taking the subfigure (b), for example, the point between adjacent regions $\mathcal{R}_1$ and $\mathcal{R}_3$ refers to the point $(r_{1,n}, r_{2,n}) \in \mathcal{G}$ with $r_{2,n} = C_{21}$. The other cases are similar.) (a) When $0 < P_n^{\mathcal{O}} < g(0, C_{21})$, curve $\mathcal{G}$ has non-empty $\mathcal{R}_1$ and empty $\mathcal{R}_2$, $\mathcal{R}_3$, $\mathcal{R}_4$. (b) When $g(0, C_{21}) < P_n^{\mathcal{O}} < g(C_{12}, 0)$, curve $\mathcal{G}$ has non-empty $\mathcal{R}_1$, $\mathcal{R}_3$ and empty $\mathcal{R}_2$, $\mathcal{R}_4$. (c) When $g(C_{12}, 0) < P_n^{\mathcal{O}} < g(C_{12}, C_{21})$, curve $\mathcal{G}$ has non-empty $\mathcal{R}_1$, $\mathcal{R}_2$, $\mathcal{R}_3$ and empty $\mathcal{R}_4$. (d) When $P_n^{\mathcal{O}} > g(C_{12}, C_{21})$, curve $\mathcal{G}$ has non-empty $\mathcal{R}_2$, $\mathcal{R}_3$, $\mathcal{R}_4$ and empty $\mathcal{R}_1$.

At the point $(r_{1,n}, r_{2,n})$ between two adjacent regions, the tangent line is not unique, and ϕ satisfies

$$\phi \in \begin{cases} \left[-1 - \frac{h_2}{h_1 2^{r_{2,n}+C_{12}}}, -1\right] \\ \qquad \text{if } (r_{1,n}, r_{2,n}) \text{ is between } \mathcal{R}_1 \text{ and } \mathcal{R}_2 \\ \left[-1, -1 + \frac{h_1}{h_1 + h_2 2^{r_{1,n}+C_{21}}}\right] \\ \qquad \text{if } (r_{1,n}, r_{2,n}) \text{ is between } \mathcal{R}_1 \text{ and } \mathcal{R}_3 \\ \left[-1 + \frac{h_1^2 - h_2^2}{h_1^2 + h_1 h_2 2^{C_{12}+C_{21}}}, -1 + \frac{h_1}{h_1 + h_2 2^{C_{12}+C_{21}}}\right] \\ \qquad \text{if } (r_{1,n}, r_{2,n}) \text{ is between } \mathcal{R}_3 \text{ and } \mathcal{R}_4 \\ \left[-1 - \frac{h_2}{h_1 2^{C_{21}+C_{12}}}, -\left(1 + \frac{(h_1^2 - h_2^2)2^{C_{12}-r_{1,n}}}{h_1 h_2 2^{C_{12}+C_{21}} + h_2^2}\right)^{-1}\right] \\ \qquad \text{if } (r_{1,n}, r_{2,n}) \text{ is between } \mathcal{R}_2 \text{ and } \mathcal{R}_4. \end{cases} \tag{3.25}$$

Remark 3.2 For $(r_{1,n}, r_{2,n}) \in \mathcal{R}_2$, ϕ is only determined by $r_{2,n}$, i.e. it is independent of $P_n^{\mathcal{O}}$, and increases as $r_{2,n}$ increases. For $(r_{1,n}, r_{2,n}) \in \mathcal{R}_3$, ϕ is only determined by $r_{1,n}$, i.e. it is independent of $P_n^{\mathcal{O}}$, and decreases as $r_{1,n}$ increases. For $(r_{1,n}, r_{2,n}) \in \mathcal{R}_4$, (i) if $h_1 > h_2$, ϕ is only determined by $r_{1,n}$, i.e. it is independent of $P_n^{\mathcal{O}}$, and decreases as $r_{1,n}$ increases with $\lim_{r_{1,n}\to\infty} \frac{dr_{2,n}}{dr_{1,n}} = -1$; (ii) if $h_1 = h_2$, $\phi = -1$ for any point belongs to $\mathcal{R}_4$.

Then, we obtain the optimal solution of problem (3.22)–(3.23) for $n = 1, \dots, N$ by finding the point on $\mathcal{G}$, which has a $\phi = -\frac{\mu_1}{\mu_2}$. The optimal solution $(r_{1,n}^{\mathcal{O}}, r_{2,n}^{\mathcal{O}})$ is concluded as follows; and then by applying (3.10)–(3.13), we complete the whole transmission scheme.

(1) If $-\frac{\mu_1}{\mu_2} < -1$, as indicated by (3.24) and (3.25), the optimal point can only possibly be in $\mathcal{R}_2$, between $\mathcal{R}_1$ and $\mathcal{R}_2$, between $\mathcal{R}_2$ and $\mathcal{R}_4$, or on the horizontal axis. Letting $-1 - \frac{h_2}{h_1 2^{r_{2,n}+C_{12}}}$ (the second term of (3.24)) equal $-\frac{\mu_1}{\mu_2}$, we have

$$r_{2,n} = \min\left\{ \left(\log \frac{h_2\mu_2}{h_1(\mu_1 - \mu_2)} - C_{12} \right)^+, C_{21} \right\} \doteq R_2. \tag{3.26}$$

(a) If $0 \le P_n^{\mathcal{O}} \le g(C_{12}, 0)$, the optimal rate allocation is given as $r_{1,n}^{\mathcal{O}} = C((h_1 + h_2)P_n^{\mathcal{O}})$ and $r_{2,n}^{\mathcal{O}} = 0$.

(b) If $g(C_{12}, 0) < P_n^{\mathcal{O}} \le g(C_{12}, R_2)$, the point between $\mathcal{R}_1$ and $\mathcal{R}_2$ satisfies $r_{2,n} < R_2$ and thus has a $\phi = -\frac{\mu_1}{\mu_2}$, which can be easily checked by (3.25) and the fact that $-1 - \frac{h_2}{h_1 2^{r_{2,n}+C_{12}}}$ decreases as $r_{2,n}$ decreases. Therefore, this point is the optimal solution, and the optimal rate allocation is given as $r_{1,n}^{\mathcal{O}} = C_{12}$ and $r_{2,n}^{\mathcal{O}} = C((h_1 + h_2)P_n^{\mathcal{O}}) - C_{12}$.

(c) If $P_n^{\mathcal{O}} > g(C_{12}, R_2)$, the optimal rate allocation is given as

$$r_{1,n}^{\mathcal{O}} = \log \frac{(h_1 + h_2)(1 + h_1 P_n^{\mathcal{O}})}{h_2 2^{-C_{12}} + h_1 2^{R_2}} > C_{12} \tag{3.27}$$

and $r_{2,n}^{\mathcal{O}} = R_2$, which can be verified directly from the definition of R_2.

It can be seen that R_2 is the maximum possible transmission rate for transmitter 2, regardless of the sum power value $P_n^{\mathcal{O}}$; thus we name it as the capping rate at transmitter 2.

(2) If $-\frac{\mu_1}{\mu_2} > -1$, similar to the above case, the optimal solution to problem (3.22)–(3.23) can be obtained as follows:

(a) If $0 \le P_n^{\mathcal{O}} \le g(0, C_{21})$, the optimal rate allocation is given as $r_{1,n}^{\mathcal{O}} = 0$ and $r_{2,n}^{\mathcal{O}} = C((h_1 + h_2)P_n^{\mathcal{O}})$.

(b) If $g(0, C_{21}) < P_n^{\mathcal{O}} \le g(R_1, C_{21})$, where R_1 is defined in

$$R_1 = \begin{cases} \min\left\{ \left(\log\left(\frac{h_1}{h_2}\left(\frac{\mu_2}{\mu_2-\mu_1} - 1\right)\right) - C_{21}\right)^+, C_{12} \right\} & \text{if } -\dfrac{\mu_1}{\mu_2} \ge -1 + \dfrac{h_1^2 - h_2^2}{h_1^2 + h_1 h_2 2^{C_{12}+C_{21}}} \\ \log\left(\dfrac{\mu_1(h_2^2 - h_1^2)}{(\mu_1 - \mu_2)h_2(h_1 2^{C_{12}+C_{21}} + h_2)}\right) + C_{12} & \text{otherwise,} \end{cases} \tag{3.28}$$

the optimal rate allocation is given as follows:

(i) If $-\frac{\mu_1}{\mu_2} \ge -1 + \frac{h_1^2-h_2^2}{h_1^2+h_1h_2 2^{C_{12}+C_{21}}}$, $r_{2,n}^{\mathcal{O}} = C_{21}$ and $r_{1,n}^{\mathcal{O}} = C((h_1 + h_2)P_n^{\mathcal{O}}) - C_{21}$.

(ii) Otherwise, $r_{2,n}^{\mathcal{O}} = C_{21}$ and

$$r_{1,n}^{\mathcal{O}} = \begin{cases} C((h_1 + h_2)P_n^{\mathcal{O}}) - C_{21} & \text{if } P_n^{\mathcal{O}} \le g(C_{12}, C_{21}) \\ \log \dfrac{(h_1 + h_2)(1 + h_1 P_n^{\mathcal{O}})}{h_2 2^{-C_{12}} + h_1 2^{C_{21}}} & \text{if } P_n^{\mathcal{O}} > g(C_{12}, C_{21}). \end{cases}$$

(c) If $P_n^{\mathcal{O}} > g(R_1, C_{21})$, the optimal rate allocation is given as follows:

(i) If $-\frac{\mu_1}{\mu_2} \ge -1 + \frac{h_1^2-h_2^2}{h_1^2+h_1h_2 2^{C_{12}+C_{21}}}$, $r_{1,n}^{\mathcal{O}} = R_1$ and $r_{2,n}^{\mathcal{O}} = \log \frac{(1+h_2 P_n^{\mathcal{O}})(h_1+h_2)}{h_1 2^{-C_{21}} + h_2 2^{R_1}}$.

(ii) Otherwise, $r_{1,n}^{\mathcal{O}} = R_1$ and $r_{2,n}^{\mathcal{O}} = \log \frac{(h_2 P_n^{\mathcal{O}}+1)h_1(h_1+h_2)}{h_2^2 2^{R_1-C_{12}-C_{21}} + h_1 h_2 2^{R_1} + (h_1^2-h_2^2)2^{-C_{21}}}$.

It can be seen that R_1 is the maximum possible transmission rate for transmitter 1, regardless of the sum power value $P_n^{\mathcal{O}}$; thus we name it as the capping rate at transmitter 1.

(3) If $-\frac{\mu_1}{\mu_2} = -1$, it follows that:

(a) If $P_n^{\mathcal{O}} \le g(C_{12}, C_{21})$, the optimal rate allocation is not unique, and any rate pair $(r_{1,n}^{\mathcal{O}}, r_{2,n}^{\mathcal{O}})$ satisfying $r_{1,n}^{\mathcal{O}} + r_{2,n}^{\mathcal{O}} = C((h_1 + h_2)P_n^{\mathcal{O}})$, $0 \le r_{1,n}^{\mathcal{O}} \le C_{12}$, and $0 \le r_{2,n}^{\mathcal{O}} \le C_{21}$, is optimal.

(b) If $P_n^{\mathcal{O}} > g(C_{12}, C_{21})$,

(i) when $h_1 > h_2$, the optimal rate allocation is given as

$$r^{\mathcal{O}}_{1,n} = \log \frac{(h_1 + h_2)(1 + h_1 P^{\mathcal{O}}_n)}{h_2 2^{-C_{12}} + h_1 2^{C_{21}}} \tag{3.29}$$

and $r^{\mathcal{O}}_{2,n} = C_{21}$, which is the point between $\mathcal{R}_2$ and $\mathcal{R}_4$. In this case, the capping rate at transmitter 2 is equal to C_{21}.

(ii) when $h_1 = h_2$, the optimal rate allocation is not unique, and any rate pair $(r^{\mathcal{O}}_{1,n}, r^{\mathcal{O}}_{2,n})$ satisfying $r^{\mathcal{O}}_{1,n} + r^{\mathcal{O}}_{2,n} = \log \frac{2+2h_1 P^{\mathcal{O}}_n}{1+2^{-C_{12}-C_{21}}}$, $r^{\mathcal{O}}_{1,n} \geq C_{12}$, and $r^{\mathcal{O}}_{2,n} \geq C_{21}$ is optimal.

3.1.5 The Online Scheme

In this section, we study the online scheme under the assumption of no priori information about the future energy arrivals. First, we introduce the competitive analysis and define the competitive ratio. Then, the greedy scheme is examined, and its corresponding competitive ratios against the optimal offline scheme for various values of μ_1 and μ_2 are derived.

3.1.5.1 Competitive Analysis

Competitive analysis is a widely adopted approach for the performance measurement of online solutions against the optimal offline one [13]. Under such a framework, the performance of an online scheme is measured by the worst (maximum) ratio, i.e. the competitive ratio, between the profits achieved by the offline and online schemes over all possible input sequences.

Given the input sequence $\boldsymbol{\omega}$ and the weighting factors μ_1, μ_2, the profits (or weighted sum throughput) obtained by the optimal offline scheme $\mathcal{O}$ and an online scheme, denoted by $\mathcal{A}$, are defined as

$$B^{\mathcal{O}}(\boldsymbol{\omega}) = \mu_1 T \sum_{n=1}^{N} r^{\mathcal{O}}_{1,n} + \mu_2 T \sum_{n=1}^{N} r^{\mathcal{O}}_{2,n}, \tag{3.30}$$

$$B^{\mathcal{A}}(\boldsymbol{\omega}) = \mu_1 T \sum_{n=1}^{N} r^{\mathcal{A}}_{1,n} + \mu_2 T \sum_{n=1}^{N} r^{\mathcal{A}}_{2,n}, \tag{3.31}$$

respectively, where $r^{\mathcal{A}}_{i,n}$, $i = 1, 2$, is the rate of transmitter i at the nth slot achieved by online scheme $\mathcal{A}$.

Definition 3.2 An online scheme $\mathcal{A}$ is called ρ-competitive or has a competitive ratio of ρ, if for all possible input sequences $\boldsymbol{\omega} \in \boldsymbol{\Omega}$ we have

$$\max_{\boldsymbol{\omega} \in \boldsymbol{\Omega}} \frac{B^{\mathcal{O}}(\boldsymbol{\omega})}{B^{\mathcal{A}}(\boldsymbol{\omega})} \leq \rho, \tag{3.32}$$

where ρ is a constant independent of the input sequences.

3.1.5.2 The Greedy Scheme

Here, we consider an online scheme $\mathcal{A}$ described as follows: the transmitters always consume up the harvested energy available at each slot. Given an arbitrary input sequence $\boldsymbol{\omega} \in \boldsymbol{\Omega}$, the sum power allocated to the nth slot by $\mathcal{A}$ is given as $P_n^{\mathcal{A}} = \frac{E_n}{T}$, $n = 1, \dots, N$. Scheme $\mathcal{A}$ determines the rates at slot n, denoted by $(r_{1,n}^{\mathcal{A}}, r_{2,n}^{\mathcal{A}})$, by applying the optimal solution of problem (3.22)–(3.23). The corresponding power allocation for the private and common messages of the two transmitters at slot n can be obtained by (3.10)–(3.13).

Before deriving the competitive ratios of the greedy scheme, we present two preliminary results that give the upper bound of the profit obtained by $\mathcal{O}$ and the lower bound of the profit obtained by $\mathcal{A}$, respectively.

Lemma 3.1 *For an arbitrary input sequence $\boldsymbol{\omega} \in \boldsymbol{\Omega}$, define its corresponding enhanced input sequence as $\tilde{\boldsymbol{\omega}} = \left(\sum_{n=1}^{N} E_n, 0, \dots, 0 \right)$. The profit obtained by $\mathcal{O}$ for serving $\boldsymbol{\omega}$ is upper bounded by that for serving $\tilde{\boldsymbol{\omega}}$, i.e. $B^{\mathcal{O}}(\boldsymbol{\omega}) \le B^{\mathcal{O}}(\tilde{\boldsymbol{\omega}})$.*

Proof: Suppose that the profit earned by $\mathcal{O}$ for serving $\tilde{\boldsymbol{\omega}}$ is smaller than that for $\boldsymbol{\omega}$, i.e. $B^{\mathcal{O}}(\boldsymbol{\omega}) > B^{\mathcal{O}}(\tilde{\boldsymbol{\omega}})$. Without violating the EH constraints, the enhanced input sequence $\tilde{\boldsymbol{\omega}}$ can always be served with the same power/rate profile that $\mathcal{O}$ serves $\boldsymbol{\omega}$ with. The resulting profits obtained by $\mathcal{O}$ for serving the two input sequences would be the same, which contradicts with the assumption that $B^{\mathcal{O}}(\boldsymbol{\omega}) > B^{\mathcal{O}}(\tilde{\boldsymbol{\omega}})$. Thus, this proposition is proved. □

As indicated by the solution of problem (3.22)–(3.23) in Section 3.1.4.1, with $\mathcal{A}$, the rate profile may have different structures over the transmission period (for example, if $-\frac{\mu_1}{\mu_2} < -1$, only transmitter 1 has nonzero rate when the sum power is small, whereas both transmitters can have nonzero rates when the sum power is larger than certain value). Consider a "lazy" version of $\mathcal{A}$, denoted as $\tilde{\mathcal{A}}$, with which the rate profile has a unified structure for all sum power levels. For an arbitrary input sequence $\boldsymbol{\omega} \in \boldsymbol{\Omega}$, $\tilde{\mathcal{A}}$ determines the sum power allocation $P_n^{\tilde{\mathcal{A}}}$ at slot n, $n = 1, \dots, N$, as $P_n^{\tilde{\mathcal{A}}} = P_n^{\mathcal{A}} = \frac{E_n}{L}$, where the rate pair $(r_{1,n}^{\tilde{\mathcal{A}}}, r_{2,n}^{\tilde{\mathcal{A}}})$ determined by $\tilde{\mathcal{A}}$ satisfies the following: (i) if $-\frac{\mu_1}{\mu_2} \le -1$, we have $r_{2,n}^{\tilde{\mathcal{A}}} = 0$ and $g(r_{1,n}^{\tilde{\mathcal{A}}}, 0) = P_n^{\tilde{\mathcal{A}}}$; (ii) otherwise, we have $r_{1,n}^{\tilde{\mathcal{A}}} = 0$ and $g(0, r_{2,n}^{\tilde{\mathcal{A}}}) = P_n^{\tilde{\mathcal{A}}}$. Denote the profit obtained by $\tilde{\mathcal{A}}$ for serving $\boldsymbol{\omega}$ by $B^{\tilde{\mathcal{A}}}(\boldsymbol{\omega})$. Notice that with $\tilde{\mathcal{A}}$, only one of the two transmitters has nonzero rate during the transmission period, regardless of the sum power levels. From (3.9), we can derive that if $-\frac{\mu_1}{\mu_2} \le -1$,

$$
\begin{aligned}
B^{\tilde{\mathcal{A}}}(\boldsymbol{\omega}) &= \mu_1 T \sum_{n=1}^{N} \begin{cases} C\left((h_1 + h_2) P_n^{\mathcal{A}}\right) & \text{if } P_n^{\mathcal{A}} \le g(C_{12}, 0) \\ \log \dfrac{(h_1 + h_2)(1 + h_1 P_n^{\mathcal{A}})}{h_2 2^{-C_{12}} + h_1} & \text{otherwise} \end{cases} \\
&> \mu_1 T \sum_{n=1}^{N} C(h_1 P_n^{\mathcal{A}}); \qquad (3.33)
\end{aligned}
$$

otherwise,

$$B^{\tilde{\mathcal{A}}}(\boldsymbol{\omega}) = \mu_2 T \sum_{n=1}^{N} \begin{cases} C\left((h_1+h_2)P_n^{\mathcal{A}}\right) & \text{if } P_n^{\mathcal{A}} \le g(0, C_{21}) \\ \log \dfrac{(h_1+h_2)(1+h_2 P_n^{\mathcal{A}})}{h_1 2^{-C_{21}} + h_2} & \text{otherwise} \end{cases}$$

$$> \mu_2 T \sum_{n=1}^{N} C(h_2 P_n^{\mathcal{A}}). \tag{3.34}$$

Lemma 3.2 *For an arbitrary input sequence $\boldsymbol{\omega} \in \boldsymbol{\Omega}$, the profit obtained by $\mathcal{A}$ is lower bounded by that obtained with $\tilde{\mathcal{A}}$. Hence, based on (3.33) and (3.34), we obtain the following: if $-\frac{\mu_1}{\mu_2} \le -1$, we have $B^{\mathcal{A}}(\boldsymbol{\omega}) \ge B^{\tilde{\mathcal{A}}}(\boldsymbol{\omega}) > \mu_1 T \sum_{n=1}^{N} C(h_1 P_n^{\mathcal{A}})$; otherwise, we have $B^{\mathcal{A}}(\boldsymbol{\omega}) \ge B^{\tilde{\mathcal{A}}}(\boldsymbol{\omega}) > \mu_2 T \sum_{n=1}^{N} C(h_2 P_n^{\mathcal{A}})$.*

Proof: For each $n \in \{1, \dots, N\}$, it is easy to check that $\mu_1 r_{1,n}^{\mathcal{A}} + \mu_2 r_{2,n}^{\mathcal{A}} \ge \mu_1 r_{1,n}^{\tilde{\mathcal{A}}} + \mu_2 r_{2,n}^{\tilde{\mathcal{A}}}$. Hence, the summation over n slots indicates $B^{\mathcal{A}}(\boldsymbol{\omega}) \ge B^{\tilde{\mathcal{A}}}(\boldsymbol{\omega})$. □

With the upper bound of $B^{\mathcal{O}}(\boldsymbol{\omega})$ and the lower bound of $B^{\mathcal{A}}(\boldsymbol{\omega})$ obtained by Lemmas 3.1 and 3.2, we can upper bound $\frac{B^{\mathcal{O}}(\boldsymbol{\omega})}{B^{\mathcal{A}}(\boldsymbol{\omega})}$ by $\frac{B^{\mathcal{O}}(\tilde{\boldsymbol{\omega}})}{B^{\tilde{\mathcal{A}}}(\boldsymbol{\omega})}$, with which an input-independent upper bound can be found by applying certain approximation. Then, we obtain the competitive ratio of $\mathcal{A}$ as summarized in the following proposition.

Proposition 3.6 *The competitive ratio for the greedy scheme $\mathcal{A}$ is given as follows:*

(1) For $-\frac{\mu_1}{\mu_2} < -1$, $\mathcal{A}$ is ρ_1-competitive, where

$$\rho_1 \triangleq k_1 \max \left\{ \begin{array}{l} (h_1+h_2)/\ln 2 \\ (h_1+h_2)\left(\dfrac{(\mu_1-\mu_2)C_{12}}{\mu_1(2^{C_{12}}-1)} + \dfrac{\mu_2}{\mu_1 \ln 2}\right) \\ \dfrac{\mu_1 \log \frac{h_1+h_2}{h_2 2^{-C_{12}} + h_1 2^{R_2}} + \mu_2 R_2}{\mu_1 g(C_{12}, R_2)} + \dfrac{h_1}{\ln 2} \end{array} \right\}, \tag{3.35}$$

and $k_1 = \frac{\frac{E_{\max}}{T}}{C\left(h_1 \frac{E_{\max}}{T}\right)}$.

(2) For $-\frac{\mu_1}{\mu_2} = -1$, $\mathcal{A}$ is ρ_2-competitive, where

$$\rho_2 \triangleq k_1 \max \left\{ \frac{h_1+h_2}{\ln 2}, \frac{\log \frac{h_1+h_2}{h_2 2^{-C_{12}-C_{21}} + h_1}}{g(C_{12}, C_{21})} + \frac{h_1}{\ln 2} \right\}. \tag{3.36}$$

(3) For $-\frac{\mu_1}{\mu_2} > -1$, (i) if $h_1 > h_2$, $\mathcal{A}$ is ρ_3-competitive when $-1 < -\frac{\mu_1}{\mu_2} < -1 + \frac{h_1^2-h_2^2}{h_1^2+h_1 h_2 2^{C_{12}+C_{21}}}$ and ρ_4-competitive when $-\frac{\mu_1}{\mu_2} \ge -1 + \frac{h_1^2-h_2^2}{h_1^2+h_1 h_2 2^{C_{12}+C_{21}}}$; (ii) if $h_1 = h_2$, the online scheme

$\mathcal{A}$ is ρ_4-competitive for all $-\frac{\mu_1}{\mu_2} > -1$, where

$$\rho_3 \triangleq k_2 \max \left\{ \begin{array}{l} (h_1 + h_2)/\ln 2 \\ (h_1 + h_2)\left(\dfrac{(\mu_2 - \mu_1)C_{21}}{\mu_2(2^{C_{21}} - 1)} + \dfrac{\mu_1}{\mu_2 \ln 2} \right) \\ \dfrac{\mu_1 \log \frac{h_1+h_2}{h_2 2^{-C_{12}} + h_1 2^{C_{21}}} + \mu_2 C_{21}}{\mu_2 g(C_{12}, C_{21})} + \dfrac{\mu_1 h_1}{\mu_2 \ln 2} \\ \dfrac{\mu_2 \log \frac{h_1(\mu_2-\mu_1)2^{C_{21}}}{\mu_2(h_1-h_2)} + \mu_1 R_1}{\mu_2 g(R_1, C_{21})} + \dfrac{h_2}{\ln 2} \end{array} \right\}, \tag{3.37}$$

$$\rho_4 \triangleq k_2 \max \left\{ \begin{array}{l} (h_1 + h_2)/\ln 2 \\ (h_1 + h_2)\left(\dfrac{(\mu_2 - \mu_1)C_{21}}{\mu_2(2^{C_{21}} - 1)} + \dfrac{\mu_1}{\mu_2 \ln 2} \right) \\ \dfrac{\mu_1 R_1 + \mu_2 \log \frac{h_1+h_2}{h_1 2^{-C_{21}} + h_2 2^{R_1}}}{\mu_2 g(R_1, C_{21})} + \dfrac{h_2}{\ln 2} \end{array} \right\}, \tag{3.38}$$

and $k_2 = \dfrac{\frac{E_{\max}}{T}}{C\left(h_2 \frac{E_{\max}}{T}\right)}$.

Proof: See [1]. □

Remark 3.3 The competitive ratios given in Proposition 3.6 also hold for the case of finite battery capacity if we truncate $E_{\max}$ at $\mathcal{E}$ (which is equivalent to truncating E_n, $n = 1, \dots, N$, at $\mathcal{E}$). This can be easily verified by the fact that given any input sequence (truncated at $\mathcal{E}$), the profit achieved by the optimal offline scheme is upper bounded by that achieved for the case of infinite battery capacity, whereas the profit achieved by the greedy scheme is the same as that for the case of infinite battery capacity.

3.1.6 Numerical Results

In this section, we adopt $C_{12} = 0.5$ bps Hz^{-1}, $C_{21} = 0.4$ bps Hz^{-1}, $h_1 = 0.8$, $h_2 = 0.7$, $N = 15$, and $T = 10$ s. For the case of finite battery capacity, we assume $\mathcal{E} = 20$ J.

Consider the input sequence

$$\boldsymbol{\omega}_1 = (20, 17.3, 16, 16, 14.7, 16, 10.7, 4, 5.3, 2.7, 10, 13.3, 17.3, 18, 18.7)\ J.$$

Given $\boldsymbol{\omega}_1$, the sum power sequences obtained by the optimal offline scheme with infinite and finite battery capacities are given as

$$\mathbf{P}^{\mathcal{O}} = [1.206, 1.206, 1.206, 1.206, 1.206, 1.206, 1.206, 1.206, \\ 1.206, 1.206, 1.206, 1.33, 1.73, 1.8, 1.87]^T \text{ J s}^{-1},$$

and

$$\mathbf{P}^{\mathcal{O}}_{\mathcal{E}} = [1.73, 1.6, 1.6, 1.535, 1.535, 1.07, 0.8, 0.8, \\ 0.8, 0.8, 1, 1.33, 1.73, 1.8, 1.87]^T \text{ J s}^{-1},$$

respectively. The sum power sequence obtained by the greedy scheme is given as

$$\mathbf{P}^{\mathcal{A}} = [2, 1.73, 1.6, 1.6, 1.47, 1.6, 1.07, 0.4, 0.53, 0.27, 1, 1.33, 1.73, 1.8, 1.87]^T \text{ J s}^{-1}.$$

Figure 3.3 illustrates the accumulated amounts of the harvested energy (solid staircase curve) and the consumed energy by the proposed schemes (the other curves) during the course of transmissions. The slopes of the curves corresponding to the "optimal offline (infinite battery)," "optimal offline (finite battery)," and "online greedy" indicate the sum power levels. To satisfy the causal EH constraints, the above three curves must be beneath the solid staircase curve. The curve corresponding to the optimal offline scheme with finite battery must also be above the dot curve (which represents the accumulated amount of harvested energy minus the battery capacity) to avoid battery overflow. In Figure 3.4, we show the comparison of performance achieved by the optimal offline schemes, i.e. the boundary of the maximum departure region, with both infinite and finite battery capacities, against that by the online greedy scheme, given $\boldsymbol{\omega}_1$.

With $\boldsymbol{\omega}_1$, the performance of the greedy scheme is close to that of the optimal offline ones. However, we may encounter some "malicious" input sequence, with which the performance of the greedy scheme can be significantly worse than that of the optimal offline schemes. For example, consider the input sequence $\boldsymbol{\omega}_2 = (20, 0, 0, 0, 0, 0, 0, 0, 0, 0, 0, 0, 0, 0, 0)$ J. Given $\boldsymbol{\omega}_2$, the optimal offline scheme spends the 20 J energy, which is available at the beginning of the first slot, uniformly over all 15 slots (it is easy to check such power allocation holds for both infinite and finite battery cases), whereas the greedy scheme depletes all the 20 J energy in the first slot. In Figure 3.5, we plot the ratios between the weighted sum throughputs obtained by the optimal offline scheme and the greedy scheme, given $\boldsymbol{\omega}_1$ and $\boldsymbol{\omega}_2$, respectively, and

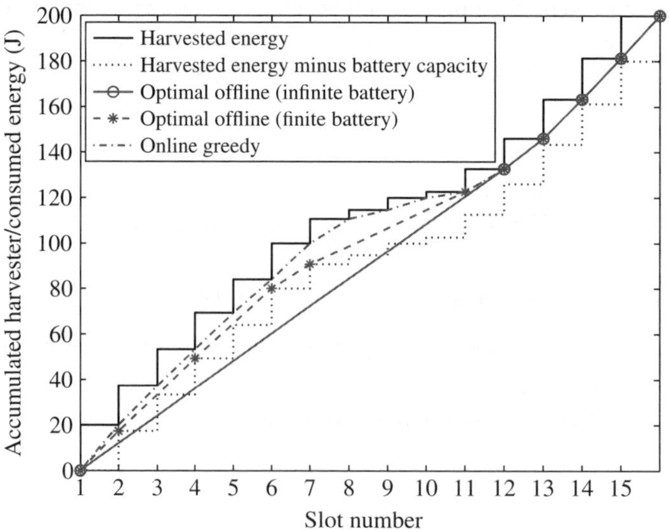

Figure 3.3 The accumulated amounts of harvested energy and energy consumed by the optimal offline scheme with infinite/finite battery capacity given $\boldsymbol{\omega}_1$.

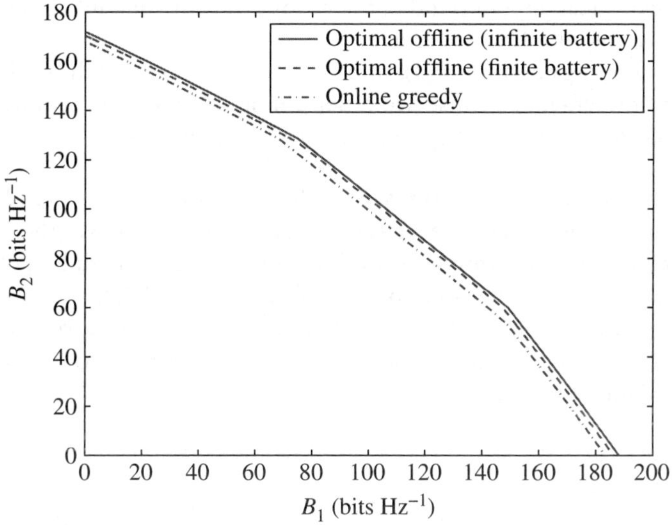

Figure 3.4 The performance achieved by the optimal offline scheme with infinite/finite battery capacity and the greedy scheme given $\boldsymbol{\omega}_1$.

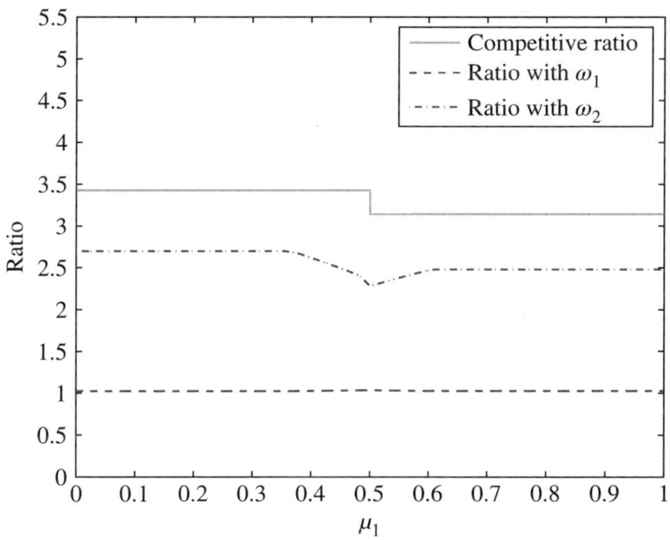

Figure 3.5 The competitive ratios and the ratios obtained with $\boldsymbol{\omega}_1$ and $\boldsymbol{\omega}_2$, with infinite battery capacity.

also the competitive ratios with $E_{max} = 20$ J and infinite battery capacity.[3] Note that the sudden change of the competitive ratio is due to the discontinuity in the profit function of $\tilde{\mathcal{A}}$ at $\mu_1 = 0.5$. As we expected, $\boldsymbol{\omega}_1$ results in ratios close to 1, whereas $\boldsymbol{\omega}_2$ results in relatively large ratios, i.e. around 2.5, due to the fact that the power profiles given by the optimal offline and the greedy schemes are dramatically different. The competitive

3 For the case of finite battery capacity, the only difference is that the ratio with ω_1 will be slightly smaller.

ratio, which serves as the theoretical upper bound of the ratios given any possible input sequence, is larger than the ratios with both $\boldsymbol{\omega}_1$ and $\boldsymbol{\omega}_2$.

Next, we consider stochastic energy arrivals and compare the performance of the optimal offline schemes, the greedy scheme, against two other online schemes that require no/partial statistical information about the EH process described as follows:

(1) *Online on–off scheme*: The two transmitters transmit with a sum power of $\frac{\tilde{E}}{T}$, where $\tilde{E}$ is the transmitters' estimation of the average energy arrival amount, whenever there is energy available; otherwise, the transmission is suspended. The rate scheduling of the two transmitters follows the optimal solution of problem (3.22)–(3.23).
(2) *Online passive scheme*: The sum power allocated in the nth slot is given as $\frac{1}{T}\sum_{i=1}^{n}\frac{E_i}{N+1-i}$. Similarly, the rate scheduling of the two transmitters follows the optimal solution of problem (3.22)–(3.23).

For the case of finite battery capacity, any energy exceeding the battery capacity will be discarded by the above two schemes.

For the purpose of exposition, we assume that the energy arrival amount, i.e. the amount of harvested energy that arrives at the beginning of a slot, follows a uniform distribution over $[0, E_{\max}]$. The average (per slot) energy arrival amount, denoted by $\bar{E}$, is thus given by $\bar{E} = \frac{E_{\max}}{2}$. For the online on–off scheme, we use $\alpha = \tilde{E}/\bar{E}$ to indicate the estimation accuracy for the average energy arrival amount (i.e. when $\alpha = 1$, the estimation is accurate; otherwise, the average energy arrival amount is either overestimated or underestimated). Figures 3.6 and 3.7 show the performance of the schemes for different average energy arrival amounts with infinite and finite battery capacities, respectively, with weighting factors $\mu_1 = 0.6$ and $\mu_2 = 0.4$. The simulation results are averaged over 2000 realizations.

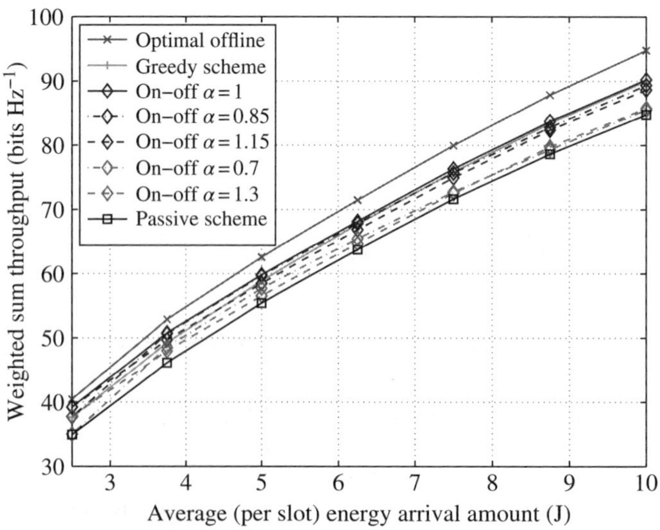

Figure 3.6 Weighted sum throughput versus average energy arrival amounts with infinite battery capacity when $\mu_1 = 0.6, \mu_2 = 0.4$.

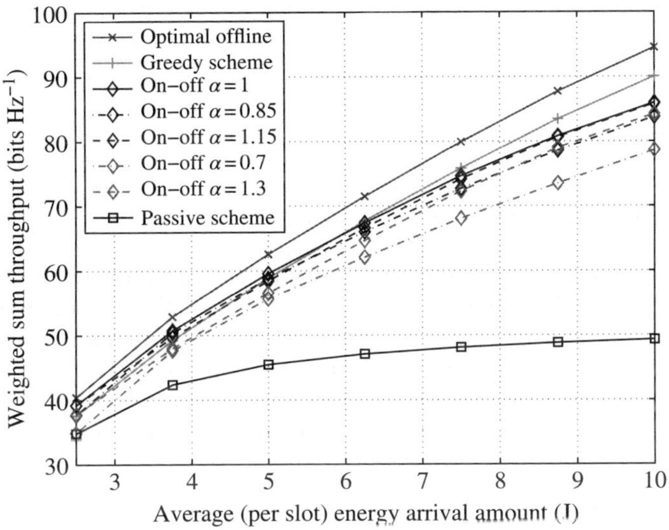

Figure 3.7 Weighted sum throughput versus average energy arrival amounts with finite battery capacity when $\mu_1 = 0.6, \mu_2 = 0.4$.

As seen from Figure 3.6, for the case of infinite battery capacity, the greedy scheme, whose weighted sum throughput is around 5% less than that of the optimal offline scheme for different average energy arrival amounts, is outperformed by the on–off scheme with accurate estimation, i.e. $\alpha = 1$. However, the performance of the on–off scheme degrades as the estimation becomes inaccurate. In particular, the performance of the on–off scheme is worse than the greedy scheme when $\alpha = 1.3$ or 0.7. The passive scheme performs the worst among all the schemes that we considered. For the case of finite battery capacity, we can observe from Figure 3.7 that the on–off scheme with accurate estimation outperforms the greedy scheme in the low average energy arrival amount regime, whereas the greedy scheme performs better than the on–off scheme in the high average energy arrival amount regime. Unlike the greedy scheme that never causes battery overflow, the on–off scheme and the passive scheme may cause battery overflow since the stored energy may not be depleted before the next energy arrival. It is worth noting that the influence of the finite battery capacity on the performance of both the on–off scheme and the passive scheme is more significant in the high average energy arrival amount regime than that in the low average energy arrival amount regime. This is due to the fact that battery overflow is more likely to happen when the energy arrival amount is large. The greedy scheme is thus recognized to enjoy robustness against the estimation error of the energy arrival process statistics and the battery capacity limitation compared with other schemes.

3.2 Relay Channels

In this section, we study the half-duplex orthogonal Gaussian relay channel with EH source and relay nodes, as shown in Figure 3.8. For the purpose of exposition, the

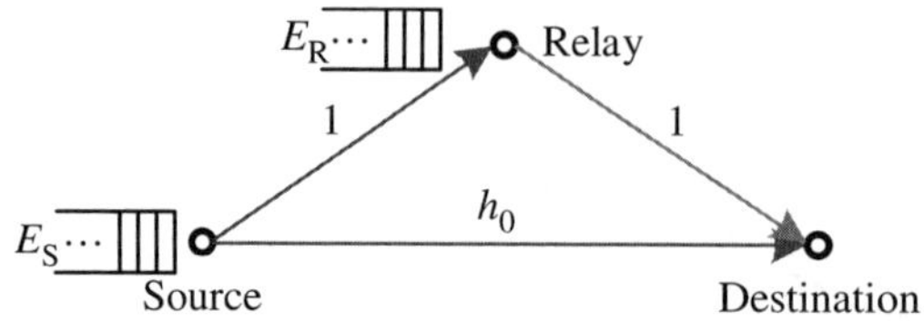

Figure 3.8 Orthogonal relay channel with energy harvesting source and relay nodes.

decode-and-forward (DF) relaying scheme is adopted to exhibit the energy cooperation between the source and relay nodes. In particular, we investigate the throughput maximization over a finite horizon of N-block transmission.[4] In each block, the source transmits a new message, which is received and decoded by the relay and then forwarded to the destination in the subsequent one or more blocks. Specifically, we consider the following two types of data traffic with different decoding delay requirements at the destination:

(1) Delay-constrained (DC) *traffic*: The destination is required to decode the ith source message, $i = 1, \dots, N$, immediately after it receives the signals from the source in the ith block and from the relay in the $(i + 1)$th block. With such a requirement, the relay needs to forward the source message received in one block to the destination immediately in the next block.
(2) No-delay-constrained (NDC) traffic: The destination can tolerate arbitrary decoding delays, provided that all source messages are decoded at the end of each N-block transmission. Consequently, the relay is allowed to store the decoded source message of the ith block and forward it to the destination in any of the remaining $(i + 1)$th, …, $(N + 1)$th blocks.

Clearly, the NDC case allows more flexible relay operations than the DC case, and in the rest of this subsection, we will rigorously prove that by allowing decoding delay at the destination, NDC can achieve higher throughput than DC and is able to exploit a new form of diversity arising from the independent source and relay energy availability over time in cooperative communications, termed "energy diversity," even with time-invariant channels.

3.2.1 System Model

We consider a classic three-node relay channel, which consists of one source–destination pair and one relay, as shown in Figure 3.8. The relay node operates in a half-duplex mode over two orthogonal frequency bands, while the source–relay and source–destination use the same band. For simplicity, we do not consider the bandwidth allocation problem for the relay and assume that the source–relay and relay–destination links operate with equal bandwidth.

We consider the DF relaying scheme, which requires the relay to successfully decode the source message. Moreover, we adopt an N-block transmission protocol described as follows: during each of the N source transmission blocks, say, the ith block, $1 \leq i \leq N$,

4 Note that in total $(N + 1)$-block time is needed for each N-block transmission due to the one-block decoding delay at the relay.

the source transmits a new message w_i with power $P_S(i)$ and rate $R(i)$; upon receiving the signal from the source, the relay decodes w_i and generates a binning index for w_i based on the "random binning" technique [14] with rate $R_B(i+1)$. In the $(i+1)$th block, the relay transmits a message v_{i+1} with power $P_R(i+1)$ and rate $C(P_R(i+1))$. It is noted that for the DC case, v_{i+1} is the binning index of source message w_i only; while for the NDC case, v_{i+1} may contain the information of binning indices for all source messages w_k's, $k \leq i$. Moreover, we assume that each block has B channel uses, where B is assumed large enough such that the channel capacity results in [14, 15] are good approximations to the communication rates in practical systems.

In addition to the block transmission model, we assume that the harvested energy arrives at the beginning of each block with known amounts $E_S(i)$ in the ith block and $E_R(i+1)$ in the $(i+1)$th block, $i = 1, 2, \ldots, N$, at the source and the relay, respectively. In this subsection, we assume that the battery capacity to store the harvested energy is infinite and the consumed energy at the source or relay other than transmission energy is small and thus negligible. Thus, the amount of energy available for each block transmission is constrained by the following source and relay EH constraints:

$$\sum_{i=1}^{k} P_S(i) \leq \frac{1}{B} \sum_{i=1}^{k} E_S(i), \tag{3.39}$$

$$\sum_{i=1}^{k} P_R(i+1) \leq \frac{1}{B} \sum_{i=1}^{k} E_R(i+1), \quad k = 1, \ldots, N. \tag{3.40}$$

For the ith source and the $(i+1)$th relay transmission blocks,[5] $i = 1, \ldots, N$, the channel input–output relationships are given as

$$y_{sr}(i) = \sqrt{h_{sr}} x_s(i) + n_r(i), \tag{3.41}$$

$$y_{sd}(i) = \sqrt{h_{sd}} x_s(i) + n_d(i), \tag{3.42}$$

$$y_{rd}(i+1) = \sqrt{h_{rd}} x_r(i+1) + w_d(i+1), \tag{3.43}$$

where $x_s(i)$ and $x_r(i+1)$ are the transmitted signals in the ith source and the $(i+1)$th relay transmission blocks with power $P_S(i)$ and $P_R(i+1)$, respectively; $y_{sr}(i)$ is the received signal at the relay; $y_{sd}(i)$ and $y_{rd}(i+1)$ are the received signals at the destination from the source and the relay, respectively; h_{sr}, h_{rd}, and h_{sd} are the constant channel power gains for the source–relay, relay–destination, and source–destination links, respectively; and $n_r(i)$, $n_d(i)$, and $w_d(i+1)$ are i.i.d. CSCG noises each with zero mean and unit variance.

With the above assumptions, the received signal-to-noise ratio (SNR) for the source–relay, source–destination, and relay–destination links are given as $\gamma_{sr}(i) = P_S(i)h_{sr}$, $\gamma_{sd}(i) = P_S(i)h_{sd}$, and $\gamma_{rd}(i+1) = P_R(i+1)h_{rd}$, respectively. Define new source–relay energy and power profiles as $\tilde{E}_S(i) = E_S(i)h_{sr}$, $\tilde{E}_R(i+1) = E_R(i+1)h_{rd}$, $\tilde{P}_S(i) = P_S(i)h_{sr}$, and $\tilde{P}_R(i+1) = P_R(i+1)h_{rd}$ and new channel gains as $\tilde{h}_{sr} = \tilde{h}_{rd} = 1$ and $\tilde{h}_{sd} = \frac{h_{sd}}{h_{sr}} \doteq h_0$. It is easy to check that with the new defined parameters, each link has the same SNR as before and the EH constraints given in (3.39) are satisfied accordingly

5 Note that the $(i+1)$th relay transmission block in fact corresponds to the ith source message in the DC case.

with the new power and energy profiles. As such, we could always determine the source and relay power profiles $\tilde{P}_S(i)$'s and $\tilde{P}_R(i+1)$'s first under the assumptions that $\tilde{h}_{sr} = \tilde{h}_{rd} = 1$ and $\tilde{h}_{sd} = \frac{h_{sd}}{h_{sr}}$ and then obtain $P_S(i)$'s and $P_R(i+1)$'s by scaling accordingly. Therefore, for notation simplicity and without loss of generality, we simplify the signal models in (3.41)–(3.43) as

$$y_{sr}(i) = x_s(i) + n_r(i), \tag{3.44}$$

$$y_{sd}(i) = \sqrt{h_0}x_s(i) + n_d(i), \tag{3.45}$$

$$y_{rd}(i+1) = x_r(i+1) + w_d(i+1), \tag{3.46}$$

by setting $h_{sr} = h_{rd} = 1$ and $h_{sd} = h_0$.

Moreover, it is assumed that $0 \leq h_0 < 1$, which means that the source–relay link is stronger than the source–destination link. Thus, the relay can always help with increasing the achievable rate from the source to the destination.

3.2.2 Problem Formulation

3.2.2.1 Delay-Constrained Case

First consider the DC case. Since in the ith source transmission block the source transmits message w_i with power $P_S(i)$ and rate $R(i)$, the relay decodes w_i reliably only if

$$R(i) \leq C(P_S(i)). \tag{3.47}$$

Then in the next block, the relay partitions w_i into bins with an equivalent rate $R_B(i+1)$ [14] and transmits the binning index in message v_{i+1} to the destination with power $P_R(i+1)$. At the destination, it first decodes v_{i+1}, if

$$R_B(i+1) \leq C(P_R(i+1)), \tag{3.48}$$

and then decodes the original message w_i if

$$R(i) \leq C(h_0 P_S(i)) + R_B(i+1) \leq C(h_0 P_S(i)) + C(P_R(i+1)), \tag{3.49}$$

where the second inequality is due to (3.48). From (3.47) and (3.49), the achievable rate for the ith source message is given by

$$R(i) = \min\{C(P_S(i)), C(h_0 P_S(i)) + C(P_R(i+1))\}. \tag{3.50}$$

Note that for the case of $h_0 = 0$, the coding scheme for the relay can be simplified to repetition coding, i.e. the source and the relay can use the same codebook.

Considering the N-block transmission, the average throughput in the unit of bits/sec/Hz (bps Hz^{-1}) is maximized by solving the following problem:

$$\max \quad \frac{1}{2(N+1)} \cdot \sum_{i=1}^{N} \min\{C(P_S(i)), C(h_0 P_S(i)) + C(P_R(i+1))\} \tag{3.51}$$

$$\text{s}\cdot\text{t}\cdot \quad (3.39), \quad P_S(i) \geq 0, \quad P_R(i+1) \geq 0, \quad i = 1, \ldots, N, \tag{3.52}$$

where the factor $\frac{1}{2}$ in (3.51) is due to half-duplex relaying and $\frac{1}{N+1}$ is due to the fact that each N-block transmission requires $(N+1)$-block duration. Next, some properties of the optimal power allocation solution for problem (3.51)–(3.52) are revealed.

Proposition 3.7 *For $0 \leq h_0 < 1$, there exist optimal power profiles $P_S^*(i)$'s and $P_R^*(i+1)$'s, which achieve the maximum throughput of problem (3.51)–(3.52) and satisfy the following inequalities:*

$$C(P_S^*(i)) \geq C(h_0 P_S^*(i)) + C(P_R^*(i+1)), \quad i = 1, \dots, N; \tag{3.53}$$

moreover, for the case of $h_0 = 0$, there exist optimal power profiles for problem (3.51)–(3.52) with

$$P_S^*(i) = P_R^*(i+1), \quad i = 1, \dots, N. \tag{3.54}$$

Proof: If (3.53) is not satisfied for any i, we can always decrease $P_R^*(i+1)$ until it is satisfied, without reducing the achievable rate of the ith source message. Similarly, if (3.54) is not satisfied at any i, we can reduce $P_S^*(i)$ if $P_S^*(i) > P_R^*(i+1)$ or $P_R^*(i+1)$ if $P_R^*(i+1) > P_S^*(i)$ until the equality holds, without reducing the rate of the ith source message. Thus, this proposition is proved. □

Remark 3.4 From Proposition 3.7, we infer that the optimal solution of problem (3.51)–(3.52) is not unique in general, e.g. when the energy harvested at the relay is excessively larger than that at the source. In the sequel, we are only interested in finding the optimal solutions for problem (3.51)–(3.52) satisfying (3.53) and (3.54) for the cases of $0 < h_0 < 1$ and $h_0 = 0$, respectively, which achieve the minimum energy consumptions at the source and relay.

By (3.53) and (3.48), we obtain that

$$\begin{aligned} C(P_S^*(i)) &\geq C(h_0 P_S^*(i)) + C(P_R^*(i+1)) \\ &\geq C(h_0 P_S^*(i)) + R_B^*(i+1), \quad i = 1, \dots, N. \end{aligned}$$

Together with (3.48), it follows that

$$R_B(i+1) = \min\{C(P_R(i+1)), C(P_S(i)) - C(h_0 P_S(i))\}, \tag{3.55}$$

$i = 1, \dots, N$. As such, if we can solve problem (3.51)–(3.52), by further applying (3.55), we can obtain the optimal relay rate allocation for the DC case.

It is easy to verify that problem (3.51)–(3.52) is convex and thus solvable by existing convex optimization techniques, e.g. the interior point method [9]. However, such an approach does not reveal any insight for the optimal solution. Thus, in this subsection, we develop an alternative method to solve problem (3.51)–(3.52) by exploiting its special structure, as will be shown later.

3.2.2.2 No-Delay-Constrained Case

For the NDC case, the relay operates the same as the DC case except that it is allowed to transmit the binning index for message w_i in messages $v_{i+1}, \dots, v_{N+1}$ instead of v_{i+1} only as in the DC case. At the destination, the binning indices for all source messages can be successfully decoded if

$$\sum_{i=1}^{N} R_{\mathrm{B}}(i+1) = \sum_{i=1}^{N} C(P_{\mathrm{R}}(i+1)), \tag{3.56}$$

$$\sum_{i=k}^{N} R_{\mathrm{B}}(i+1) \leq \sum_{i=k}^{N} C(P_{\mathrm{R}}(i+1)), \quad 2 \leq k \leq N, \tag{3.57}$$

which is equivalent to

$$\sum_{i=1}^{k} R_{\mathrm{B}}(i+1) \geq \sum_{i=1}^{k} C(P_{\mathrm{R}}(i+1)), \quad 1 \leq k \leq N-1, \tag{3.58}$$

$$\sum_{i=1}^{N} R_{\mathrm{B}}(i+1) = \sum_{i=1}^{N} C(P_{\mathrm{R}}(i+1)). \tag{3.59}$$

With the decoded binning index, the ith source message can be decoded successfully at the destination if $R(i) \leq C(h_0 P_{\mathrm{S}}(i)) + R_{\mathrm{B}}(i+1), \quad i = 1, \ldots, N$. Combining this with (3.47), the achievable rate of the ith source message is given as

$$\begin{aligned} R(i) &= \min\{C(P_{\mathrm{S}}(i)), C(h_0 P_{\mathrm{S}}(i)) + R_{\mathrm{B}}(i+1)\} \\ &= C(h_0 P_{\mathrm{S}}(i)) + R_{\mathrm{B}}(i+1), \quad i = 1, \ldots, N, \end{aligned} \tag{3.60}$$

where the second equality is due to a similar argument as Proposition 3.7. Note that for the special case of $h_0 = 0$, we have $R(i) = R_{\mathrm{B}}(i+1)$ in (3.60). In addition, (3.60) implies that $C(h_0 P_{\mathrm{S}}(i)) + R_{\mathrm{B}}(i+1) \leq C(P_{\mathrm{S}}(i)), i = 1, \ldots, N$, which leads to

$$\sum_{i=1}^{k} C(P_{\mathrm{S}}(i)) - C(h_0 P_{\mathrm{S}}(i)) \geq \sum_{i=1}^{k} R_{\mathrm{B}}(i+1), \quad k = 1, \ldots, N. \tag{3.61}$$

From (3.58) and (3.61), we obtain

$$\sum_{i=1}^{k} C(h_0 P_{\mathrm{S}}(i)) + C(P_{\mathrm{R}}(i+1)) \leq \sum_{i=1}^{k} C(P_{\mathrm{S}}(i)), \quad k = 1, \ldots, N. \tag{3.62}$$

Using (3.60) and (3.62), the average throughput for the NDC case is maximized by solving the following problem:

$$\max \quad \frac{1}{2(N+1)} \sum_{i=1}^{N} C(h_0 P_{\mathrm{S}}(i)) + C(P_{\mathrm{R}}(i+1)) \tag{3.63}$$

$$\text{s.t.} \quad \sum_{i=1}^{k} C(h_0 P_{\mathrm{S}}(i)) + C(P_{\mathrm{R}}(i+1)) \leq \sum_{i=1}^{k} C(P_{\mathrm{S}}(i)), \quad (3.52). \tag{3.64}$$

Problem (3.63)–(3.64) is non-convex due to the first constraint in (3.64) and thus difficult to solve at a first glance. We will derive the optimal solution for this problem based on a separate source and relay power allocation strategy in Section 3.2.4. We conclude this section by the following two propositions regarding problems (3.51)–(3.52) and (3.63)–(3.64).

Proposition 3.8 *The maximum value of problem (3.63)–(3.64) is no smaller than that of problem (3.51)–(3.52).*

Proof: By Proposition 3.7, it follows that problem (3.51)–(3.52) has the same maximum value as the following problem:

$$\max \quad \frac{1}{2(N+1)} \sum_{i=1}^{N} C(h_0 P_S(i)) + C(P_R(i+1)) \tag{3.65}$$

$$\text{s.t.} \quad C(h_0 P_S(i)) + C(P_R(i+1)) \leq C(P_S(i)), \quad (3.52). \tag{3.66}$$

It is easy to see that any solution that satisfies (3.66) will also satisfy (3.64) for problem (3.63)–(3.64), but not vice versa. As such, the feasible set of problem (3.63)–(3.64) contains that of problem (3.65)–(3.66), which implies that the maximum value of problem (3.63)–(3.64) is no smaller than that of problem (3.65)–(3.66) or (3.51)–(3.52). The proof is thus completed. □

Proposition 3.9 *For any optimal power profiles $P_S^*(i)$'s and $P_R^*(i+1)$'s for problem (3.51)–(3.52) or (3.63)–(3.64), if $0 < h_0 < 1$, the constraint $B\sum_{i=1}^{N} P_S^*(i) \leq \sum_{i=1}^{N} E_S(i)$ must be satisfied with equality; if $h_0 = 0$, at least one of two constraints $B\sum_{i=1}^{N} P_S^*(i) \leq \sum_{i=1}^{N} E_S(i)$ and $B\sum_{i=1}^{N} P_R^*(i+1) \leq \sum_{i=1}^{N} E_R(i+1)$ must be satisfied with equality.*

Proof: Supposing that the above equalities do not hold, we can increase the source or relay transmission power without violating the energy constraints to further improve the throughput, which contradicts the fact that $P_S^*(i)$'s and $P_R^*(i+1)$'s are optimal solutions of problem (3.51)–(3.52) or (3.63)–(3.64). Thus, this proposition is proved. □

3.2.3 Optimal Solution for the Delay-Constrained Case

In this section, we solve problem (3.51)–(3.52) for the DC case. We first present a monotonic property for the optimal power allocation in problem (3.51)–(3.52), upon which we then develop an efficient algorithm to solve this problem.

3.2.3.1 Monotonic Power Allocation

Since the optimal solution of problem (3.51)–(3.52) may not be unique (cf. Remark 3.4), we are interested in finding one optimal solution for this problem that leads to the minimum power consumption at the source and relay. For such an optimal solution, we have the following monotonic property.

Proposition 3.10 *The optimal solution of problem (3.51)–(3.52), satisfying (3.53) for the case of $0 < h_0 < 1$ or (3.54) for the case of $h_0 = 0$, is non-decreasing over i, i.e. $P_S^*(i) \leq P_S^*(i+1)$, $P_R^*(i+1) \leq P_R^*(i+2)$, $i = 1, 2, \ldots, N-1$.*

Proof: Denote the optimal source and relay power profiles of problem (3.51)–(3.52) as $P_S^*(i)$'s and $P_R^*(i+1)$'s, respectively. For any consecutive power pair, consider the following three cases:

(1) $P_S^*(i) > P_S^*(i+1)$ and $P_R^*(i+1) > P_R^*(i+2)$: Define a new power allocation profile as $\tilde{P}_S(i) = \tilde{P}_S(i+1) = \frac{P_S^*(i)+P_S^*(i+1)}{2}$ and $\tilde{P}_R(i+1) = \tilde{P}_R(i+2) = \frac{P_R^*(i+1)+P_R^*(i+2)}{2}$. It is easy to check that the new power allocation profile still satisfies the energy constraint in (3.52); moreover, since the objective function in (3.51) is concave, it follows that the new allocation leads to a larger sum rate over these two blocks. Thus, this case cannot happen.

(2) $P_S^*(i) > P_S^*(i+1)$ and $P_R^*(i+1) \le P_R^*(i+2)$: First, we prove that the ith constraint in (3.53) is not achieved with equality by contradiction as follows. Assuming that its equality is achieved, it is then observed that $P_R^*(i+1) \le P_R^*(i+2)$ is contradicted with the following results:

$$\log(1 + P_R^*(i+1)) = \log\left(\frac{1+P_S^*(i)}{1+h_0P_S^*(i)}\right) \tag{3.67}$$

$$> \log\left(\frac{1+P_S^*(i+1)}{1+h_0P_S^*(i+1)}\right) \tag{3.68}$$

$$\ge \log(1 + P_R^*(i+2)), \tag{3.69}$$

where (3.68) is due to the fact that $\log\left(\frac{1+x}{1+h_0x}\right)$ is strictly increasing over $x > 0$ for any fixed h_0 with $0 \le h_0 < 1$ and (3.69) is by Proposition 3.7.
Thus, there must exist δ, $0 < \delta \le \frac{P_S^*(i)-P_S^*(i+1)}{2}$, such that

$$C(P_S^*(i) - \delta) \ge C(h_0(P_S^*(i) - \delta)) + C(P_R^*(i+1)). \tag{3.70}$$

Define a new power allocation profile as $\tilde{P}_S(i) = P_S^*(i) - \delta$, $\tilde{P}_S(i+1) = P_S^*(i+1) + \delta$, $\tilde{P}_R(i+1) = P_R^*(i+1)$, and $\tilde{P}_R(i+2) = P_R^*(i+2)$. It is easy to check that the new allocation satisfies the energy constraints in (3.52) and yields a larger sum rate over these two blocks, i.e.

$$\begin{aligned}\tilde{R}(i) + \tilde{R}(i+1) &\ge C(h_0\tilde{P}_S(i)) + C(P_R^*(i+1)) \\ &\quad + C(h_0\tilde{P}_S(i+1)) + C(P_R^*(i+2)) \end{aligned}\tag{3.71}$$

$$\begin{aligned}&> C(h_0P_S^*(i)) + C(P_R^*(i+1)) \\ &\quad + C(h_0P_S^*(i+1)) + C(P_R^*(i+2)) \end{aligned}\tag{3.72}$$

$$= R^*(i) + R^*(i+1),$$

where (3.71) is due to (3.70) and the following fact: since $C(x+\delta) - C(x) \ge C(h_0(x+\delta)) - C(h_0x)$ for $0 < h_0 < 1$, it follows that

$$\begin{aligned}&C(\tilde{P}_S(i+1)) - (C(h_0\tilde{P}_S(i+1)) + C(P_R^*(i+2))) \\ &\ge C(P_S^*(i+1)) - C(h_0P_S^*(i+1)) - C(P_R^*(i+2)) \ge 0;\end{aligned}$$

and (3.72) is due to the concavity of $\log(x)$ and $\delta \le \frac{P_S^*(i)-P_S^*(i+1)}{2}$. Therefore, this case cannot happen.

(3) $P_S^*(i) \le P_S^*(i+1)$ and $P_R^*(i+1) > P_R^*(i+2)$: Since $\log\left(\frac{1+x}{1+h_0 x}\right)$ is strictly increasing over $x > 0$ for fixed h_0, $0 \le h_0 < 1$, it follows that

$$\log\left(\frac{1+P_S^*(i+1)}{1+h_0 P_S^*(i+1)}\right) \ge \log\left(\frac{1+P_S^*(i)}{1+h_0 P_S^*(i)}\right) \ge \log(1+P_R^*(i+1)),$$

where the last inequality is due to Proposition 3.7. Thus, it is obtained that

$$\begin{aligned} C(P_S^*(i+1)) &\ge C(h_0 P_S^*(i+1)) + C(P_R^*(i+1)) \\ &> C(h_0 P_S^*(i+1)) + C(\tilde{P}_R(i+2)), \end{aligned} \tag{3.73}$$

where $\tilde{P}_R(i+1) = \tilde{P}_R(i+2) = \frac{P_R^*(i+1)+P_R^*(i+2)}{2}$. By keeping $P_S^*(i)$ and $P_S^*(i+1)$ unchanged and updating the relay power values as $\tilde{P}_R(i+1)$ and $\tilde{P}_R(i+2)$, it is observed that the relay energy constraints are still satisfied and the sum rate is improved, i.e.

$$\begin{aligned} \tilde{R}(i) + \tilde{R}(i+1) &= C(h_0 P_S^*(i)) + C(\tilde{P}_R(i+1)) \\ &\quad + C(h_0 P_S^*(i+1)) + C(\tilde{P}_R(i+2)) &(3.74) \\ &> C(h_0 P_S^*(i)) + C(P_R^*(i+1)) \\ &\quad + C(h_0 P_S^*(i+1)) + C(P_R^*(i+2)) &(3.75) \\ &= R^*(i) + R^*(i+1), \end{aligned}$$

where (3.74) is due to (3.73) and the fact that $C(P_S^*(i)) \ge C(h_0 P_S^*(i)) + C(P_R^*(i+1)) > C(h_0 P_S^*(i)) + C(\tilde{P}_R(i+1))$ and (3.75) is due to the concavity of $\log(x)$. Thus, this case cannot happen. To summarize, since all the above three cases cannot be true, the only remaining case of $P_S^*(i) \le P_S^*(i+1)$ and $P_R^*(i+1) \le P_R^*(i+2)$ must be true. Proposition 3.10 is thus proved. □

With the above monotonic properties, we next address the optimal solution of problem (3.51)–(3.52) for the cases with and without the direct source–destination link, respectively.

3.2.3.2 The Case with Direct Link

For the case of $0 < h_0 < 1$, we consider the Lagrangian of problem (3.51)–(3.52), which is given by

$$\begin{aligned} &\mathcal{L}(P_S(i), P_R(i+1), \mu_k, \lambda_k, \gamma_i, \eta_{i+1}) \\ &= \frac{1}{2(N+1)} \sum_{i=1}^{N} \min\{C(P_S(i)), C(h_0 P_S(i)) + C(P_R(i+1))\} \\ &\quad - \sum_{k=1}^{N} \mu_k \left(\sum_{i=1}^{k} BP_S(i) - E_S(i) \right) - \sum_{k=1}^{N} \lambda_k \left(\sum_{i=1}^{k} BP_R(i+1) - E_R(i+1) \right) \\ &\quad + \sum_{i=1}^{N} \gamma_i P_S(i) + \sum_{i=1}^{N} \eta_{i+1} P_R(i+1), \end{aligned} \tag{3.76}$$

where μ_k, λ_k, γ_i, and η_{i+1} are the nonnegative Lagrangian multipliers. By taking the derivative over $P_{\rm S}(i)$ and $P_{\rm R}(i+1)$, it is obtained that

$$\frac{\partial \mathcal{L}}{\partial P_{\rm S}(i)} = \begin{cases} \dfrac{1}{4(N+1)} \times \dfrac{1}{1+P_{\rm S}(i)} & P_{\rm R}(i+1) \geq \dfrac{(1-h_0)P_{\rm S}(i)}{1+h_0 P_{\rm S}(i)} \\ \dfrac{1}{4(N+1)} \times \dfrac{h_0}{1+h_0 P_{\rm S}(i)} & \text{otherwise} \end{cases}$$
$$- B \sum_{k=i}^{N} \mu_k + \gamma_i, \tag{3.77}$$

$$\frac{\partial \mathcal{L}}{\partial P_{\rm R}(i+1)} = \begin{cases} 0 & P_{\rm R}(i+1) \geq \dfrac{(1-h_0)P_{\rm S}(i)}{1+h_0 P_{\rm S}(i)} \\ \dfrac{1}{4(N+1)} \times \dfrac{1}{1+P_{\rm R}(i+1)} & \text{otherwise} \end{cases}$$
$$- B \sum_{k=i}^{N} \lambda_k + \eta_{i+1}. \tag{3.78}$$

By letting $\frac{\partial \mathcal{L}}{\partial P_{\rm S}(i)} = 0$ and $\frac{\partial \mathcal{L}}{\partial P_{\rm R}(i+1)} = 0$, the optimal solutions for problem (3.51)–(3.52) are obtained as follows:

(1) *Case I*: If $P_{\rm R}^*(i+1) \geq \frac{(1-h_0)P_{\rm S}^*(i)}{1+h_0 P_{\rm S}^*(i)}$,

$$\begin{cases} P_{\rm S}^*(i) = \left(\dfrac{1}{4(N+1) \sum_{k=i}^{N} \mu_k} - 1 \right)^+ \\ P_{\rm R}^*(i+1) = \dfrac{(1-h_0)P_{\rm S}^*(i)}{1+h_0 P_{\rm S}^*(i)} \end{cases}. \tag{3.79}$$

(2) *Case II*: If $P_{\rm R}^*(i+1) \leq \frac{(1-h_0)P_{\rm S}^*(i)}{1+h_0 P_{\rm S}^*(i)}$,

$$\begin{cases} P_{\rm S}^*(i) = \left(\dfrac{1}{4(N+1) \sum_{k=i}^{N} \mu_k} - \dfrac{1}{h_0} \right)^+ \\ P_{\rm R}^*(i+1) = \left(\dfrac{1}{4(N+1) \sum_{k=i}^{N} \lambda_k} - 1 \right)^+ \end{cases}. \tag{3.80}$$

Remark 3.5 From the above expressions, it is observed that the source and relay power profiles need to be jointly optimized, since the achievable rate for the ith source message is limited by the available source power in Case I, but by the available relay power in Case II.

From the KKT optimality conditions of problem (3.51)–(3.52), it follows that λ_k and μ_k are strictly positive only when their corresponding relay and source energy constraints are satisfied with equality. Thus, it follows that the optimal source power can change the value from one block to another only when the harvested source energy is exhausted at the current block or there is a transition between the two values given by (3.79) and (3.80). The latter case is due to the fact that $h_0 < 1$, and thus the source power values given by (3.79) and (3.80) are different even when the source energy constraint is not active. By further considering the result of Proposition 3.10, we know that changing source power values from (3.79) to (3.80) is not possible, and only transitions from (3.80) to (3.79) can occur. Thus, we have the following proposition for the optimal source power allocation.

Proposition 3.11 *Consider the optimal source power $P_S^*(i)$'s for problem (3.51)–(3.52), which satisfy Proposition 3.10. For any two successive source energy exhausting blocks, k_i and k_j with $k_i < k_j$, i.e. the source energy constraints $\sum_{i=1}^{k_i} P_S^*(i) \leq \frac{1}{B} \sum_{i=1}^{k_i} E_S(i)$ and $\sum_{i=1}^{k_j} P_S^*(i) \leq \frac{1}{B} \sum_{i=1}^{k_j} E_S(i)$ are active, while the other constraints $\sum_{i=1}^{j} P_S^*(i) \leq \frac{1}{B} \sum_{i=1}^{j} E_S(i)$, $j = k_i + 1, \ldots, k_j - 1$, are all inactive, the optimal source power values from the $(k_i + 1)$th to the k_jth blocks can only be one of the following three cases:*

(1) Scenario I: $P_S^*(i)$, $i = k_i + 1, \ldots, k_j$, are identical and given by (3.79).
(2) Scenario II: $P_S^*(i)$, $i = k_i + 1, \ldots, k_j$, are identical and given by (3.80).
(3) Scenario III: There exists k_0 with $k_i < k_0 < k_j$ such that $P_S^*(i) = P_0$, $i = k_i + 1, \ldots, k_0$, and $P_S^*(i) = P_0 + \frac{1}{h_0} - 1$, $i = k_0 + 1, \ldots, k_j$, where P_0 is jointly determined by (3.79) and (3.80). Define the k_0th block as the source power transition block for this scenario.

Based on Proposition 3.11, we know that if we could identify all blocks at which the source energy gets exhausted and furthermore all the scenarios corresponding to Proposition 3.11, the optimal source and relay power profiles for problem (3.51)–(3.52) can be obtained accordingly from (3.79) and (3.80). Therefore, we propose Algorithm 3.1 to solve problem (3.51)–(3.52), whose optimality proof can be seen in [2]. The main procedure of this proposed algorithm is described as follows.

Starting from the first block, Algorithm 3.1 implements a forward searching for the optimal power allocation until the Nth block is reached. Suppose that the $(i-1)$th block is an energy exhausting block for the source and the optimal power allocations for the source and relay have been obtained up to this block, denoted by $P_S^*(n)$ and $P_R^*(n+1)$, respectively, $n = 1, \ldots, i-1$. Then, starting from the ith source and $(i+1)$th relay transmission blocks, we first compute $i_{s,0}$ and $i_{r,0}$ (defined in (3.81)), corresponding to the next possible source and relay energy exhausting blocks, respectively, and the source and relay power values $\tilde{P}_S^{i,0}$ (from the ith to $i_{s,0}$th blocks) and $\tilde{P}_R^{i+1,0}$ (from the $(i+1)$th to

Algorithm 3.1: Compute the optimal solution of problem (3.51)–(3.52) for the case of $0 < h_0 < 1$.

Input: $i = 1, N$
Output: $P_S^*(i), P_R^*(i+1)$

repeat
  Compute $i_{s,0}$ and $i_{r,0}$ by (3.81), and $\widetilde{P}_S^{i,0}$ and $\widetilde{P}_R^{i+1,0}$ by (3.82)
  if $\widetilde{P}_R^{i+1,0} \geq \frac{(1-h_0)\widetilde{P}_S^{i,0}}{1+h_0\widetilde{P}_S^{i,0}}$ **then**
    Compute $P_S^*(i)$ and $P_R^*(i+1)$ by (3.83), and set $i = i_{s,0} + 1$
    if $\widetilde{P}_R^{i+1,0} < \frac{(1-h_0)\widetilde{P}_S^{i,0}}{1+h_0\widetilde{P}_S^{i,0}}$ **then**
      Compute $i_{s,k}$, $i_{r,p}$, $\widetilde{P}_S^{i,k}$ and $\widetilde{P}_R^{i+1,p}$, $k, p \geq 1$ by (3.84)–(3.86), and check
      if these exists k_0 satisfying (3.87) **then**
        Compute j_s and $\widehat{P}_S^i$ by (3.88) and (3.89), respectively, and check
        if $\widehat{P}_S^i$ satisfies (3.90) **then**
          Compute $P_S^*(i)$ and $P_R^*(i+1)$ by (3.91) and (3.92)
          Set $i = j_s + 1$
        else
          Compute $P_S^*(i)$ and $P_R^*(i+1)$ by (3.93)
          Set $i = \min\{k_0, i_{s,0}\} + 1$
        end if
      else
        Compute $P_S^*(i)$ and $P_R^*(i+1)$ by (3.93)
        Set $i = i_{s,0} + 1$
      end if
    end if
  end if
until $i > N$

($i_{r,0}$ + 1)th blocks), respectively, which are given as

$$\begin{cases} i_{s,0} = \arg\min_{i \leq j \leq N} \left\{ \frac{\sum_{k=i}^{j} E_S(k)}{(j-i+1)B} \right\}, \\ i_{r,0} = \arg\min_{i \leq j \leq N} \left\{ \frac{\tilde{E}_R(i+1) + \sum_{k=i}^{j} E_R(k+1)}{(j-i+1)B} \right\}, \end{cases} \tag{3.81}$$

with $\tilde{E}_R(i+1)$ denoting the relay energy left before the $(i+1)$th relay transmission block, i.e. $\tilde{E}_R(2) = 0$ and $\tilde{E}_R(i+1) = \sum_{k=1}^{i-1} E_R(k+1) - BP_R^*(k+1)$, $i = 2, \ldots, N$, and

$$\tilde{P}_S^{i,0} = \frac{\sum_{k=i}^{i_{s,0}} E_S(k)}{(i_{s,0} - i + 1)B}, \quad \tilde{P}_R^{i+1,0} = \frac{\tilde{E}_R(i+1) + \sum_{k=i}^{i_{r,0}} E_R(k+1)}{(i_{r,0} - i + 1)B}. \tag{3.82}$$

Next, by comparing $\tilde{P}_S^{i,0}$ and $\tilde{P}_R^{i+1,0}$, we determine which scenario shown in Proposition 3.11 should happen:

(1) If $\tilde{P}_{\mathrm{R}}^{i+1,0} \geq \frac{(1-h_0)\tilde{P}_{\mathrm{S}}^{i,0}}{1+h_0\tilde{P}_{\mathrm{S}}^{i,0}}$, it is claimed that Scenario I happens, and the optimal source and relay power values are given as

$$P_{\mathrm{S}}^*(n) = \tilde{P}_{\mathrm{S}}^{i,0}, \quad P_{\mathrm{R}}^*(n+1) = \frac{(1-h_0)\tilde{P}_{\mathrm{S}}^{i,0}}{1+h_0\tilde{P}_{\mathrm{S}}^{i,0}}, \quad n = i, \dots, i_{s,0}. \tag{3.83}$$

Then, we set $i = i_{s,0} + 1$ and continue the forward search.

(2) If $\tilde{P}_{\mathrm{R}}^{i+1,0} < \frac{(1-h_0)\tilde{P}_{\mathrm{S}}^{i,0}}{1+h_0\tilde{P}_{\mathrm{S}}^{i,0}}$, Scenario II or III may happen. To determine whether Scenario III happens or not, we need to compute the index k_0, $1 < k_0 < N$, of the source power transition block defined in Scenario III of Proposition 3.11.

Let the indices $i_{s,k}$ and $i_{r,p}$, $k, p \geq 1$, correspond to the source and relay energy exhausting blocks after $i_{s,k-1}$ and $i_{r,p-1}$, respectively, and $\tilde{P}_{\mathrm{S}}^{i,k}$ and $\tilde{P}_{\mathrm{R}}^{i+1,p}$ be the source and relay power values between the two successive energy exhausting blocks, i.e.

$$i_{s,k} = \arg \min_{i_{s,k-1} < j \leq N} \left\{ \frac{\sum_{k=i_{s,k-1}+1}^{j} E_{\mathrm{S}}(k)}{(j - i_{s,k-1})B} \right\}, \tag{3.84}$$

$$i_{r,p} = \arg \min_{i_{r,p-1} < j \leq N} \left\{ \frac{\sum_{k=i_{r,p-1}+1}^{j} E_{\mathrm{R}}(k+1)}{(j - i_{r,p-1})B} \right\}, \tag{3.85}$$

$$\tilde{P}_{\mathrm{S}}^{i,k} = \frac{\sum\limits_{k=i_{s,k-1}+1}^{i_{s,k}} E_{\mathrm{S}}(k)}{(i_{s,k} - i_{s,k-1})B}, \quad \tilde{P}_{\mathrm{R}}^{i+1,p} = \frac{\sum\limits_{k=i_{r,p-1}+1}^{i_{r,p}} E_{\mathrm{R}}(k+1)}{(i_{r,p} - i_{r,p-1})B}. \tag{3.86}$$

Then, we find $k_0 > i$ such that

$$\begin{aligned} \tilde{P}_{\mathrm{R}}(j+1) &< \frac{(1-h_0)\tilde{P}_{\mathrm{S}}(j)}{1+h_0\tilde{P}_{\mathrm{S}}(j)}, \quad j = i, \dots, k_0, \\ \tilde{P}_{\mathrm{R}}(k_0+2) &> \frac{(1-h_0)\tilde{P}_{\mathrm{S}}(k_0+1)}{1+h_0\tilde{P}_{\mathrm{S}}(k_0+1)}, \end{aligned} \tag{3.87}$$

where $\tilde{P}_{\mathrm{S}}(j) = \tilde{P}_{\mathrm{S}}^{i,k}$ for $i_{s,k-1} < j \leq i_{s,k}$, and $\tilde{P}_{\mathrm{R}}(j+1) = \tilde{P}_{\mathrm{R}}^{i+1,p}$ for $i_{r,p-1} < j \leq i_{r,p}$, $k, p \geq 0$, assuming $i_{s,-1} = i_{r,-1} = i - 1$.

Suppose that k_0 is found in the above. Assume that the source power values from block i to k_0 are $\hat{P}_{\mathrm{S}}^{i}$ and from block $k_0 + 1$ to the next source energy exhausting block denoted by j_s are $\hat{P}_{\mathrm{S}}^{i} + \frac{1}{h_0} - 1$, where j_s is given by

$$j_s = \arg \min_{k_0+1 \leq j \leq N} \left\{ \frac{\sum_{k=i}^{j} E_{\mathrm{S}}(k) - (j - k_0)(\frac{1}{h_0} - 1)B}{(j - i + 1)B} \right\} \tag{3.88}$$

and $\widehat{P}_{\mathrm{S}}^{i}$ is given as

$$\widehat{P}_{\mathrm{S}}^{i} = \frac{\sum_{k=i}^{j_s} E_{\mathrm{S}}(k) - (j_s - k_0)(\frac{1}{h_0} - 1)B}{(j_s - i + 1)B}. \tag{3.89}$$

Then, we further check the following conditions inspired by Proposition 3.10 and Scenario III in Proposition 3.11:

$$\begin{aligned}
&\tilde{P}_{\mathrm{S}}^{i,0} \geq \widehat{P}_{\mathrm{S}}^{i} \geq \max\left\{\frac{\tilde{P}_{\mathrm{R}}(k_0+1)}{1 - h_0 - h_0\tilde{P}_{\mathrm{R}}(k_0+1)}, P_{\mathrm{S}}^{*}(i-1)\right\},\\
&\widehat{P}_{\mathrm{S}}^{i} + \frac{1}{h_0} - 1 \leq \frac{\tilde{P}_{\mathrm{R}}(k_0+2)}{1 - h_0 - h_0\tilde{P}_{\mathrm{R}}(k_0+2)}.
\end{aligned} \tag{3.90}$$

If (3.91) is true, it is confirmed that Scenario III happens, and the optimal source and relay power values are given as

$$P_{\mathrm{S}}^{*}(n) = \begin{cases} \widehat{P}_{\mathrm{S}}^{i} & n = i, \dots, k_0 \\ \widehat{P}_{\mathrm{S}}^{i} + \dfrac{1}{h_0} - 1 & n = k_0 + 1, \dots, j_s, \end{cases} \tag{3.91}$$

$$P_{\mathrm{R}}^{*}(n+1) = \begin{cases} \tilde{P}_{\mathrm{R}}^{i+1,p} & i_{r,p-1} < n \leq i_{r,p}, \quad n \leq k_0, \quad p \geq 0 \\ \dfrac{(1-h_0)P_{\mathrm{S}}^{*}(n)}{1 + h_0 P_{\mathrm{S}}^{*}(n)} & n = k_0 + 1, \dots, j_s. \end{cases} \tag{3.92}$$

Then, we set $i = j_s + 1$ and continue the forward search.

If k_0 satisfying (3.87) cannot be found, or with such k_0 found in (3.87) but the conditions in (3.90) are not satisfied, we claim that Scenario III cannot happen and Scenario II must be true. In this case, the optimal source and relay power profiles are given as

$$P_{\mathrm{S}}^{*}(n) = \tilde{P}_{\mathrm{S}}^{i,0}, \quad P_{\mathrm{R}}^{*}(n+1) = \tilde{P}_{\mathrm{R}}^{i+1,0}, \qquad n = i, \dots, \min(k_0, i_{s,0}), \tag{3.93}$$

where $k_0 = \infty$ if no such k_0 satisfies (3.87). Then, we set $i = \min(k_0, i_{s,0}) + 1$ and continue the forward search.

Remark 3.6 According to the proof given in [2], it follows that for the case of $0 < h_0 < 1$, the optimal source power solution of problem (3.51)–(3.52) obtained using Algorithm 3.1 is unique, while this is not necessarily true for the obtained optimal relay power solution according to Proposition 3.7. However, the obtained relay power solution achieves the minimum energy consumption at the relay, since (3.53) is satisfied for each i, $i = 1, \dots, N$.

3.2.3.3 The Case Without Direct Link

Similar to the case with direct link, we obtain the following optimal power solutions for problem (3.51)–(3.52) in the case of $h_0 = 0$:

(1) If $P_{\mathrm{R}}^{*}(i+1) \geq P_{\mathrm{S}}^{*}(i)$,

$$\begin{cases} P_{\mathrm{S}}^{*}(i) = \left(\dfrac{1}{4(N+1)\sum_{k=i}^{N}\mu_k} - 1\right)^{+} \\ P_{\mathrm{R}}^{*}(i+1) = P_{\mathrm{S}}^{*}(i). \end{cases} \tag{3.94}$$

(2) If $P_{\mathrm{R}}^{*}(i+1) \leq P_{\mathrm{S}}^{*}(i)$,

$$\begin{cases} P_{\mathrm{S}}^{*}(i) = P_{\mathrm{R}}^{*}(i+1) \\ P_{\mathrm{R}}^{*}(i+1) = \left(\dfrac{1}{4(N+1)\sum_{k=i}^{N} \lambda_k} - 1 \right)^{+} \end{cases}. \tag{3.95}$$

It is worth noting that to obtain the optimal power profile with the minimum energy consumption, we set the source and relay power levels the same, while in general this is not necessary since the optimal source and relay power profiles may not be unique.

From the above solutions, it is observed that (i) the source power $P_{\mathrm{S}}^{*}(i)$ at the ith block and relay power $P_{\mathrm{R}}^{*}(i+1)$ at the $(i+1)$th block should be identical, $i = 1, \ldots, N$, and (ii) since λ_k and μ_k are strictly positive only when their corresponding energy constraints are satisfied with equality, it follows that the source–relay power changes value only when the harvested energy at either the source or the relay is exhausted.

Based on the above observations, Algorithm 3.1 for the case with direct link can be simplified to obtain the optimal source and relay power allocation for problem (3.51)–(3.52) in the case without direct link. We denote this algorithm as Algorithm 3.2. Since the optimality proof of Algorithm 3.2 is similar to that of Algorithm 3.1, it is omitted here for brevity.

Algorithm 3.2: Compute the optimal solution of problem (3.51)–(3.52) for the case of $h_0 = 0$.

Input: $i = 1, N$

Output: $P_{\mathrm{S}}^{*}(k), P_{\mathrm{R}}^{*}(k+1)$

repeat

Compute $\widetilde{P}_{\mathrm{S}}^{i}$, $\widetilde{P}_{\mathrm{R}}^{i+1}$, i_s, and i_r as follows

$$i_s = \arg\min_{i \leq j \leq N} \left\{ \frac{\widetilde{E}_{\mathrm{S}}(i) + \sum_{k=i}^{j} E_{\mathrm{S}}(k)}{(j-i+1)B} \right\},$$

$$i_r = \arg\min_{i \leq j \leq N} \left\{ \frac{\widetilde{E}_{\mathrm{R}}(i+1) + \sum_{k=i}^{j} E_{\mathrm{R}}(k+1)}{(j-i+1)B} \right\},$$

$$\widetilde{P}_{\mathrm{S}}^{i} = \frac{\widetilde{E}_{\mathrm{S}}(i) + \sum_{k=i}^{i_s} E_{\mathrm{S}}(k)}{(i_s - i + 1)B}, \quad \widetilde{P}_{\mathrm{R}}^{i+1} = \frac{\widetilde{E}_{\mathrm{R}}(i+1) + \sum_{k=i}^{i_r} E_{\mathrm{R}}(k+1)}{(i_r - i + 1)B},$$

where $\widetilde{E}_{\mathrm{S}}(1) = \widetilde{E}_{\mathrm{R}}(2) = 0$, $\widetilde{E}_{\mathrm{S}}(i) = \sum_{k=1}^{i-1} E_{\mathrm{S}}(k) - BP_{\mathrm{S}}^{*}(k)$ and $\widetilde{E}_{\mathrm{R}}(i+1) = \sum_{k=1}^{i-1} E_{\mathrm{R}}(k+1) - BP_{\mathrm{R}}^{*}(k+1)$, $i = 2, \ldots, N$, respectively. Then, the optimal source and relay power profiles are given as

if $\widetilde{P}_{\mathrm{S}}^{i} \geq \widetilde{P}_{\mathrm{R}}^{i+1}$ **then**

  Set $P_{\mathrm{S}}^{*}(k) = P_{\mathrm{R}}^{*}(k+1) = \widetilde{P}_{\mathrm{R}}^{i+1}$ for $k = i, \ldots, i_r$, and $i = i_r + 1$

else

  Set $P_{\mathrm{S}}^{*}(k) = P_{\mathrm{R}}^{*}(k+1) = \widetilde{P}_{\mathrm{S}}^{i}$ for $k = i, \ldots, i_s$, and $i = i_s + 1$

end if

until $i > N$

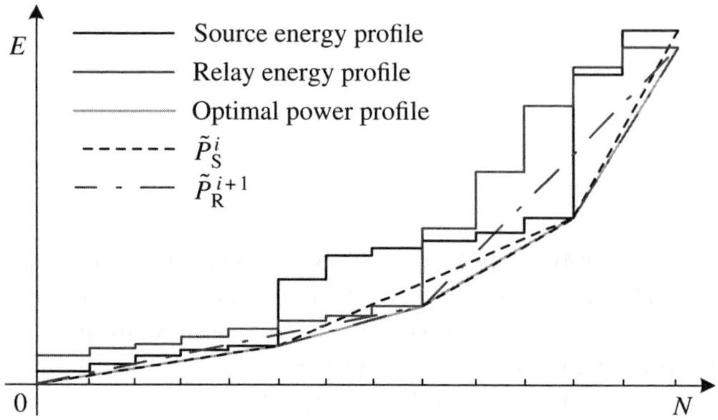

Figure 3.9 An example for the optimal source and relay power allocation for the DC case with $h_0 = 0$. Note that the slope of each dashed line corresponds to $\tilde{P}_S^i$ or $\tilde{P}_R^{i+1}$ obtained using Algorithm 3.2.

Remark 3.7 The optimal solution for problem (3.51)–(3.52) obtained using Algorithm 3.2 in the case of $h_0 = 0$ is illustrated in Figure 3.9, where we see that the optimal source–relay power profile corresponds to the shortest path from the origin to the highest possible sum energy point at the end of the N-block transmission under two stair-like source and relay energy constraints. It is worth noting that for the case of $h_0 = 0$, our relay channel model degrades to a cascade of two additive white Gaussian noise (AWGN) point-to-point channels that were studied in [11] with individual EH constraints. Thus, at each block, either the source–relay or the relay–destination link can be the performance bottleneck.

3.2.4 Optimal Solution for the No-Delay-Constrained Case

In this section, we solve problem (3.63)–(3.64) for the NDC case. We first prove that a separation principle for the source and relay power allocation problem holds, upon which problem (3.63)–(3.64) can be solved by a two-stage strategy: first obtain the optimal source power allocation by ignoring the relay, and then optimize the relay power allocation with the obtained source power solution. Since this separation principle applies to both cases with and without the direct link, we address these two cases with a unified analysis as follows.

3.2.4.1 Optimal Source Power Allocation

First, we consider the following source power allocation problem by ignoring the relay:

$$\max_{P_S(i)\geq 0} \quad \sum_{i=1}^{N} C(hP_S(i)) \tag{3.96}$$

$$\text{s.t.} \quad \sum_{i=1}^{k} P_S(i) \leq \frac{1}{B}\sum_{i=1}^{k} E_S(i), \quad k = 1, 2, \ldots, N, \tag{3.97}$$

where h is a constant with $0 < h \leq 1$. Problem (3.96)–(3.97) has been solved in [11], for which the algorithm is summarized in Algorithm 3.3 for the sake of completeness. Note

Algorithm 3.3: Compute the optimal solution for problem (3.96)–(3.97).

Input: $i = 1, N$

Output: i_s, P^i_{S}

1: **repeat**

2: Compute

$$i_s = \arg\min_{i \le j \le N} \left\{ \frac{\sum_{k=i}^{j} E_{\mathrm{S}}(k)}{(j - i + 1) B} \right\}, \quad P^i_{\mathrm{S}} = \frac{\sum_{k=i}^{i_s} E_{\mathrm{S}}(k)}{(i_s - i + 1) B}$$

3: The optimal source power profile is given as $P^*_{\mathrm{S}}(n) = P^i_{\mathrm{S}},\ n = i, \dots, i_s$

4: Set $i = i_s + 1$

5: **until** $i > N$

that the optimal source power profile $P^*_{\mathrm{S}}(i)$'s of problem (3.96)–(3.97) are non-decreasing over i [11].

It is easy to see that the optimal source power profile of problem (3.96)–(3.97) maximizes the average throughput of both the source–relay and source–destination links. Moreover, since, for the NDC case, the relay can store the binning indices of the decoded source messages with arbitrary delay before forwarding them to the destination with best effort transmissions, the relay power profile intuitively should have no effect on the optimal source power profile. This conjecture is affirmed by the following proposition.

Proposition 3.12 *For the NDC case, the optimal source power solution for problem (3.96)–(3.97) is also globally optimal for problem (3.63)–(3.64).*

Proof: Define $f(x) = \frac{1}{2}\log\left(\frac{1+x}{1+h_0 x}\right)$ over $x > 0$ with fixed $0 \le h_0 < 1$. Since $f''(x) = -\frac{1}{2}(1+x)^{-2} + \frac{1}{2}h_0^2(1+h_0 x)^{-2}$ and $\frac{h_0}{1+h_0 x} < \frac{1}{1+x}$, it follows that $f''(x) < 0$ and $f(x)$ is concave. Moreover, it is easy to check that $f(x)$ is increasing over $x > 0$. Then, we obtain the following lemma. □

Lemma 3.3 *For the case of $0 < h_0 < 1$, the optimal source power profile $P^*_{\mathrm{S}}(i)$'s of problem (3.63)–(3.64) are non-decreasing over i; for $h_0 = 0$, there exist optimal source power profile $P^*_{\mathrm{S}}(i)$'s, which are non-decreasing over i, $i = 1, \dots, N$.*

Proof: Denote the optimal solution of problem (3.63)–(3.64) as $P^\star_{\mathrm{S}}(i)$'s and $P^\star_{\mathrm{R}}(i+1)$'s. For the case of $0 < h_0 < 1$, consider any consecutive source–relay power pair corresponding to the ith and the $(i+1)$th source messages, $i = 1, \dots, N-1$, and the following two cases:

(1) $P^\star_{\mathrm{S}}(i) > P^\star_{\mathrm{S}}(i+1)$, $P^\star_{\mathrm{R}}(i+1) > P^\star_{\mathrm{R}}(i+2)$: Define $r^\star(i+1) = C(P^\star_{\mathrm{R}}(i+1))$. Since $\log(x)$ is increasing, it follows that $r^\star(i+1) > r^\star(i+2)$. Then, the constraint in (3.64) is equivalent to $\sum_{k=1}^{i} r^\star(k+1) \le \sum_{k=1}^{i} f(P^\star_{\mathrm{S}}(k))$, which is convex. Thus, similar to Case (1) of the proof of Proposition 3.10, this case is not optimal for problem (3.63)–(3.64).

(2) $P_S^\star(i) > P_S^\star(i+1)$, $P_R^\star(i+1) \leq P_R^\star(i+2)$: It is first proved that the constraint $\sum_{k=1}^{i} C(P_R^\star(k+1)) \leq \sum_{k=1}^{i} f(P_S^\star(k))$ is not satisfied with equality in this case by contradiction as follows. Suppose that the above rate inequality is satisfied with equality. From the $(i-1)$th and the $(i+1)$th constraints in (3.64), it follows that $C(P_R^\star(i+1)) \geq f(P_S^\star(i))$ and $C(P_R^\star(i+2)) \leq f(P_S^\star(i+1))$, and together with the assumption that $P_R^\star(i+1) \leq P_R^\star(i+2)$, it follows that $f(P_S^\star(i)) \leq C(P_R^\star(i+1)) \leq C(P_R^\star(i+2)) \leq f(P_S^\star(i+1))$. Since $f(x)$ is an increasing function, it follows that $P_S^\star(i) \leq P_S^\star(i+1)$, which contradicts the assumption that $P_S^\star(i) > P_S^\star(i+1)$.

Then, there must exist $\delta > 0$, with which we define a new power allocation as $\tilde{P}_S(i) = P_S^\star(i) - \delta$ and $\tilde{P}_S(i+1) = P_S^\star(i+1) + \delta$, while keeping $P_R^\star(i+1)$ and $P_R^\star(i+2)$ unchanged. It is easy to check that the new power allocation still satisfies the constraints in problem (3.96)–(3.97) and increases the sum rate over these two blocks. Thus, Case (2) is not optimal for problem (3.63)–(3.64).

In conclusion, for the case of $0 < h_0 < 1$, only the case $P_S^\star(i) \leq P_S^\star(i+1)$ and $P_R^\star(i+1) \leq P_R^\star(i+2)$ can be optimal. For the case of $h_0 = 0$, similar argument can be applied to show the existence of such a non-decreasing optimal solution, which is omitted for brevity. Thus, this lemma is proved. □

Next, we prove Proposition 3.12. First, consider the case of $0 < h_0 < 1$. For the optimal source power $P_S^\star(i)$'s of problem (3.63)–(3.64) and $P_S^*(i)$'s obtained using Algorithm 3.3, we assume that $P_S^\star(j) = P_S^*(j), j = 1, \ldots, i-1$, and $P_S^\star(i) \neq P_S^*(i)$. Consider the following two cases:

(1) $P_S^\star(i) > P_S^*(i)$: Similar to the Case (Ia) in the proof of Algorithm 3.1, it follows that this case violates the source energy constraints, since $P_S^\star(i)$'s are non-decreasing and thus cannot be true.

(2) $P_S^\star(i) < P_S^*(i)$: First, we claim the following two results, which will be proved later: (i) $P_S^\star(i)$'s reduce the optimal value of problem (3.63)–(3.64) without considering the relay power allocation. (ii) By further considering the relay power allocation and the constraint (3.64), $P_S^\star(i)$'s shrink the feasible set of $P_R(i+1)$'s. If (i) and (ii) are both true, it follows that $P_S^\star(i) < P_S^*(i)$ cannot be true.

For the proof of (i), from [11], it follows that $P_S^*(i)$'s maximize $\sum_{k=1}^{N} C(h_0 P_S(k))$ subject to the constraint (3.97); in other words, for the case of $P_S^\star(i) < P_S^*(i)$, this value will be strictly decreased. Thus, (i) is proved.

For the proof of (ii), it is equivalent to prove that $\sum_{k=1}^{j} f(P_S^\star(k)) \leq \sum_{k=1}^{j} f(P_S^*(k))$, $i \leq j \leq N$. When $j = i$, by the assumption that $P_S^\star(i) < P_S^*(i)$, it follows that $\sum_{k=1}^{j} f(P_S^\star(k)) \leq \sum_{k=1}^{j} f(P_S^*(k))$. For $j = i+1$, if $\sum_{k=1}^{j} f(P_S^\star(k)) > \sum_{k=1}^{j} f(P_S^*(k))$, it follows that $f(P_S^\star(j)) > f(P_S^*(j))$ and thus $P_S^\star(j) > P_S^*(j)$ since $f(x)$ is monotonically increasing. For $P_S^*(i)$'s, denote the index of the next source energy exhausted block after the jth block as j_s. It is then obtained that $P_S^\star(k) \geq P_S^\star(j) > P_S^*(j) = P_S^*(k)$, $j \leq k \leq j_s$, since by Lemma 3.3, $P_S^\star(i)$'s are non-decreasing over i. Moreover, by Yang and Ulukus [11], since $f(x)$ is concave, it can be shown that with the same energy budget over the first to the jth blocks, if $\sum_{k=1}^{j} f(P_S^\star(k)) > \sum_{k=1}^{j} f(P_S^*(k))$, it follows that

$\sum_{k=1}^{j} P_S^{\star}(k) > \sum_{k=1}^{j} P_S^{*}(k)$. Then, it is easy to check that $P_S^{\star}(j)$'s violate the source energy constraint at the j_sth block, i.e. $B\sum_{k=1}^{j_s} P_S^{\star}(k) > B\sum_{k=1}^{j_s} P_S^{*}(k) = \sum_{k=1}^{j_s} E_S(k)$. Thus, it is obtained that $\sum_{k=1}^{j} f(P_S^{\star}(k)) \le \sum_{k=1}^{j} f(P_S^{*}(k))$ for $j = i+1$. By using the mathematical induction method, it can be shown that $\sum_{k=1}^{j} f(P_S^{\star}(k)) \le \sum_{k=1}^{j} f(P_S^{*}(k))$, $1 \le j \le N$, which suggests that (ii) is true.

From the above analysis, it is proved that for the case of $0 < h_0 < 1$, $P_S^{*}(i) = P_S^{\star}(i)$, $i = 1, \dots, N$, and thus $P_S^{*}(i)$'s are optimal for problem (3.63)–(3.64). For the case of $h_0 = 0$, since there always exists one optimal source power profile that is non-decreasing over i (by Lemma 3.3), without loss of generality, we can assume that $P_S^{\star}(i)$'s are still non-decreasing over i. With the same argument as the case of $0 < h_0 < 1$, it is observed that claim (ii) is still true, which means that the maximum value with $P_S^{\star}(i)$'s is not larger than that with $P_S^{*}(i)$'s. In conclusion, this proposition is proved. □

This proposition implies that the separation principle for the source and relay power allocation problems is optimal for problem (3.63)–(3.64). Thus, even though problem (3.63)–(3.64) is non-convex, we can still find its globally optimal solution efficiently.

Remark 3.8 It is shown as the proof of Proposition 3.12 that for the case of $0 < h_0 < 1$, the optimal source power profile $P_S^{*}(i)$'s of problem (3.63)–(3.64) given by Algorithm 3.3 is unique. However, for the case of $h_0 = 0$, this result is not true in general, since the source energy may not necessarily be exhausted at the end of each N-block transmission for problem (3.63)–(3.64) (cf. Proposition 3.9).

3.2.4.2 Optimal Relay Power Allocation

With the optimal source power profile $P_S^{*}(i)$'s obtained using Algorithm 3.3, the optimal relay power profile can be determined by the following problem:

$$\max_{P_R(i+1)\ge 0} \quad \sum_{i=1}^{N} C(P_R(i+1)) \tag{3.98}$$

$$\text{s.t.} \quad \sum_{i=1}^{k} C(P_R(i+1)) \le \sum_{i=1}^{k} C(P_S^{*}(i)) - \sum_{i=1}^{k} C(h_0 P_S^{*}(i)), \tag{3.99}$$

$$\sum_{i=1}^{k} P_R(i+1) \le \frac{1}{B}\sum_{i=1}^{k} E_R(i+1), \quad k = 1, \dots, N. \tag{3.100}$$

This problem is non-convex due to the constraint in (3.99). However, by letting $r(i+1) = C(P_R(i+1))$, problem (3.98)–(3.100) can be rewritten as

$$\max_{r(i+1)\ge 0} \quad \sum_{i=1}^{N} r(i+1) \tag{3.101}$$

$$\text{s.t.} \quad \sum_{i=1}^{k} r(i+1) \le \sum_{i=1}^{k} C(P_S^{*}(i)) - \sum_{i=1}^{k} C(h_0 P_S^{*}(i)), \tag{3.102}$$

$$\sum_{i=1}^{k} (2^{2r(i+1)} - 1) \le \frac{1}{B}\sum_{i=1}^{k} E_R(i+1), \tag{3.103}$$

where $k = 1, \dots, N$. It can be shown that problem (3.101)–(3.103) is convex over $r(i+1)$'s. By the KKT optimality conditions, we obtain the optimal solution for problem (3.101)–(3.103) as

$$r^*(i+1) = \left(\frac{1}{2} \log \frac{1 - \sum_{k=i}^{N} \lambda_k}{2 \ln 2 \cdot \sum_{k=i}^{N} \gamma_k} \right)^+, \quad i = 1, \dots, N, \tag{3.104}$$

where λ_k and γ_k are the nonnegative Lagrangian multipliers corresponding to the kth constraint in (3.102)–(3.103), respectively. Problem (3.101)–(3.103) can be solved by a forward-search algorithm, denoted by Algorithm 3.4, for which the optimality proof is similar to that of Algorithm 3.2 and is thus omitted.

Algorithm 3.4: Compute the optimal solution for problem (3.101)–(3.103).

Input: $i = 1, N$

Output: $i_1, i_2, \widetilde{r}_1, \widetilde{r}_2$

1: **repeat**

2: Compute

$$i_1 = \arg\min_{i \le j \le N} \left\{ \frac{\widetilde{C}_i + \sum_{k=i}^{j} C\left(P_S^*(k)\right) - C\left(h_0 P_S^*(k)\right)}{(j - i + 1)\, B} \right\},$$

$$i_2 = \arg\min_{i \le j \le N} \left\{ \frac{\widetilde{E}_{i+1} + \sum_{k=i}^{j} E_S(k+1)}{(j - i + 1)\, B} \right\},$$

$$\widetilde{r}_1 = \frac{\widetilde{C}_i + \sum_{k=i}^{i_1} C\left(P_S^*(k)\right) - C\left(h_0 P_S^*(k)\right)}{(i_1 - i + 1)\, B},$$

$$\widetilde{r}_2 = C\left(\frac{\widetilde{E}_{i+1} + \sum_{k=i}^{i_2} E_S(k+1)}{(i_2 - i + 1)\, B} \right),$$

where $\widetilde{C}_1 = \widetilde{E}_2 = 0$, $\widetilde{C}_i = \sum_{k=1}^{i-1} C\left(P_S^*(k)\right) - C\left(h_0 P_S^*(k)\right) - r^*(k)$, and $\widetilde{E}_{i+1} = \sum_{k=1}^{i-1} \left(E_S(k+1) - 2^{2r^*(k+1)} - 1\right)$, $i = 2, \dots, N$. Let $j_0 = \arg\min_{j=1,2} \{\widetilde{r}_j\}$. Set $r^*(j+1) = \widetilde{r}_{j_0}$, $j = i, \dots, i_{j_0}$, and $i = i_{j_0} + 1$

3: **until** $i > N$

Remark 3.9 It is worth noting that from (3.104), we observe that the optimal relay transmission rate $r^*(i+1)$ is non-decreasing over i and strictly increases when any one of the constraints (3.102)–(3.103) is satisfied with equality. Thus, the optimal relay power profile $P_R^*(i+1)$'s of problem (3.63)–(3.64) with $P_R^*(i+1) = 2^{2r^*(i+1)} - 1$, $i = 1, \dots, N$, are also non-decreasing over i. Furthermore, it can be shown that the obtained $P_R^*(i+1)$'s achieve the minimum energy consumption at the relay.

Based on the analysis in the above two subsections, we obtain the following proposition for the monotonic properties for $P_S^*(i)$'s and $P_R^*(i+1)$'s for the NDC case, similar to the DC case (cf. Proposition 3.10).

Proposition 3.13 *For problem (3.63)–(3.64), there exist optimal source power solution $P_S^*(i)$'s and relay power solution $P_R^*(i+1)$'s that are non-decreasing over i, $i = 1, \dots, N$.*

Remark 3.10 It is worth noting that problems of similar structures as (3.98)–(3.100) have been solved independently in prior works [3, 6].

3.2.4.3 Optimal Rate Scheduling

It has been shown in Section 3.2.3 that for the DC case, the relay binning rate $R_B(i+1)$ for each source message can be directly computed by (3.55) given the optimal source and relay power profiles. However, for the NDC case, it requires additional effort to obtain $R_B(i+1)$, since we need to consider two sets of different constraints in (3.58) and (3.61) (as will be shown next). Supposing that $R_B(i+1)$ is obtained with the optimal source and relay power profiles $P_S^*(i)$'s and $P_R^*(i+1)$'s, the source transmission rates $R(i)$'s can be determined from (3.60). Then, with the obtained $R(i)$'s, we can update the obtained source power profile to achieve the same maximum throughput but with the minimum energy consumption.

To compute $R_B(i+1)$'s, the following observations are first drawn. If $C(P_R^*(i+1)) > C(P_S^*(i)) - C(h_0 P_S^*(i))$, $\forall i \in \{1, \dots, N\}$, the relay should transmit not only the binning index of the ith source message at the $(i+1)$th block but also those of source messages $1 \le j < i$. Moreover, due to the constraint in (3.64), it follows that if $C(P_R^*(i+1)) > C(P_S^*(i)) - C(h_0 P_S^*(i))$, $\forall i$, there must exist j with $1 \le j < i$, such that $C(P_R^*(j+1)) < C(P_S^*(j)) - C(h_0 P_S^*(j))$. The above observations imply that to obtain $R_B(i+1)$'s, we need to find all i's with $C(P_R^*(i+1)) > C(P_S^*(i)) - C(h_0 P_S^*(i))$ and then use their surplus rates to transmit the binning indices of source messages $j \le i$. In Figure 3.10, we show an example for the relationship between $C(P_R^*(i+1))$ and $C(P_S^*(i)) - C(h_0 P_S^*(i))$, where Case A stands for the case of $C(P_R^*(i+1)) < C(P_S^*(i)) - C(h_0 P_S^*(i))$ and Case B for the case with a reversed inequality. Then, a geometric interpretation for the algorithm of computing $R_B(i+1)$'s is given as follows: use the surplus rate area in Case B to fill the deficient rate area in Case A. Obviously, if the area to be filled in Case A is larger than the filling area in Case B, the corresponding values for $R_B(i+1)$'s are not unique.

Thus, we now develop a backward-search algorithm (Algorithm 3.5) to obtain one of the feasible solutions for $R_B(i+1)$'s. The main procedure of this algorithm is described as follows. First, $R_B(i+1)$'s are initialized as the minimum values between $C(P_R^*(i+1))$

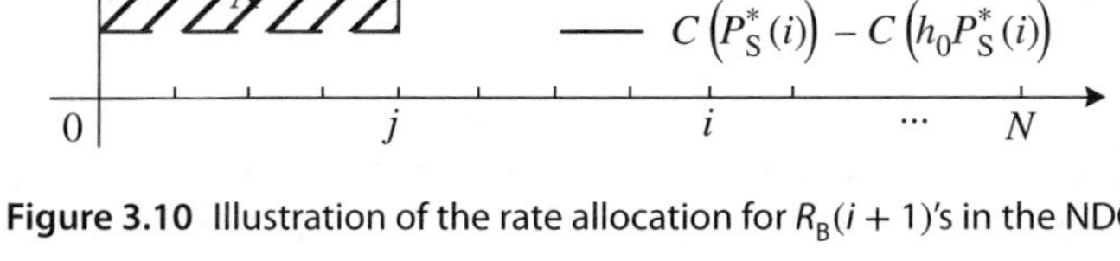

Figure 3.10 Illustration of the rate allocation for $R_B(i+1)$'s in the NDC case.

and $C(P_S^*(i)) - C(h_0 P_S^*(i))$ for all i's, and a parameter t (which, e.g. corresponds to the filling area in Case B of Figure 3.10) is set to be 0. The algorithm then searches the values for $R_B(i+1)$'s in a backward way from $i = N$ to 1. For any ith block, the algorithm computes temp $= C(P_R^*(i+1)) - (C(P_S^*(i)) - C(h_0 P_S^*(i)))$. Then, if temp > 0, Case B occurs, and temp is added to t; if temp < 0, Case A occurs, and $R_B(i+1)$ is increased by min($-$temp, t), and this amount is then subtracted from t.

Algorithm 3.5: Compute $R_B(i+1)$'s for the NDC case.

Input: $P_S^*(i)$'s by Algorithm 3.3 and $P_R^*(i+1)$s by Algorithm 3.4
Output: $R_B(i+1)$
1: Initialize

$$R_B(i+1) = \min\left\{C\left(P_R^*(i+1)\right), C\left(P_S^*(i)\right) - C\left(h_0 P_S^*(i)\right)\right\},$$

where $i = 1, \dots, N$, and $t = 0$
2: From $i = N$ to 1, compute

$$\text{temp} = C\left(P_R^*(i+1)\right) - C(P_S^*(i)) + C(h_0 P_S^*(i))$$

3: **if** temp > 0 **then**
4: Set $t = t + \text{temp}$
5: **else**
6: Set $R_B(i) = R_B(i) + \min(-\text{temp}, t)$ and $t = (t + \text{temp})^+$
7: **end if**

Remark 3.11 It is recalled in Remark 3.8 that in the case of $h_0 = 0$, the optimal source power profile $P_S^*(i)$'s given by Algorithm 3.3 may not achieve the minimum energy consumption. In this case, from (3.60), it follows that $R(i) = R_B(i+1)$, $i = 1, \dots, N$, with $R_B(i+1)$'s obtained by using Algorithm 3.5. In order to achieve the minimum energy consumption at the source, the optimal source power solution of problem (3.63)–(3.64) can be updated as $P_S^*(i) = 2^{2R_B(i+1)} - 1$. It is worth noting that the above obtained source power profile is still non-decreasing over i (thus in accordance with Proposition 3.13), since $R_B(i+1)$'s obtained using Algorithm 3.5 are non-decreasing over i.

3.2.4.4 Throughput Comparison: DC Versus NDC

As shown by Proposition 3.8, the throughput of the NDC case is no smaller than that of the DC case. To further compare these two cases, the following proposition shows when the NDC case is strictly better than the DC case in terms of the average throughput.

Proposition 3.14 *The average throughput of the NDC case is strictly larger than that of the DC case if and only if there exists* $i \in \{1, \dots, N\}$ *such that* $C(P_R^*(i+1)) > C(P_S^*(i)) - C(h_0 P_S^*(i))$, *where* $P_S^*(i)$*'s and* $C(P_R^*(i+1))$*'s are the optimal solutions to problems (3.96)–(3.97) and (3.101)–(3.103) obtained by Algorithms 3.3 and 3.4, respectively.*

Proof: Denote $P_S^\star(i)$'s and $P_R^\star(i+1)$'s as the optimal solution for problem (3.51)–(3.52) and $P_S^*(i)$'s and $P_R^*(i+1)$'s as the optimal solution for problem (3.63)–(3.64). First, we prove that the condition given in Proposition 3.14 is sufficient. Assume that $C(P_R^*(k+1)) \leq C(P_S^*(k)) - C(h_0P_S^*(k)), 1 \leq k \leq i$, and $C(P_R^*(i+2)) > C(P_S^*(i+1)) - C(h_0P_S^*(i+1))$. Due to constraint (3.64), without loss of generality, we further assume that $C(P_R^*(i+1)) < C(P_S^*(i)) - C(h_0P_S^*(i))$. Then, for $P_S^{\star}(i)$'s and $P_R^{\star}(i+1)$'s, it is easy to check that only Scenario II or III given in Proposition 3.11 can happen at the ith block, which is further discussed as follows:

(1) Scenario III happens for the ith source message in problem (3.51)–(3.52), i.e. $P_S^{\star}(i) = P_0$, $P_S^{\star}(i+1) = P_0 + \frac{1}{h_0} - 1$, and thus the source energy constraint at the ith block is not satisfied. Then, there exists $\delta > 0$, such that the newly defined source power allocation $\tilde{P}_S(i) = P_0 + \delta, \tilde{P}_S(i+1) = P_0 + \frac{1}{h_0} - 1 - \delta$ satisfies the source energy constraint at the ith block. Moreover, as for the NDC case, we can increase the binning rate of ith source message with the amount $C(P_S^{\star}(i+1)) - C(\tilde{P}_S(i+1))$ (note that this operation is possible since this amount is less than $C(\tilde{P}_S(i)) - C(P_S^{\star}(i))$) and transmit it at the $(i+1)$th relay transmission. With the above scheme, it is easy to check that the sum rate over these two blocks is strictly improved.
(2) Scenario II happens at the ith block for problem (3.51)–(3.52). Consider two subcases: (i) If the relay energy is not exhausted at the end of each N-block transmission, as for the NDC case, we can increase $R_B(i+1)$ and use the available relay energy at the $(N+1)$th block to transmit the increased part in $R_B(i+1)$, which strictly improves the throughput of the DC case. (ii) If the relay energy is exhausted, there must exists $k > i+2$, such that $P_R^{\star}(k) > P_R^{\star}(i+2)$. Thus, there exists $0 < \delta < \frac{P_R^{\star}(k) - P_R^{\star}(i+2)}{2}$. Define a new relay power allocation satisfying the relay energy constraint as $\tilde{P}_R(i+2) = P_R^{\star}(i+2) + \delta$ and $\tilde{P}_R(k) = P_R^{\star}(k) - \delta$. By increasing the binning rate of the ith message with the amount $C(\tilde{P}_R(i+2)) - C(P_R^{\star}(i+2))$ and decreasing that of the kth message with $C(P_R^{\star}(k)) - C(\tilde{P}_R(k))$, it can be shown that the new scheme improves the sum rate of these two blocks, since $C(\tilde{P}_R(i+2)) - C(P_R^{\star}(i+2)) > C(P_R^{\star}(k)) - C(\tilde{P}_R(k))$. Based on (1) and (2), the "if" part is proved.

Next, we prove that the condition in Proposition 3.14 is also necessary. Assume that $C(P_R^*(i+1)) \leq C(P_S^*(i)) - C(h_0P_S^*(i)), \forall i \in \{1, \dots, N\}$. It is easy to check that with the power allocation $P_S^*(i)$'s and $P_R^*(i+1)$'s of problem (3.63)–(3.64), the achievable rate of the ith block for problem (3.51)–(3.52) is given as $R(i) = \min\{C(P_S^*(i)), C(P_R^*(i+1)) + C(h_0P_S^*(i))\} = C(P_R^*(i+1)) + C(h_0P_S^*(i))$, which equals that of the NDC case. By searching over the whole feasible set, the throughput of the DC case will be no smaller than that of the NDC case. Together with Proposition 3.8, it is obtained that the throughput of the two cases is identical, and thus the "only if" part is proved. □

3.2.5 Numerical Results

In this section, we present some numerical results to validate our theoretical results. For the purpose of exposition, we assume a periodic energy profile model for some predictable EH sources. Specifically, the source and relay energy profiles are given as

$$E_S(i) = A_S \sin\left(\frac{i-1}{N}2\pi + \frac{\pi}{2}\right) + A_S,$$
$$E_R(i+1) = A_R \sin\left(\frac{i-1}{N}2\pi + \theta\right) + A_R, \quad 1 \leq i \leq N,$$

respectively, where $A_S, A_R > 0$ are the amplitudes of the sinusoidal energy profiles at the source and relay, respectively, and θ is the phase shift between these two energy profiles. Here, we choose $B = 100$, $N = 40$, $\theta = \frac{5}{4}\pi$, and $A_S = A_R = 200$.

We compare our proposed algorithms with a greedy power allocation strategy. Here, we adopt a nontrivial greedy algorithm by assuming that both the source and relay know the harvested energy amounts up to the current block prior to transmissions. The transmission rate of the ith source message in the greedy algorithm is given as

$$R_G(i) = \min\left\{ C\left(\frac{\tilde{E}_S(i)}{B}\right), C\left(h_0\frac{\tilde{E}_S(i)}{B}\right) + C\left(\frac{\tilde{E}_R(i+1)}{B}\right)\right\}, \tag{3.105}$$

where $\tilde{E}_S(i) = \sum_{k=1}^{i} E_S(k) - B\sum_{k=1}^{i-1} P_S(k)$, and $\tilde{E}_R(i+1) = \sum_{k=1}^{i} E_R(k+1) - B\sum_{k=1}^{i-1} P_R(k+1)$, with $P_S(i) = 2^{2R_G(i)} - 1$ and $P_R(i+1) = 2^{2\left(R_G(i) - C\left(h_0\frac{\tilde{E}_S(i)}{B}\right)\right)} - 1$, $i = 1, \dots, N$. Note that in the above greedy algorithm, both the source and relay consume as much available power as possible at two successive blocks to maximize the instantaneous throughput, thus achieving the minimum delay as for the DC case.

In Figure 3.11, we show the average throughputs versus the direct link channel gain h_0 for the proposed power allocation algorithms and the greedy algorithm. It is observed that as the direct link becomes stronger, i.e. h_0 increases, there is a throughput limit of 0.387 bps Hz^{-1}. For the NDC case, this throughput limit is achieved even for very small h_0 around 0.05. In contrast, for the DC case, the throughput increases almost linearly and achieves the throughput limit when h_0 exceeds 0.75. Thus, the throughput gain of NDC over DC cases by relaxing the decoding delay is present only in the regime of small direct link gains. This is due to the fact that with small h_0, the condition in Proposition 3.14 is more likely to be satisfied. For the greedy algorithm, it is observed that the throughput loss can be large, especially when h_0 is small, as compared with the proposed algorithm for the DC case.

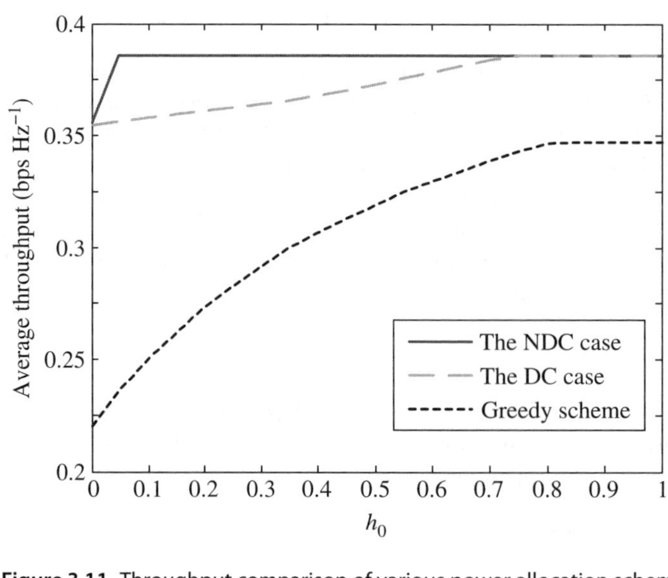

Figure 3.11 Throughput comparison of various power allocation schemes for the relay channel with energy harvesting constraints.

3.3 Large Relay Networks

In conventional energy-constrained wireless communication systems such as wireless sensor networks (WSNs), sensors are equipped with fixed energy supply devices, e.g. batteries, which have limited operation time. When thousands of sensors are deployed in a large area, recharging or replacing batteries becomes inconvenient. In such scenarios, harvesting energy from the environment at each node could be a solution to provide almost unlimited energy supply [16, 17]. Compared to the case with conventional constant energy suppliers, however, relays powered by energy harvesters with possibly random power levels are subject to the new EH constraints. In this section, we study the large fading relay networks with EH relay nodes. We aim to develop some efficient and yet simple-to-implement power control algorithms to maximize the system throughput.

3.3.1 System Model and Assumptions

We consider a large relay network, which consists of one source–destination pair and N EH relays. The relay node operates in a half-duplex mode during each communication block, which is composed of two successive time slots: the source transmits during the first time slot to all the relays, and the relays transmit simultaneously during the second time slot to the destination. We consider the case that there is no direct link between the source and the destination.

The channel input–output relationship of the considered network is given as follows. During the first time slot of the ith communication block, the received signal $y_{i,n}$ at the nth relay is given as

$$y_{i,n} = \sqrt{P_i^{\mathrm{S}}} h_{i,n} x_i + u_{i,n}, \tag{3.106}$$

for $n = 1, \ldots, N, \quad i = 1, \ldots, T$, where x_i is the transmit signal in the ith communication block, P_i^{S} is the transmit power at the source node, $h_{i,n}$ is the channel coefficient from the source to the nth relay in the ith transmission period, and $u_{i,n}$ is the i.i.d. CSCG noise with zero mean and unit variance. The source node is powered by a conventional power supply, and thus P_i^{S} is constant over time, i.e. $P_i^{\mathrm{S}} = P^{\mathrm{S}}$. Similarly, we define the input–output relationship in the second time slot over the relay-to-destination links as

$$z_i = \sum_{n=1}^{N} g_{i,n} t_{i,n} + w_i, \tag{3.107}$$

where z_i is the received signal at the destination, $g_{i,n}$ is the channel coefficient of the nth relay-to-destination link, $t_{i,n}$ is the transmit signal at the nth relay, and w_i is the i.i.d. CSCG noise with zero mean and unit variance. In this subsection, we consider the case that $h_{i,n}$'s are i.i.d. random variables across i and n, with known probability density function (PDF) of $|h_{i,n}|$ and $|h_{i,n}|^2$, denoted as $p_{h1}(x)$ and $p_{h2}(x)$, respectively. For $g_{i,n}$, we adopt the same i.i.d. assumption and define $p_{g1}(x)$ and $p_{g2}(x)$ similarly. Moreover, $\mathbb{E}(|h_{i,n}|)$, $\mathbb{E}(|h_{i,n}|^2)$, $\mathbb{E}(\frac{1}{|h_{i,n}|^2})$, $\mathbb{E}(|g_{i,n}|)$, and $\mathbb{E}(|g_{i,n}|^2)$ are all finite.

For the purpose of exposition, we consider the amplify-and-forward (AF) relaying scheme in the sequel, which has already been shown in [18] to achieve the optimal throughput scaling order. Moreover, the backward and forward link gains $h_{i,n}$ and $g_{i,n}$ are perfectly known at each relay. As such, the transmit signal $t_{i,n}$ at the relay is given by

$$t_{i,n} = \frac{h^*_{i,n} g^*_{i,n}}{|h_{i,n}||g_{i,n}|} \cdot \frac{\sqrt{P^{\mathrm{R}}_{i,n}} y_{i,n}}{\sqrt{P^{\mathrm{S}}_i |h_{i,n}|^2 + 1}}, \tag{3.108}$$

where $P^{\mathrm{R}}_{i,n}$ is the transmit power at the nth relay. In this subsection, we are interested in developing some distributed power allocation algorithms for implementation simplicity, which requires only local information about channel state and harvested energy rate.

Based on the above assumptions, the received signal at the destination for the ith communication block is given as

$$z_i = \sum_{n=1}^{N} \frac{\sqrt{P^{\mathrm{S}}_i P^{\mathrm{R}}_{i,n}} |h_{i,n}||g_{i,n}|}{\sqrt{P^{\mathrm{S}}_i |h_{i,n}|^2 + 1}} x_i + \sum_{n=1}^{N} \frac{\sqrt{P^{\mathrm{R}}_{i,n}} |g_{i,n}| h^*_{i,n}}{|h_{i,n}| \sqrt{P^{\mathrm{S}}_i |h_{i,n}|^2 + 1}} u_{i,n} + w_i. \tag{3.109}$$

We denote $\mathcal{A}_i = \sum_{n=1}^{N} \mathcal{A}_{i,n}$ as the amplitude of the received signal with $\mathcal{A}_{i,n} = \frac{\sqrt{P^{\mathrm{S}}_i P^{\mathrm{R}}_{i,n}} |h_{i,n}||g_{i,n}|}{\sqrt{P^{\mathrm{S}}_i |h_{i,n}|^2+1}}$ and $\mathcal{B}_i = \sum_{n=1}^{N} \mathcal{B}_{i,n}$ as the relay noise with $\mathcal{B}_{i,n} = \frac{\sqrt{P^{\mathrm{R}}_{i,n}} |g_{i,n}| h^*_{i,n} u_{i,n}}{|h_{i,n}| \sqrt{P^{\mathrm{S}}_i |h_{i,n}|^2+1}}$.

In addition to the block transmission model, the EH process at each relay is also period based: within one EH period, the harvested energy rate r_n at the nth relay is random but constant, while it may change across different EH periods. It is worth pointing out that in most practical scenarios, the communication and EH are operating at different time scales. Specifically, if we denote the length of the EH period (which could be roughly set equal to the coherent time of the energy arrive process) and that of the communication block as τ_{E}[6] and τ_{C}, respectively, we usually have $\frac{\tau_{\mathrm{E}}}{\tau_{\mathrm{C}}} \gg 1$. For example, the harvested energy rate may keep constant over several seconds, which is the case for EH from the wind or solar sources, while the communication block may only last several milliseconds, which is the case for most modern wireless communication systems. As such, the harvested energy rate r_n at each relay could be treated as identical over thousands of communication blocks. Furthermore, the EH processes are i.i.d. across the N relays, and during each EH period, the harvested energy rate r_n at the nth relay is a discrete-value random variable, with the distribution $\Pr\{r_n = \lambda_k\} = \nu_k, \quad k = 1, \ldots, K$.

In this subsection, we focus on the average performance optimization during one EH period and leave the more general cases with resource allocation over multiple EH periods to our future studies. Let $T = \left\lfloor \frac{\tau_{\mathrm{E}}}{\tau_{\mathrm{C}}} \right\rfloor$ and T be large enough such that the asymptotic result over many communication blocks could be obtained. Within one EH period, the transmit power at each relay should satisfy the following causal EH constraints:

$$\sum_{i=1}^{t} P^{\mathrm{R}}_{i,n} \le t r_n, \quad t = 1, \ldots, T. \tag{3.110}$$

Finally, we consider the case that the battery capacity to store the harvested energy rate at each relay is infinite. Moreover, the consumed energy at the relays other than the

6 For simplicity, the lengths of the EH periods at different relays are identical.

transmission energy is negligible, or they could be modeled as constants that can be easily absorbed into the EH constraints.

3.3.2 Average Throughput for Threshold-Based Transmissions

In this section, we derive the average throughput with general power allocation at the relays and then introduce a simple threshold-based transmission scheme. First, we discuss the following asymptotic results for the statistics of the received signal at the destination in (3.109). In particular, since the considered power allocation algorithms are only determined by local information about the channel state and harvested energy rate, it is easy to see that $P^{\mathrm{R}}_{i,n}$ is i.i.d. across different relays for a given i.

Proposition 3.15 *For the case with i.i.d. $h_{i,n}$, $g_{i,n}$, and $P^{\mathrm{R}}_{i,n}$, $\frac{1}{\sqrt{N}}\mathcal{A}_i$ and $\frac{1}{\sqrt{N}}\mathcal{B}_i$ are asymptotically Gaussian as N goes to infinity, with the distributions given by [19]*

$$\frac{1}{\sqrt{N}}\mathcal{A}_i \sim \mathcal{N}\left(\sqrt{N}\mathbb{E}(\mathcal{A}_{i,n}), Var(\mathcal{A}_{i,n})\right), \tag{3.111}$$

$$\frac{1}{\sqrt{N}}\mathcal{B}_i \sim \mathcal{CN}(0, Var(\mathcal{B}_{i,n})), \tag{3.112}$$

where $Var(\mathcal{A}_{i,n})$ is a finite value, i.e. not a function of N, since $\mathbb{E}(|h_{i,n}|^2)$ and $\mathbb{E}(|g_{i,n}|^2)$ are bounded, and

$$\mathbb{E}(\mathcal{A}_{i,n}) = \mathbb{E}\left(\frac{\sqrt{P^{\mathrm{S}}_i P^{\mathrm{R}}_{i,n}}|h_{i,n}||g_{i,n}|}{\sqrt{P^{\mathrm{S}}_i|h_{i,n}|^2+1}}\right), \tag{3.113}$$

$$Var(\mathcal{B}_{i,n}) = \mathbb{E}\left(\frac{P^{\mathrm{R}}_{i,n}|g_{i,n}|^2}{P^{\mathrm{S}}_i|h_{i,n}|^2+1}\right), \tag{3.114}$$

with the expectation $\mathbb{E}(\cdot)$ taken over the random variables $h_{i,n}$, $g_{i,n}$, and r_n (note that $P^{\mathrm{R}}_{i,n}$ is a function of these random variables).

Then with Gaussian signaling, the achievable rate for the considered network at the ith communication block is approximately given as $\frac{1}{2}\log\left(1+N\frac{\mathbb{E}^2(\mathcal{A}_{i,n})}{\mathrm{Var}(\mathcal{B}_{i,n})}\right)$. Therefore, the average throughput of the large relay network adopting the AF relaying scheme could be calculated as

$$R_{\mathrm{AF}} = \lim_{T\to\infty}\frac{1}{T}\sum_{i=1}^{T}\frac{1}{2}\log\left(1+N\frac{\mathbb{E}^2(\mathcal{A}_{i,n})}{\mathrm{Var}(\mathcal{B}_{i,n})}\right). \tag{3.115}$$

3.3.2.1 Threshold-Based Transmission

As shown in (3.115), it is quite difficult to develop an optimal online relay power allocation algorithm to maximize R_{AF}, with only the current $h_{i,n}$, $g_{i,n}$, r_n, and battery state, since the problem is non-convex and involves many design variables for each possible r_n and battery state. In this subsection, we propose a distributed (in the sense of requiring only local information about the channel state and harvested energy rate) and efficient threshold-based suboptimal transmission scheme. The proposed scheme is inspired by the following two observations:

(1) Consider the terms $\frac{|h_{i,n}|}{\sqrt{P_i^S|h_{i,n}|^2+1}}$ and $\frac{1}{P_i^S|h_{i,n}|^2+1}$ in (3.113) and (3.114) for the signal and noise power parts, respectively. It is easy to observe that for larger $|h_{i,n}|$, the signal becomes strengthened and the noise is weakened simultaneously. As such, it is natural to require that $|h_{i,n}|^2 \geq H_n$ in the threshold-based transmission scheme, where H_n is a constant determined by the harvested energy rate r_n and the channel fading statistics.
(2) For the term $|g_{i,n}|$, it is easy to check that for a larger value of this term, we obtain a larger received SNR $\frac{\mathbb{E}^2(\mathcal{A}_{i,n})}{\text{Var}(\mathcal{B}_{i,n})}$. As such, we also require that $|g_{i,n}|^2 \geq G_n$, where G_n is a constant also determined by the harvested energy rate r_n and the channel fading statistics.

Based on the above observations, we design the threshold-based transmission scheme as follows: each relay transmits with all its available energy if and only if $|h_{i,n}|^2 \geq H_n$ and $|g_{i,n}|^2 \geq G_n$ are satisfied simultaneously; otherwise, the relay saves energy for future transmission. Note that there are other threshold-based transmission schemes, e.g. setting threshold for only one of the two links or letting the relay transmit with a constant power, which will be shown to have similar Markov property to this scheme.

It is easy to check that under the proposed threshold-based transmission scheme, the EH constraints defined in (3.108) are satisfied. Since during one EH period, the harvested energy rate at any given relay is a constant, the stored energy level at the battery is within the collection $\mathcal{E}_n = \{s_{n,j}, j = 1, \dots, \infty\}$, with $s_{n,j} = jr_n\tau_C, j = 1, 2, \dots, \infty$.

3.3.2.2 Markov Property of the Transmission Scheme

In this subsection, we first show that under the proposed threshold-based relaying scheme, the stored energy level in each relay battery actually forms a Markov chain over time. By investigating the properties of this Markov chain, we could derive the asymptotical average throughput of the considered large relay network.

We denote the stored energy state at the nth relay and the ith communication block as $S_{i,n}$. Based on the proposed threshold-based transmission scheme, the state transition probability from state $S_{i,n} = s_{n,j}$ to state $S_{i+1,n}$ is given as

$$\begin{aligned} p_{j,1} &= \Pr\{S_{i+1,n} = s_{n,1} | S_{i,n} = s_{n,j}\} \\ &= \Pr\{|h_{i,n}|^2 \geq H_n\} \Pr\{|g_{i,n}|^2 \geq G_n\}, \end{aligned} \tag{3.116}$$

$$\begin{aligned} p_{j,j+1} &= \Pr\{S_{i+1,n} = s_{n,j+1} | S_{i,n} = s_{n,j}\} \\ &= 1 - p_{j,1}. \end{aligned} \tag{3.117}$$

It is easy to see that for different j, $p_{j,j+1}$, and $p_{j,1}$ remain the same over j. Thus, we denote $p_0 = p_{j,1}$ and $p_1 = p_{j,j+1}$.

Remark 3.12 It is observed that the transmission power over time at each relay actually forms a first-order Markov chain [19]. Since the channel coefficients $h_{i,n}$ and $g_{i,n}$ are i.i.d. across different relays, the overall relay transmission power vector across the whole network also forms a Markov chain.

Denote the probability that the stored energy state of each relay equals to $s_{n,j}$ as q_j, $j = 1, \dots, \infty$. Thus, when the system approaches to its corresponding stationary state, q_j could be obtained by solving the following equations:

$$\begin{cases} \sum_{j=1}^{\infty} p_0 q_j = q_1 \\ p_1 q_1 = q_2 \\ \quad \vdots \\ p_1 q_{j-1} = q_j \\ \quad \vdots \\ \sum_{j=1}^{\infty} q_j = 1, \quad q_j \geq 0, \quad j = 1, \dots, \infty. \end{cases} \tag{3.118}$$

It can be proved that the solution of (3.118) is given as

$$q_j = \frac{p_1^{j-1}}{\sum_{k=1}^{\infty} p_1^{k-1}}, \quad j = 1, \dots, \infty. \tag{3.119}$$

Given the Markov property of the proposed transmission scheme, the AF rate defined in (3.115) could be evaluated when the system approaches to the stationary state [19], i.e. when T goes to infinity. Therefore, the optimal transmission thresholds for maximizing (3.115) are also given by solving the following normalized received SNR maximization problem at the stationary state of the Markov transmission process:

$$\max_{\{H_k, G_k\}} \quad \frac{\mathcal{A}_0^2}{\mathcal{B}_0} \tag{3.120}$$

$$\text{s.t.} \qquad H_k \geq 0, \quad G_k \geq 0, \tag{3.121}$$

where $\mathcal{A}_0$ and $\mathcal{B}_0$ can be computed as

$$\mathcal{A}_0 = \sum_{k=1}^{K} \sqrt{\lambda_k} v_k \mathcal{F}(H_k, G_k), \tag{3.122}$$

$$\mathcal{B}_0 = \sum_{k=1}^{K} \lambda_k v_k \mathcal{G}(H_k, G_k), \tag{3.123}$$

with

$$\mathcal{F}(H_k, G_k) = \left(\sum_{j=1}^{\infty} \sqrt{j q_j} \right) \int_{x \geq \sqrt{H_k}, y \geq \sqrt{G_k}} \frac{xy p_{h1}(x) p_{g1}(y) \mathrm{d}x \, \mathrm{d}y}{\sqrt{P^{\mathrm{S}} x^2 + 1}}, \tag{3.124}$$

$$\mathcal{G}(H_k, G_k) = \left(\sum_{j=1}^{\infty} j q_j \right) \int_{x \geq H_k, y \geq G_k} \frac{y p_{h2}(x) p_{g2}(y)}{P^{\mathrm{S}} x + 1} \mathrm{d}x \, \mathrm{d}y. \tag{3.125}$$

With a little abuse of notation, here we use H_k and G_k to denote the transmission thresholds for the kth harvested energy rate, while H_n and G_n for the nth relay. In problem (3.120)–(3.121), it actually implies that the relays with the same harvested energy rate should be assigned with identical thresholds; and for each possible harvested energy rate λ_k, $k = 1, \dots, K$, we need to assign a pair of values $(\mathcal{F}(H_k, G_k), \mathcal{G}(H_k, G_k))$, which are functions of the transmission thresholds H_k and G_k, to maximize the normalized SNR defined in (3.120). In general, problem (3.120)–(3.121) is non-convex with arbitrary fading distributions of $h_{i,n}$ and $g_{i,n}$. Moreover, it is worth pointing out that the $\mathcal{F}$ and $\mathcal{G}$ functions are actually independent of the harvested energy rate λ_k.

Remark 3.13 To solve problem (3.120)–(3.121), we need to know the statistics of the channel fading and the EH process over space (i.e. the distribution of r_n across n). In practice, we usually collect the harvested energy rate information r_n from the N relays at the beginning of each EH period, based on which we can generate an estimate for the distribution of the harvested energy rate r_n by treating r_n's i.i.d. It is easy to see that our approach is valid even though the harvested energy rate r_n's are correlated among different relays in the considered EH period, since our optimization is within each EH period. On the opposite, power control algorithms considering multi-period joint optimization, e.g. the dynamic programming algorithm for solving the Markov decision process problem in [20], need to consider both the time-domain and space-domain correlation of the EH processes among the relays, where such information is usually difficult to be obtained.

3.3.3 Transmission Threshold Optimization

In this section, we show how to solve problem (3.120)–(3.121) with state-dependent threshold and also present a more efficient but heuristic scheme, where the transmission thresholds are treated independent of the harvested energy rate.

3.3.3.1 Convexification via Randomization

Before reformulating problem (3.120)–(3.121) into a solvable form, we present the following result.

Proposition 3.16 *At the optimal point of problem (3.120)–(3.121), denoted as* $\mathcal{F}(H_k^*, G_k^*)$ *and* $\mathcal{G}(H_k^*, G_k^*)$ *for* $k = 1, \dots, K$, *the following relationship holds:*

$$\mathcal{G}(H_k^*, G_k^*) = \min_{H_k, G_k} \quad \mathcal{G}(H_k, G_k) \tag{3.126}$$

$$\text{s.t.} \quad \mathcal{F}(H_k, G_k) = \mathcal{F}(H_k^*, G_k^*), \quad H_k \geq 0, G_k \geq 0. \tag{3.127}$$

The proof is straightforward and based on contradiction.

Based on the above proposition, we define a mapping $\mathcal{Q} : \tilde{F} \to \tilde{G}$ as follows:

$$\tilde{G} = \mathcal{Q}(\tilde{F}) = \min_{H,G} \quad \mathcal{G}(H, G) \tag{3.128}$$

$$\text{s.t.} \quad \mathcal{F}(H, G) = \tilde{F}, \quad H \geq 0, G \geq 0, \tag{3.129}$$

where the domain of $\tilde{F}$ is given by the set $\{\mathcal{F}(H, G) : H \geq 0, G \geq 0\}$.

Remark 3.14 It is worth noting that the function $\mathcal{Q}$ defined above is also independent of the harvested energy rate λ_k and only determined by the distributions of $h_{i,n}$ and $g_{i,n}$.

Moreover, for an arbitrary pair $(\tilde{F}, \tilde{G})$ satisfying the mapping $\mathcal{Q}$, it can be achieved by at least one transmission threshold pair (H, G), whose relationship is given as follows:

$$\begin{cases} \mathcal{F}(H,G) = \tilde{F} \\ \mathcal{G}(H,G) = \tilde{G}. \end{cases} \tag{3.130}$$

In general, function $\mathcal{Q}$ is non-convex for arbitrary channel distributions. Here we present an example for the Rayleigh fading case, as shown in Figure 3.12, where all the link coefficients $h_{i,n}$ and $g_{i,n}$ are complex Gaussian random variables with zero mean and unit variance. It can be verified that (i) the domain of the function $\mathcal{Q}$ is bounded, i.e. $\tilde{F} \in [0, B]$, with B being a constant, and (ii) the function $\mathcal{Q}$ is concave over $[0, A]$ and convex over $[A, B]$, where A is a constant. Due to the non-convexity of the function $\mathcal{Q}$, the original problem (3.120)–(3.121) is also non-convex and thus difficult to obtain its global optimal solution.

Next, we define the convex hull of function $\mathcal{Q}$ as

$$\mathcal{Q}_0 = \text{Conv}(\mathcal{Q}), \tag{3.131}$$

where Conv$(\cdot)$ denotes the convex-hull operation. To obtain the function $\mathcal{Q}_0$, we adopt the following convexification technique by introducing randomization in the threshold-based transmission: by Proposition 3.15, it is easy to see that for a given $\tilde{F}$, a smaller $\tilde{G}$ leads to a larger value of the objective function in problem (3.120)–(3.121). As such, for the case of $\tilde{F} \in [0, A]$, we could make function $\mathcal{Q}$ convex within this regime by transmitting under the thresholds corresponding to the two pairs $(0, 0)$ and $(A, \mathcal{Q}(A))$ with certain probabilities, respectively. Note that each point on the curve (the solid curve in Figure 3.12) corresponding to the function $\mathcal{Q}$ can be achieved by letting the relays transmit with certain thresholds given by (3.130). Therefore, the point $(C, \mathcal{Q}_0(C))$, $0 \le C \le A$, on the segment from $(0, 0)$ to $(A, \mathcal{Q}(A))$ can be achieved by letting the relays transmit under the thresholds corresponding to $(0, 0)$ with probability $\frac{A-C}{A}$ and under

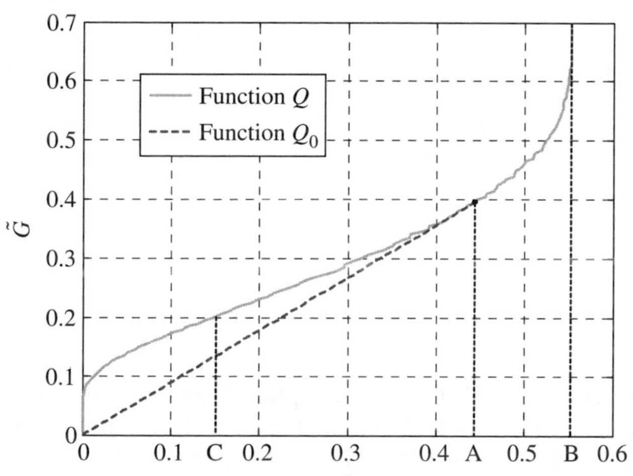

Figure 3.12 The relationship between $\tilde{F}$ and $\tilde{G}$ for the Rayleigh fading case.

those corresponding to $(A, \mathcal{Q}(A))$ with probability $\frac{C}{A}$. With the function $\mathcal{Q}_0$ instead of $\mathcal{Q}$, we may solve the threshold optimization problem, which will be shown later.

Remark 3.15 Since the functions $\mathcal{Q}$ and $\mathcal{Q}_0$ are only determined by the channel statistics of $h_{i,n}$ and $g_{i,n}$, and independent of the harvested energy rate r_n at each relay, they could be computed offline prior to the transmissions. Due to the non-convexity of the optimization problem in (3.128)–(3.129), it is difficult to derive the closed form of the function $\mathcal{Q}$ for general cases. Instead, we would build a look-up table offline and compute $\mathcal{Q}$ over a finite set $\{a_m\}$, $m = 1, \dots, M$, where we obtain the corresponding values $b_m = \mathcal{Q}(a_m)$ by exhaustive search for the general channel fading cases, and meanwhile, for each point (a_m, b_m), we also obtain the corresponding transmission thresholds H and G during the search process. Accordingly, the offline look-up table for the function $\mathcal{Q}_0$ could be obtained by the convex-hulling operation over $\{(a_m, \mathcal{Q}(a_m))\}$, and the corresponding transmission thresholds corresponding to each point in $\{(a_m, \mathcal{Q}_0(a_m))\}$ can be obtained by further considering the convexification operation discussed above. Based on the obtained look-up table, we then approximate the function $\mathcal{Q}_0$ as follows: connect the two adjacent points $(a_m, \mathcal{Q}_0(a_m))$ and $(a_{m+1}, \mathcal{Q}_0(a_{m+1}))$, $m = 1, \dots, M-1$ with straight line, and use the segment between $(a_m, \mathcal{Q}_0(a_m))$ and $(a_{m+1}, \mathcal{Q}_0(a_{m+1}))$ to approximate the function $\mathcal{Q}_0$ over the regime $[a_m, a_{m+1}]$. Note that the points on the obtained piecewise linear approximation function are also achievable for (3.130) with some transmission thresholds H and G, since the function $\mathcal{Q}_0$ is convex. Moreover, the approximation error diminishes as M goes to infinity and $\{a_m\}$ are properly chosen.

3.3.3.2 State-Dependent Threshold Optimization

In this subsection, we show how to solve problem (3.120)–(3.121) in a more generalized form, based on the results in the previous subsection. By changing variables as $\tilde{F}_k = \mathcal{F}(H_k, G_k)$ and $\tilde{G}_k = \mathcal{G}(H_k, G_k)$ and using the relationship $\tilde{G}_k = \mathcal{Q}_0(\tilde{F}_k)$ obtained in the previous subsection, we reformulate problem (3.120)–(3.121) as the following problem:

$$\max_{\{\tilde{F}_k, \tilde{G}_k\}} \quad \frac{\tilde{\mathcal{A}}_0^2}{\tilde{\mathcal{B}}_0} \tag{3.132}$$

$$\text{s.t.} \quad \tilde{G}_k = \mathcal{Q}_0(\tilde{F}_k), \quad k = 1, \dots, K, \tag{3.133}$$

$$\tilde{F}_k \geq 0, \quad \tilde{G}_k \geq 0, k = 1, \dots, K, \tag{3.134}$$

where

$$\tilde{\mathcal{A}}_0 = \sum_{k=1}^{K} \sqrt{\lambda_k \nu_k \tilde{F}_k}, \tag{3.135}$$

$$\tilde{\mathcal{B}}_0 = \sum_{k=1}^{K} \lambda_k \nu_k \tilde{G}_k. \tag{3.136}$$

It is easy to check that problem (3.132)–(3.134) has a larger optimal value than problem (3.120)–(3.121), since the constraints in (3.133) lead to a smaller $\tilde{G}_k$ with a given $\tilde{F}_k$ than the function $\mathcal{Q}$ derived in (3.128)–(3.129) for problem (3.120)–(3.121).

Remark 3.16 It is worth pointing out that with the convexification technique proposed in the previous subsection, we actually change the original problem

(3.120)–(3.121) into a more general problem (3.132)–(3.134). Specifically, in the original threshold-based transmission scheme, the relays with the same harvested energy rate are only allowed to transmit with the same thresholds. However, after the convexification, relays under the same harvested energy rate may transmit with one of the two different groups of thresholds.

It is easy to see that if we obtain the optimal solution of problem (3.132)–(3.134), we could use the look-up table discussed in Remark 3.15 to find the corresponding thresholds H_k and G_k for each possible harvested energy rate. To solve problem (3.132)–(3.134), we further introduce a new variable $\theta \doteq \frac{-\tilde{\mathcal{B}}_0}{\tilde{\mathcal{A}}_0^2}$, and it is easy to check that problem (3.132)–(3.134) has the same optimal point $\{\tilde{F}_k, \tilde{G}_k\}$ as the following problem:

$$\max_{\{\theta, \tilde{F}_k, \tilde{G}_k\}} \quad \theta \tag{3.137}$$

$$\text{s.t.} \quad \left(\sum_{k=1}^{K} \sqrt{\lambda_k} \nu_k \tilde{F}_k\right)^2 \cdot \theta + \sum_{k=1}^{K} \lambda_k \nu_k \tilde{G}_k \le 0, \tag{3.138}$$

$$\tilde{G}_k = \mathcal{Q}_0(\tilde{F}_k), \quad k = 1, \dots, K, \tag{3.139}$$

$$\tilde{F}_k \ge 0, \quad \tilde{G}_k \ge 0, k = 1, \dots, K. \tag{3.140}$$

It is easy to see that for an arbitrarily given θ, problem (3.137)–(3.140) is convex. Therefore, the global optimal solution of problem (3.137)–(3.140) can be obtained by bisection search [9] over the parameter θ, where, in each iteration, we solve the following convex feasibility problem with an updated θ.

$$\text{Find} \quad \{\tilde{F}_k\} \tag{3.141}$$

$$\text{s.t.} \quad \left(\sum_{k=1}^{K} \sqrt{\lambda_k} \nu_k \tilde{F}_k\right)^2 \cdot \theta + \sum_{k=1}^{K} \lambda_k \nu_k \mathcal{Q}_0(\tilde{F}_k) \le 0, \tag{3.142}$$

$$\tilde{F}_k \ge 0, \quad k = 1, \dots, K. \tag{3.143}$$

Remark 3.17 In problem (3.120)–(3.121), the number of the design variables is $2K$, where K is the number of the nonzero harvested energy rates with nonzero probability. Since problem (3.120)–(3.121) is non-convex, we need to search over these $2K$ variables to obtain the global optimal solution in general if we want to solve problem (3.120)–(3.121) directly. However, as shown in this section, we can solve problem (3.132)–(3.134) by bisection search, with the help of an offline look-up table built by the nonlinear equation groups defined in (3.130).

3.3.3.3 State-Oblivious Transmission Threshold

In the previous subsection, we examined the state-dependent threshold, which leads to a manageable but relatively large computation burden. In this subsection, we propose a more efficient heuristic suboptimal scheme to derive the transmission thresholds, which are treated independent of the harvested energy rate at each relay.

When $\mathcal{F}(H_k, G_k)$ and $\mathcal{G}(H_k, G_k)$ are independent of the harvested energy rate λ_k, the transmission thresholds for problem (3.120)–(3.121) are given by solving the following problem:

$$\max_{\{H,G\}} \frac{\mathcal{F}^2(H,G)}{\mathcal{G}(H,G)}, \tag{3.144}$$

where $\mathcal{F}(H,G)$ and $\mathcal{G}(H,G)$ are defined the same as (3.124) and (3.125), respectively. Similar to the analysis in the previous subsection, it follows that the optimal point of problem (3.144) also satisfies the function $\mathcal{Q}_0$ defined in (3.131), and thus problem (P5) can be solved by searching over the domain of function $\mathcal{Q}_0$ with the look-up table discussed in Remark 3.15.

3.3.4 Numerical Results

In this section, we present some numerical results to validate our theoretical analysis. For the purpose of exposition, we assume i.i.d. Rayleigh fading for all the links, and the harvested energy rate at each relay can be of the following three values: 0, 1, and 2, with the probabilities 0.1, 0.45, and 0.45, respectively. We compare the proposed schemes with the constant power allocation scheme, for which we set $H_n = G_n = 0$.

In Figure 3.13, we plot the normalized received SNR at the destination as a function of the source transmission power for the optimal threshold-based, state-oblivious threshold-based, and the constant power allocation schemes. Here the normalized received SNR is defined as the received SNR contributed by all the relays, i.e. $\frac{\mathbb{E}^2(\mathcal{A})}{\text{Var}(\mathcal{B})}$ in problem (3.120)–(3.121). It can be observed that there is about 2.54 dB power gain for the optimal threshold-based transmission scheme against the state-oblivious one. Compared with the constant power allocation scheme, the optimal threshold-based scheme has a relatively larger performance gain in the high source transmission power regime than that in the low-power regime.

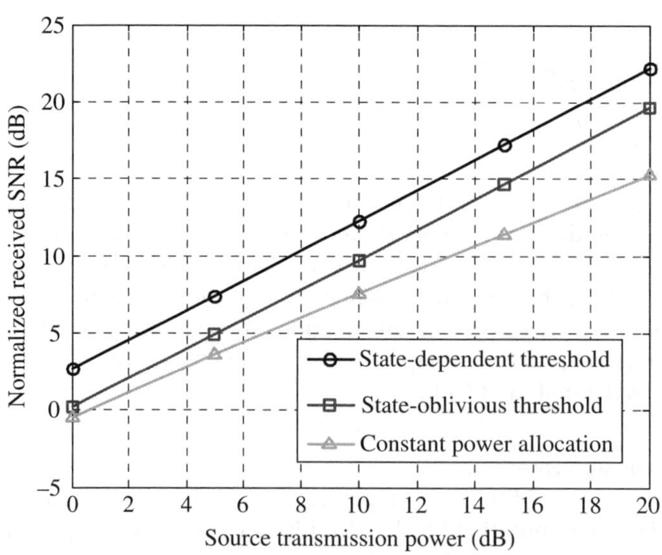

Figure 3.13 Comparison of the normalized received SNR among various transmission schemes.

3.4 Summary

In this chapter, we first study the resource allocation schemes for the two-user Gaussian MAC with a shared energy harvester and a bidirectional conferencing link. Both the noncausal and causal EH models were considered. For the noncausal EH model, we develop the optimal offline scheme that achieves the boundary of the maximum departure region by investigating the structure of the optimal sum power allocation and then deriving the optimal rate scheduling over the two transmitters. In particular, we show that there exists a capping rate at one of the two transmitters. For the causal EH model, one greedy scheme is considered under the assumption of no priori information about the future energy arrivals, and competitive analysis is used to measure the performance of the online scheme against the optimal offline scheme.

Then, we study the throughput maximization problem for the orthogonal relay channel with EH source and relay nodes, assuming a deterministic EH model. For both the cases with and without delay constraints at the destination, we examine the structures of the optimal source and relay power profiles over time and develop algorithms to efficiently compute these optimal power profiles.

Finally, we study the optimal threshold-based transmission scheme for a large relay network powered by renewable energy sources. The scheme is distributed with efficient power control, only requiring local information about the channel state and harvested energy rate. By utilizing the convexification technique by randomization, we derive a more general threshold-based transmission strategy and show that the newly obtained received SNR maximization problem can be solved via bisection search, combined with some offline computation. Beyond the optimal threshold determining scheme, we propose a suboptimal scheme, which requires less computation.

References

1 Zhao, D., Huang, C., Chen, Y. et al. (2015). Resource allocation for multiple access channel with conferencing links and shared renewable energy sources. *IEEE Journal on Selected Areas in Communications* 33 (3): 423–437. https://doi.org/10.1109/JSAC.2015.2391791.

2 Huang, C., Zhang, R., and Cui, S. (2013). Throughput maximization for the relay channel with energy harvesting nodes. *IEEE Journal on Selected Areas in Communications* 13 (10): 1469–1479. https://doi.org/10.1109/JSAC.2013.130811.

3 Huang, C., Zhang, J., Pin, Z., and Cui, S. (2013). Threshold-based transmissions for large relay networks powered by renewable energy. In: *Proceedings of IEEE Global Communications Conference (GLOBECOM)*, 1921–1926. IEEE. https://doi.org/10.1109/GLOCOM.2013.6831355.

4 Willems, F. (1983). The discrete memoryless multiple access channel with partially cooperating encoders (Corresp.). *IEEE Transactions on Information Theory* 29 (3): 441–445. https://doi.org/10.1109/TIT.1983.1056660.

5 Bross, S., Lapidoth, A., and Wigger, M. (2008). The Gaussian MAC with conferencing encoders. In: *Proceedings of IEEE International Symposium on Information Theory (ISTI)*, 2702–2706. IEEE. https://doi.org/10.1109/ISIT.2008.4595483.

6 Bross, S., Lapidoth, A., and Wigger, M. (2012). Dirty-paper coding for the Gaussian multiaccess channel with conferencing. *IEEE Transactions on Information Theory* 58 (9): 5640–5668. https://doi.org/10.1109/TIT.2012.2202210.

7 Maric, I., Yates, R.D., and Kramer, G. (2007). Capacity of interference channels with partial transmitter cooperation. *IEEE Transactions on Information Theory* 53 (10): 3536–3548. https://doi.org/10.1109/TIT.2007.904792.

8 Simeone, O., Gunduz, D., Poor, H.V. et al. (2009). Compound multiple-access channels with partial cooperation. *IEEE Transactions on Information Theory* 55 (6): 2425–2441. https://doi.org/10.1109/TIT.2009.2018343.

9 Boyd, S. and Vandenberghe, L. (2004). *Convex Optimization.* Cambridge: Cambridge University Press.

10 Yang, J., Ozel, O., and Ulukus, S. (2012). Broadcasting with an energy harvesting rechargeable transmitter. *IEEE Transactions on Wireless Communications* 11 (2): 571–583. https://doi.org/10.1109/TWC.2011.120911.101813.

11 Yang, J. and Ulukus, S. (2012). Optimal packet scheduling in an energy harvesting communication system. *IEEE Transactions on Communications* 60 (1): 220–230. https://doi.org/10.1109/TCOMM.2011.112811.100349.

12 Tutuncuoglu, K. and Yener, A. (2012). Optimum transmission policies for battery limited energy harvesting nodes. *IEEE Transactions on Wireless Communications* 11 (3): 1180–1189. https://doi.org/10.1109/TWC.2012.012412.110805.

13 Borodin, A. and ElYaniv, R. (1998). *Online Computation and Competitive Analysis.* Cambridge: Cambridge University Press.

14 Liang, Y. and Veeravalli, V. (2005). Gaussian orthogonal relay channels: optimal resource allocation and capacity. *IEEE Transactions on Information Theory* 51 (9): 3284–3289. https://doi.org/10.1109/TIT.2005.853305.

15 Cover, T. and Gamal, A.E. (1979). Capacity theorems for the relay channel. *IEEE Transactions on Information Theory* 25 (5): 572–584. https://doi.org/10.1109/TIT.1979.1056084.

16 Niyato, D., Hossain, E., and Fallahi, A. (2007). Sleep and wakeup strategies in solar-powered wireless sensor/mesh networks: performance analysis and optimization. *IEEE Transactions on Mobile Computing* 6 (2): 221–236. https://doi.org/10.1109/TMC.2007.30.

17 Kansal, A., Hsu, J., Zahedi, S., and Srivastava, M.B. (2007). Power management in energy harvesting sensor networks. *ACM Transactions in Embedded Computing Systems* 6 (4): (Art. No.: 32). https://doi.org/10.1145/1274858.1274870.

18 Wang, B. and Zhang, J. (2006). Throughput scaling of wideband sensory relay networks: cooperative relaying, power allocation and achievable rates. In: *Proceedings of the 25th IEEE International Conference on Computer Communications (INFOCOM)*, 1–12. IEEE. https://doi.org/10.1109/INFOCOM.2006.148.

19 Gallager, R.G. (1996). *Discrete Stochastic Process.* Boston, MA: Springer. https://doi.org/10.1007/978-1-4615-2329-1.

20 Ho, C.K. and Zhang, R. (2012). Optimal energy allocation for wireless communications with energy harvesting constraints. *IEEE Transactions on Signal Processing* 60 (9): 4808–4818. https://doi.org/10.1109/TSP.2012.2199984.

4

Cross-Layer Design for Energy Harvesting Links

4.1 Introduction

Although energy harvesting (EH) makes external renewable energy available to communication systems, it also brings instability of the system performance due to its irregularity and randomness, which requires extra efforts on the design of communication systems. This chapter will deal with system design across different layers concerning EH characteristics.

While EH brings uncertainty to communication systems, quality of service (QoS) is highly valued in most scenarios and requires strict guarantee. In the first part of the chapter, power control schemes to minimize the delay in the radio interface are discussed. For a single transmission task, the main concern is how to complete the transmission in as short a time as possible. To this end, completion time minimization problem has been thoroughly considered, mainly in the offline context, corresponding to the case with noncausal energy state information (ESI), such that transmission scheme can be optimized exploiting the knowledge about future energy arrivals *a priori* [1–3]. In [3], the relationship between a throughput maximization problem and a completion time minimization problem is identified. In a finite time horizon, the optimal policy for completion time minimization is identical to a throughput maximization policy, with the maximum departure curve representing the maximum number of bits that can be delivered before a given deadline. For online cases with causal ESI, i.e. when the system only knows the historical energy arrivals, Ref. [4, 5] investigate the performance bound of an online algorithm by competitive analysis. Specifically, in [5], it is shown that an online algorithm can only ensure the completion time less than at least twice of its offline optimum. Optimal algorithms achieving the performance bound are given. Besides completion time, average delay of an EH queuing system is analyzed in an offline way [6] and in the case where statistical information is available [7].

In addition to renewable energy, grid power can be integrated for the energy supply of wireless communications to guarantee the QoS, and this also leads to interesting trade-off between grid power consumption and QoS [8]. This is particularly important for system-level analysis, e.g. on cellular networks with base stations (BSs) jointly powered by renewable energy and power grid. Based on measured data, the statistics of harvested energy profile [9] has been studied. A practical way to optimize the resource allocation is to exploit the statistical information of the harvested energy, for instance, the average arrival rate or the statistical distribution. For the infinite time horizon case, Ref. [10]

Energy Harvesting Wireless Communications, First Edition.
Chuan Huang, Sheng Zhou, Jie Xu, Zhisheng Niu, Rui Zhang and Shuguang Cui.

considers a cross-layer resource allocation problem to maximize the total system utility, as a function of the average transmission rate, using a Markov decision process (MDP) approach [11]. The packet dropping and blocking probabilities are analyzed with different sleep and wake-up strategies in sensor networks powered by solar energy [12]. In [13], it is shown that the wireless link is strongly influenced by the renewable energy profile, and system parameters, e.g. the transmit power and packet durations, should be accordingly tuned to improve the performance. Throughput optimal routing scheme is proposed in [14] for rechargeable sensor networks. The closed-form maximum stable throughput is studied and derived in cognitive radio networks [15] and cooperative networks [16], respectively. Besides, scheduling mechanisms for throughput maximization have been extended to finite time horizon cases [17, 18]. While most of the existing work focuses on link-level analysis, some recent studies look into resource allocation strategies for cellular systems with both renewable energy sources and power grid access [8, 19, 20], which will be the focus of the second part of this chapter.

This chapter is organized as follows. We first illustrate the online way of minimizing the completion time and the delay in Section 4.2 via competitive analysis for completion time minimization in Sections 4.2.1–4.2.3 and via MDP for average delay minimization in Section 4.2.4, respectively. Then in Section 4.3, we show how to coordinate the renewable energy utilization, wireless resource allocation, and BS sleeping to minimize the grid power while guaranteeing the blocking probability of users.

4.2 Completion Time and Delay Minimization

Delay performance has always been a major criterion for QoS evaluation. In an EH wireless link, reducing latency is more challenging compared with grid-powered or purely battery-powered systems, since energy shortage might occur, and this leads to a pressing need for cross-layer design for EH links. In this section, completion time of a single packet and average delay of multiple packets are studied.

4.2.1 Completion Time Minimization

Consider a time-slotted EH wireless link as shown in Figure 4.1. Each slot is of the same length, and thus we use energy and power interchangeably without causing confusion. At the first time slot $t = 0$, the transmitter is requested to deliver a packet of D bits to the receiver. The only available power source of the transmitter is an EH module that gathers

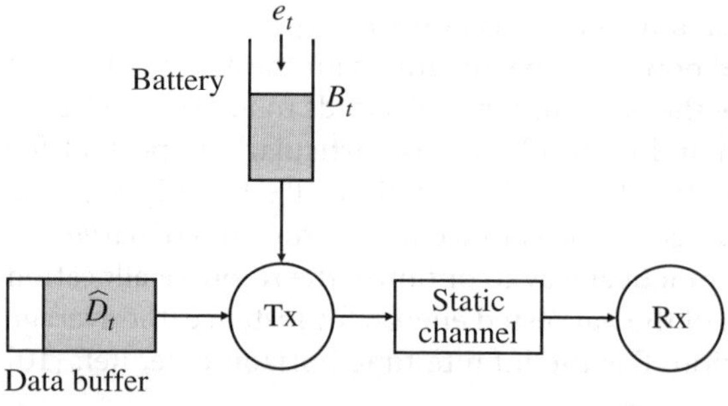

Figure 4.1 Energy harvesting system with random energy arrivals and an infinite-sized battery.

energy from renewable sources. At each time slot $t \in \mathbb{N}$, the harvested energy, denoted by e_t, is stored in an infinite-sized battery. Here, assume that the energy consumption for auxiliary actions at the transmitter side, such as calculation and decision-making, is negligible compared with that for transmissions over the air, and thus only the transmit power, denoted by p_t, is considered in the sequel. The transmitter transmits over a time-invariant channel, and the corresponding rate-power mapping $g : \mathbb{R}_{\geq 0} \mapsto \mathbb{R}_{\geq 0}$ from transmit power to channel rate, or the number of bits transmitted in a slot, is assumed to have the following properties:

(1) $g(\cdot)$ is monotonically increasing.
(2) $g(\cdot)$ is strictly concave.
(3) $g(0) = 0$, and $g(\cdot)$ is unbounded.
(4) There exists a constant α, such that $\alpha = \lim_{x \to 0} g'(x)$.

Some remarks are as follows. In the first two assumptions, monotonicity and concavity of the rate function together indicate the law of diminishing marginal utility, referring to the decrease in marginal increment of channel rate as transmit power rises. This property suggests a uniform allocation of energy over different slots in order to maximize the short-term throughput. The third assumption admits that an extremely high transmit power can produce any desirable channel rate. The fourth assumption forbids the exchange of high latency for excessively low energy consumption with the finite α.

Here are some examples supporting the rationality of the above assumptions. For static AWGN channels with transmit power p and noise power n, the classic Shannon capacity $C = \frac{1}{2}\log(1 + p/n)$ satisfies all four conditions. Further, note that the single receiver in Figure 4.1 does not limit our model to a single-input–single-output (SISO) channel. In fact, the analysis applies to all channel types with a qualified sum-rate function, such as the two-user multiple-access channel with uncoordinated coding [4], which multiplies the channel rate and the transmit power in Shannon capacity by a scale factor and therefore does not affect the concavity. Similar extensions can be made to multiple-receiver scenarios.

Due to the causality constraint on the energy usage, the allocated transmit power p_t cannot exceed currently available energy B_t in the battery, i.e.

$$p_t \leq B_t, \tag{4.1}$$

where

$$B_t = B_{t-1} + e_{t-1} - p_{t-1}. \tag{4.2}$$

4.2.1.1 Offline Optimum

Given an energy arrival profile E and data size D, the *offline* minimum completion time is the solution to the following optimization problem:

$$\max_{p_\tau} \quad T \tag{4.3}$$

$$\text{s.t.} \quad \sum_{\tau=1}^{t} p_\tau \leq \sum_{\tau=0}^{t-1} e_\tau, \quad \forall t \in \{1, 2, \dots, T\} \tag{4.4}$$

$$\sum_{\tau=1}^{T} g(p_\tau) \geq D. \tag{4.5}$$

Constraint (4.4), which is equivalent to (4.1), stands for the causality constraint in using energy before the transmission is completed. Upon the termination of transmission, at least D bits are delivered to the receiver, which explains (4.5).

An offline completion time minimization problem is similar to the finite-horizon throughput maximization, which can be written as

$$\max_{p_\tau} \quad \sum_{\tau=1}^{T} g(p_\tau) \tag{4.6}$$
$$\text{s.t.} \quad \sum_{\tau=1}^{t} p_\tau \leq \sum_{\tau=0}^{t-1} e_\tau, \quad \forall t \in \{1, 2, \dots, T\}.$$

Theorem 2 in [2] summarizes some insights into the relationship between completion time minimization and throughput maximization. It states that the power allocation schemes of the two problems are identical for correspondent data size (throughput) D and time length T. This property suggests that a throughput maximization algorithm can be easily transformed into a completion time minimization algorithm. The optimal offline algorithm for transmission completion time minimization for a system with infinite-sized battery and static channels is given in [1]. The main idea of the algorithm is summarized in three lemmas. First, the transmit power does not decrease in the optimal policy. This property is straightforward, since constant transmit power produces the largest throughput, while energy usage is upper bounded by energy arrivals. Therefore, energy insufficiency becomes the only possible reason that transmit power is not identical throughout the transmission process. This gives the rest two lemmas, which are that transmit power only changes when energy arrives and that when transmit power changes, previously harvested energy is completely consumed. Details can be found in [1].

4.2.1.2 Online Settings

Consider the online case where only historical energy arrivals are known to the transmitter, or, namely, the causal case. In every time slot t, the transmitter observes the size of the remaining data in the buffer, denoted by $\hat{D}_t$, historical energy arrivals as denoted by E_{t-1}, and historical power allocations p_{t-1}, where we let

$$E_t = (e_0, e_1, \dots, e_t),$$

and

$$P_t = (p_1, p_2, \dots, p_t).$$

Further denote the whole energy arrival process by $E = \{e_t, t \in \mathbb{N}\}$.

The transmitter then selects a certain power level for transmission. For brevity, the size of unfinished bits at the beginning of slot t is denoted by $\hat{D}_t$, such that

$$\hat{D}_t = D - \sum_{\tau=1}^{t-1} g(p_\tau), \quad \forall t \in \mathbb{N}^+. \tag{4.7}$$

The goal is to minimize the completion time T of the transmission of data with size D, i.e. to find an online algorithm, denoted by π, which depends only on P_{t-1}, E_{t-1}, and D, such that algorithm π leads to a relatively small completion time $T(\pi, E)$ satisfying $\hat{D}(T(\pi, E) + 1) = 0$ and Eqs. (4.1) and (4.2).

The deficiency of future information brings a dilemma at the transmitter. On one hand, the optimal offline algorithm suggests a relatively conservative power allocation scheme, i.e. spending less energy at the very beginning and increasing the transmit power when more energy arrives, which might lead to a long completion time. On the other hand, an aggressive algorithm can result in the shortage of available energy for subsequent transmissions, even possibly in failing to deliver the required D bits of data.

4.2.1.3 Preliminaries on Competitive Analysis

Since neither prior probability distributions nor statistical characteristics of the energy arrival process E are assumed, a distributional or average-case analysis is impractical. Here, competitive analysis is adopted to study completion time minimization from a worst-case viewpoint to evaluate the performance of an online algorithm. Relevant definitions in [21] and some notations are reviewed as follows.

Denoting the offline solution of (4.3) by $T_{\text{off}}(E)$, an online algorithm π is said to be strictly c-competitive if for all the energy arrivals E that the offline completion time $T_{\text{off}}(E)$ is finite, the completion time $T(\pi, E)$ is upper bounded as

$$T(\pi, E) \leq cT_{\text{off}}(E).$$

The *competitive ratio* of π is defined as the infimum of all the c that π is called c-competitive. Equivalently, the competitive ratio of π is denoted as $\lambda(\pi)$, and the following equation holds:

$$\lambda(\pi) = \sup_E \frac{T(\pi, E)}{T_{\text{off}}(E)}. \tag{4.8}$$

Further, an online algorithm π is called *competitive* if it attains a finite competitive ratio, i.e. $\lambda(\pi) < +\infty$.

Since the performance of any online algorithm is always no better than its offline optimum counterpart, the relationship

$$\lambda(\pi) \geq 1, \quad \forall \pi,$$

which means that the competitive ratio of online algorithms is lower bounded, always holds. The goal is to find the tight bound of competitive ratio of online algorithms for completion time minimization with a general rate-power function $g(\cdot)$:

$$\inf_\pi \lambda(\pi) = \inf_\pi \sup_E \frac{T(\pi, E)}{T_{\text{off}}(E)}.$$

Further, an online algorithm π^* is called *optimal* if it attains the minimum competitive ratio $\inf_\pi \lambda(\pi)$.

Other than seeking for an optimal algorithm after obtaining the lower bound, a simple 2-competitive algorithm is first proposed in the next subsection, and then its optimality is proved in Section 4.2.3.

4.2.2 A 2-Competitive Online Algorithm

Here we propose a 2-competitive online algorithm called linger-on-then-keep-invariable (LOKI), which works as shown in Algorithm 4.1.

Algorithm 4.1: LOKI for completion time minimization.

Input: D

Output: power level at each slot

$t \leftarrow 0$
$B \leftarrow 0$
repeat
    Harvest an energy packet of size e_t
    $B \leftarrow B + e_t$
    $t \leftarrow t + 1$
until $g\left(\frac{B}{t}\right) t \geq D$
$P_0 \leftarrow \frac{B}{t}$
repeat
    Transmit with power level P_0 at slot t
    $t \leftarrow t + 1$
until transmission completed

Algorithm LOKI is rather intuitive. It postpones the transmission until adequate energy and time, with which the data can be delivered, is gained. With energy arrival process E, the allocated transmit power of each time slot can be given as

$$p_t = \begin{cases} 0 & t = 1, 2, \dots, T_1 - 1 \\ P_0 & t = T_1, T_1 + 1, \dots, T_{\text{LOKI}}(E), \end{cases}$$

where T_1 is the time slot in which the transmission begins, i.e. the smallest integer that satisfies

$$g\left(\frac{\sum_{\tau=0}^{T_1-1} e_\tau}{T_1}\right) T_1 \geq D,$$

and

$$P_0 = \frac{\sum_{\tau=0}^{T_1-1} e_\tau}{T_1}.$$

It is easy to show that LOKI does not violate the causality constraint in (4.1). Since the transmission can be completed with a constant power P_0 in T_1 slots, the second repeat-until iteration terminates in T_1 slots; therefore the energy accumulated before $t = T_1$ is enough for the transmission.

Although LOKI might seem slightly pessimistic and conservative, it actually attains a simple form as well as a good performance. Additionally, it will show great importance in subsequent analysis on the infimum of competitive ratio in Section 4.2.3. The following lemma gives the performance of LOKI.

Lemma 4.1 *LOKI attains a competitive ratio of 2.*

Proof: First, prove that the offline minimum completion time T_{off} is no less than T_1 by *reductio ad absurdum.*

According to the algorithm, T_1 is the result of the following optimization problem:

$$\begin{aligned} \min \quad & t \\ \text{s.t.} \quad & g\left(\frac{\sum_{\tau=0}^{t-1} e_\tau}{t}\right) t \geq D \\ & t \in \mathbb{N}. \end{aligned}$$

Thus we have

$$g\left(\frac{\sum_{\tau=0}^{t-1} e_\tau}{t}\right) t < D, \quad \forall t = 1, 2, \ldots, T_1 - 1. \tag{4.9}$$

Assuming that the offline optimal solution is $\{p_t^*\}_{t>0}$ and it has the offline completion time of T_{off},

$$\sum_{\tau=1}^{T_{\text{off}}} g(p_\tau^*) \geq D \tag{4.10}$$

with one of the causal constraints

$$\sum_{\tau=1}^{T_{\text{off}}} p_\tau^* \leq \sum_{\tau=0}^{T_{\text{off}}-1} e_\tau.$$

Due to the concavity and monotonicity of the rate-power function $g(\cdot)$,

$$\sum_{\tau=1}^{T_{\text{off}}} g(p_\tau^*) \leq g\left(\frac{\sum_{\tau=1}^{T_{\text{off}}} p_\tau^*}{T_{\text{off}}}\right) T_{\text{off}} \leq g\left(\frac{\sum_{\tau=0}^{T_{\text{off}}-1} e_\tau}{T_{\text{off}}}\right) T_{\text{off}}.$$

According to (4.9), if $T_{\text{off}} < T_1$, then

$$\sum_{\tau=1}^{T_{\text{off}}} g(p_\tau^*) \leq g\left(\frac{\sum_{\tau=0}^{T_{\text{off}}-1} e_\tau}{T_{\text{off}}}\right) T_{\text{off}} < D,$$

which contradicts (4.10). Therefore, the following inequality holds:

$$T_{\text{off}} \geq T_1. \tag{4.11}$$

Second, show that the competitive ratio of algorithm LOKI is no larger than 2. Since

$$T_1 g(P_0) = g\left(\frac{\sum_{\tau=0}^{T_1-1} e_\tau}{T_1}\right) T_1 \geq D,$$

the transmitter stays silent for the first $T_1 - 1$ slots and needs at most T_1 time slots after slot T_1 to fulfill the transmission, so that

$$T_{\texttt{LOKI}}(E) \leq 2T_1 - 1 \leq 2T_{\text{off}} - 1.$$

Thus, the competitive ratio of the LOKI algorithm satisfies

$$\lambda_{\texttt{LOKI}} = \sup_{E} \frac{T_{\texttt{LOKI}}(E)}{T_{\text{off}}(E)} \leq 2.$$

Third, show that the competitive ratio is no less than 2. Consider an energy arrival process $E = \{e_t, t \in \mathbb{N}\}$ as follows:

$$e_t = \begin{cases} g^{-1}\left(\frac{D}{T_0}\right) & t = 0, 1, \dots, T_0 - 1 \\ 0 & \text{otherwise.} \end{cases}$$

The optimal offline strategy is to transmit with a constant power level $g^{-1}\left(\frac{D}{T_0}\right)$ during the first T_0 slots, and

$$T_{\text{off}}(E) = T_0, \tag{4.12}$$

while LOKI suggests to transmit with the same power level from slot T_0, and

$$T_{\texttt{LOKI}}(E) = 2T_0 - 1. \tag{4.13}$$

Dividing both sides of (4.13) by the corresponding side of (4.12) and taking supremum over $T_0 \in \mathbb{N}$,

$$\lambda_{\texttt{LOKI}} \geq \sup_{T_0} \frac{T_{\texttt{LOKI}}(E)}{T_{\text{off}}(E)} = 2.$$

Therefore, the competitive ratio of algorithm LOKI is 2. □

4.2.3 Game-Theoretic Analysis on the Completion Time Minimization

Up to now, the online algorithms are required to follow only the causality constraint of energy, while no description of the energy arrival process is provided. To simplify the problem, potential candidates of energy arrival processes and power allocation algorithms are visited before introducing the game-theoretic formulation and analysis of the completion time minimization problem. Hereafter, the set of all potential energy arrival processes is referred to as the action set of the nature, and the set of all candidate power allocation algorithms as the action set of the transmitter.

4.2.3.1 The Action Set of the Nature

Due to the fourth assumption about the rate function $g(\cdot)$, a disproportionate amount of energy might result in the failure of transmission, even in an offline setting. In order to make this problem reasonable, the total harvested energy must be adequately large. The following lemma states the condition for an energy arrival process being proper for the completion time minimization problem.

Lemma 4.2 *For an energy arrival process $E = \{e_t, t \in \mathbb{N}\}$, the transmission of data with size D can be completed in a finite length of time if and only if $D < \alpha \sum_{\tau=0}^{+\infty} e_\tau$.*

Proof:

Sufficiency: For any power allocation scheme $\{p_t\}_{t>0}$, the amount of data transmitted in the first t slots, denoted by D_t, satisfies

$$\begin{aligned} D_t &= \sum_{\tau=1}^{t} g(p_\tau) \\ &< g\left(\frac{\sum_{\tau=1}^{t} p_\tau}{t}\right) t \\ &\leq g\left(\frac{\sum_{\tau=0}^{+\infty} e_\tau}{t}\right) t. \end{aligned}$$

The first inequality holds because of the concavity of $g(\cdot)$, and the second is due to the causality constraint of energy. Taking the limit of the last expression as t approaches infinity,

$$\begin{aligned} D_t &\leq \lim_{\to+\infty} g\left(\frac{\sum_{\tau=0}^{+\infty} e_\tau}{t}\right) t \\ &= \lim_{x\to 0} g'(x) \sum_{\tau=0}^{+\infty} e_\tau \\ &= \alpha \sum_{\tau=0}^{+\infty} e_\tau. \end{aligned}$$

Thus, any transmission to be fulfilled in a finite number of slots must have a data size strictly less than $\alpha \sum_{\tau=0}^{+\infty} e_\tau$.

Necessity: Let $\epsilon = \frac{-D+\alpha\sum_{\tau=0}^{+\infty} e_\tau}{2} > 0$. Since $D + 2\epsilon = \alpha \sum_{\tau=0}^{+\infty} e_\tau$, there exists a positive integer T_2, such that

$$\alpha \sum_{\tau=0}^{T_2-1} e_\tau > D + \epsilon. \tag{4.14}$$

Meanwhile, because

$$\lim_{t\to+\infty} g\left(\frac{\sum_{\tau=0}^{T_2-1} e_\tau}{t}\right) t = \alpha \sum_{\tau=0}^{T_2-1} e_\tau,$$

there exists another positive integer T_3, such that

$$g\left(\frac{\sum_{\tau=0}^{T_2-1} e_\tau}{T_3}\right) T_3 > \alpha \sum_{\tau=0}^{T_2-1} e_\tau - \epsilon. \tag{4.15}$$

Allocating the energy as

$$\begin{cases} p_t = 0 & t = 1, \ldots, T_2 - 1 \\ p_t = \dfrac{\sum_{\tau=0}^{T_2-1} e_\tau}{T_3} & t = T_2, \ldots, T_2 + T_3 - 1, \end{cases}$$

and substituting (4.14) into (4.15), the total size of data that can be transmitted in $T_2 + T_3 - 1$ slots is

$$g\left(\frac{\sum_{\tau=0}^{T_2-1} e_\tau}{T_3}\right) T_3 > D.$$

Therefore, the transmission can be fulfilled in a finite time interval. □

The action set $\mathcal{I}$ of the nature is defined as the set of all energy arrival processes under which it is possible to fulfill the transmission in an finite time length. Therefore, the action set is written as

$$\mathcal{I} = \left\{ \{e_t, t \in \mathbb{N}\} \middle| \sum_{\tau=0}^{+\infty} e_\tau > \frac{D}{\alpha} \right\}.$$

4.2.3.2 The Action Set of the Transmitter

Combined with the causality constraint, the action set Π of the minimizer is defined to be the set of all competitive and causal strategies, i.e.

$$\Pi = \{\pi | \lambda(\pi) < +\infty, p_t \leq B_t, \forall t\}. \tag{4.16}$$

Denoting Algorithm 4.1 by π_{LOKI}, since the competitive ratio $\lambda(\pi_{\mathrm{LOKI}}) = 2$, π_{LOKI} is both causal and competitive, and thus $\pi_{\mathrm{LOKI}} \in \Pi$. The necessary condition for an online algorithm being both competitive and causal is given in the following lemma.

Lemma 4.3 *(Reliability guarantee) For any strategy $\pi \in \Pi$, the transmit power p_t at each time slot t satisfies*

$$\alpha p_t - g(p_t) < \max(\alpha B_t - \hat{D}_t, 0). \tag{4.17}$$

Proof: In order to guarantee the completion of transmission in a finite time interval, according to Lemma 4.2, the transmit power must follow the constraint

$$\hat{D}_t - g(p_t) < \alpha[B_t - p_t], \quad \forall t \in \mathbb{N}^+, \tag{4.18}$$

as long as there is enough energy stored in the battery, i.e.

$$\hat{D}_t < \alpha B_t.$$

The left-hand side of (4.18) equals the residual data at the beginning of slot $t + 1$, and the right-hand side is the upper bound of the size of data that can be transmitted, assuming no further energy arrivals after t. Adding $\alpha p - \hat{D}_t$ to both sides of (4.18), (4.17) holds when $\hat{D}_t < \alpha B_t$.

If $\hat{D}_t \geq \alpha B_t$, any power allocation $p_t > 0$ will lead to the failure in completion and thus bring an infinite competitive ratio. Therefore, the transmitter should be silent, and (4.17) is also satisfied. □

From Lemma 4.3, the following corollary can be directly obtained.

Corollary 4.1 *If* $\alpha \sum_{\tau=0}^{t_0-1} e_\tau < D$, *then* $\forall \pi \in \Pi, \forall t \leq t_0, p_t = 0$.

Corollary 4.1 provides a hint about the structure of π. It indicates that for any competitive and causal strategy, the transmitter will remain silent before the total amount of harvested energy reaches a level that is associated with the data size D.

4.2.3.3 Two-Person Zero-Sum Game

From a game-theoretic viewpoint, the original problem, which focuses on finding an online algorithm with a minimum competitive ratio, forms a two-person zero-sum game in pure strategy. The kernel function of the game is given by

$$J(\pi, E) = \frac{T(\pi, E)}{T_{\text{off}}(E)}, \tag{4.19}$$

and

$$\lambda(\pi) = \sup_E J(\pi, E). \tag{4.20}$$

In this game, the transmitter, or the strategy designer, decides on the online algorithm and acts as the minimizer, while the nature seeks an energy arrival process E to maximize the kernel function.

In accordance with [22], the upper value of this two-person zero-sum infinite game is defined by

$$\bar{V} = \inf_\pi \sup_E J(\pi, E), \tag{4.21}$$

while the lower value is

$$\underline{V} = \sup_E \inf_\pi J(\pi, E). \tag{4.22}$$

If the upper value equals the lower value, i.e. $V = \bar{V} = \underline{V}$, then the game is said to have a value of V.

To illustrate the relationship between the two-person zero-sum game and the competitive analysis of the completion time minimization, substitute (4.20) into (4.21), and get

$$\bar{V} = \inf_\pi \lambda(\pi),$$

which suggests that the upper value of the game equals exactly the infimum of the competitive ratio of online algorithms. Therefore, if the game has a value, it can be simpler to obtain the tight lower bound of competitive ratio.

However, it is evident that the original game does not have a value, since $\underline{V} = 1$ and $\bar{V} > 1$. But with the reduced action set from the transmitter and the nature $\Pi \times \mathcal{I}$, the game has a value of 2, i.e.

$$\sup_{E \in \mathcal{I}} \inf_{\pi \in \Pi} J(\pi, E) = \inf_{\pi \in \Pi} \sup_{E \in \mathcal{I}} J(\pi, E) = 2. \tag{4.23}$$

The above relation is supported by the following lemma.

Lemma 4.4 *For any positive value* ϵ, *this game has an* ϵ *saddle point. More explicitly,* $\forall \epsilon > 0, \exists (\pi^*, E^*) \in \Pi \times \mathcal{I}$, *s.t.* $\forall (\pi, E) \in \Pi \times \mathcal{I}$,

$$J(\pi^*, E) - \epsilon \leq J(\pi^*, E^*) \leq J(\pi, E^*) + \epsilon. \tag{4.24}$$

Proof: Construct an energy arrival process $E^*_{T_0} = \{e^*_t, t \in \mathbb{N}\}$ as

$$e^*_t = \begin{cases} g^{-1}\left(\frac{D}{T_0}\right) & t = 0, 1, \ldots, T_0 - 1 \\ 0 & \text{otherwise,} \end{cases}$$

and let $l(T_0)$ represent the integer satisfying

$$\frac{D}{\alpha g^{-1}\left(\frac{D}{T_0}\right)} - 1 \leq l(T_0) < \frac{D}{\alpha g^{-1}\left(\frac{D}{T_0}\right)}. \tag{4.25}$$

Dividing (4.25) by T_0, the inequality becomes

$$\frac{\frac{D}{T_0}}{\alpha g^{-1}\left(\frac{D}{T_0}\right)} - \frac{1}{T_0} \leq \frac{l(T_0)}{T_0} < \frac{\frac{D}{T_0}}{\alpha g^{-1}\left(\frac{D}{T_0}\right)}. \tag{4.26}$$

Taking limit as T_0 approaches infinity at both sides of (4.26),

$$\lim_{T_0 \to +\infty} \frac{\frac{D}{T_0}}{\alpha g^{-1}\left(\frac{D}{T_0}\right)} = \lim_{T_0 \to +\infty} \frac{l(T_0)}{T_0} = 1.$$

Therefore,

$$\lim_{T_0 \to +\infty} \frac{l(T_0)}{T_0} = 1. \tag{4.27}$$

It is obvious that the optimal offline strategy under $E^*_{T_0}$ is to transmit with a constant power level $g^{-1}\left(\frac{D}{T_0}\right)$ in the first T_0 slots, and

$$T_{\text{off}}(E^*_{T_0}) = T_0.$$

For online strategies, according to Corollary 4.1, any strategy $\pi \in \Pi$ will not choose to transmit before $l(T_0) + 1$, since

$$\sum_{\tau=0}^{l(T_0)-1} e^*_\tau = l(T_0) g^{-1}\left(\frac{D}{T_0}\right) < \frac{D}{\alpha}.$$

Also notice that all algorithms need at least T_0 slots for transmission; thus

$$T(\pi, E^*_{T_0}) \geq l(T_0) + T_0, \quad \forall \pi \in \Pi. \tag{4.28}$$

Dividing both sides of (4.28) by offline optimum T_0,

$$J(\pi, E^*_{T_0}) \geq 1 + \frac{l(T_0)}{T_0}.$$

It is clear that (4.27) can be equivalently interpreted as that $\forall \epsilon > 0$, $\exists N_0(\epsilon) \in \mathbb{N}^+$, s.t. $\forall T_0 \geq N_0(\epsilon)$,

$$\frac{l(T_0)}{T_0} \geq 1 - \epsilon;$$

therefore

$$J(\pi, E^*_{T_0}) \geq 2 - \epsilon.$$

Recall algorithm LOKI proposed in Section 4.2.2, which gives

$$J(\pi_{\text{LOKI}}, E) \leq \lambda(\pi_{\text{LOKI}}) = 2, \quad \forall E \in \mathcal{I}.$$

For any $\epsilon > 0$, let $T_0 = \max\left(N_0(\epsilon), \lceil \frac{1}{\epsilon} \rceil\right)$ and $\pi^* = \pi_{\text{LOKI}}$, then $\forall \pi \in \Pi$

$$\begin{aligned} J(\pi, E^*_{T_0}) + \epsilon &\geq 2 \\ &\geq J(\pi^*, E^*_{T_0}). \end{aligned}$$

And $\forall E \in \mathcal{I}$,

$$\begin{aligned} J(\pi^*, E) - \epsilon &\leq 2 - \epsilon \\ &\leq 2 - \frac{1}{T_0} \\ &\leq J(\pi^*, L^*_{T_0}). \end{aligned}$$

As a result, inequality (4.24) holds. □

According to Theorem 4.1 in [22], a two-person zero-sum game has an ϵ saddle point for every positive value ϵ if and only if this game has a finite value and the value equals the limit of the Cauchy sequence $\{J_{\epsilon_k}(\pi^*, E^*)\}_k$ with $\epsilon_{k+1} < \epsilon_k$ and $\lim_{k\to+\infty}\epsilon_k = 0$, where $J_{\epsilon_k}(\pi^*, E^*)$ represents the middle item of (4.24) with $\epsilon = \epsilon_k$. Therefore, the following theorem is obtained.

Theorem 4.1 With an infinite-sized battery and a time-invariant rate-power function satisfying the four assumptions in Section 4.2.1, the infimum of competitive ratio of online algorithms for completion time minimization is 2:

$$\inf_{\pi} \sup_{E} \frac{T(\pi, E)}{T_{\text{off}(E)}} = 2. \tag{4.29}$$

Proof: First, prove that the two-person zero-sum game has a value of 2 in pure strategy. Applying the ϵ saddle points found in the proof of Lemma 4.4, the limit of the Cauchy sequence is

$$\begin{aligned} \lim_{k\to+\infty} J_{\epsilon_k}(\pi^*, E^*) &= \lim_{\epsilon\to 0} J(\pi^*, E^*_{T_0(\epsilon)}) \\ &= \lim_{T_0\to+\infty} \frac{2T_0 - 1}{T_0} \\ &= 2. \end{aligned}$$

Therefore,

$$\sup_{E\in\mathcal{I}} \inf_{\pi\in\Pi} J(\pi, E) = \inf_{\pi\in\Pi} \sup_{E\in\mathcal{I}} J(\pi, E) = 2$$

holds.

Substituting the kernel function (4.19) into (4.23), the equation is given by $\inf_{\pi\in\Pi}\sup_{E\in\mathcal{I}}\frac{T(\pi,E)}{T_{\text{off}}(E)} = 2$, which is quite similar to the left-hand side of (4.29). The only difference is that the action sets in the zero-sum game are restricted to all competitive and causal algorithms and all reasonable energy arrival processes. Thus, any causal algorithms that are not in π must have a competitive ratio of infinity, so the above equation gives

$$\inf_{\pi}\sup_{E\in\mathcal{I}}\frac{T(\pi,E)}{T_{\text{off}}(E)} = 2.$$

With all energy arrival processes that might lead to a finite completion time included in $\mathcal{I}$, it is sufficient to conclude that the infimum of competitive ratio of online algorithms for completion time minimization is 2. □

4.2.3.4 Discussions

Another intuitive algorithm that achieves the optimality is the generalized lazy online (GLO) algorithm proposed in [4]. In the algorithm, the transmitter waits till enough energy is gathered and then transmits with the power that is the solution for the offline completion time minimization assuming no forthcoming energy arrivals. An energy arrival instance and its allocation schemes under the offline optimum, LOKI, and GLO in [4] are illustrated in Figure 4.2. The solid line represents the total amount of harvested energy up to the corresponding time slot, while the short-dashed line stands for the strategy that leads to the shortest completion time, the dot-dashed line for the policy suggested by LOKI, and the long-dashed line for the policy by GLO. In the example, it is assumed that the rate-power function is $g(p) = \log(1 + p)$ and the data size $D = 4$, and thus the transmission can be completed with four units of energy and four time slots, i.e. $4g(1) = D$. Since the available energy at the first slot is less than one unit, the transmission cannot be completed in four slots. The optimal strategy is to use up all the initial energy at the first slot and then set the transmit power to a constant level during the following four slots, which gives a total completion time of five slots. With algorithm LOKI, the transmitter waits for three slots and then allocates the harvested

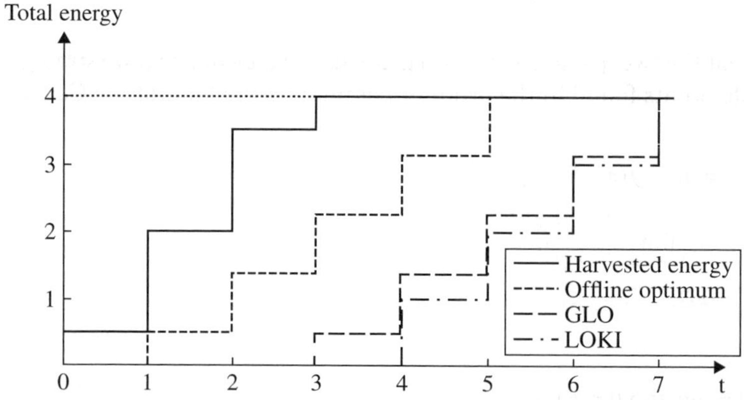

Figure 4.2 The energy allocation policies of offline optimum and LOKI under an example energy arrival realization.

energy equally to the following four slots. Under GLO algorithm, the transmission does not start until the third slot as well because of energy shortage. It first transmits with 0.5 unit of energy at the third slot and increases the power to 0.875 at the fourth slot. For both LOKI and GLO algorithms, the completion time is seven slots, which is less than twice of the offline minimum.

Theorem 4.1 gives the tight lower bound of the competitive ratio. It is proved that both GLO in [4] and LOKI reach the minimum competitive ratio of 2, which suggests that they are both optimal from the worst-case viewpoint. With GLO implemented, the controller needs to set the transmit power to a reasonably higher level at each energy arrival, which might require excessive computational steps. Instead, LOKI shows better potential for facilitating computation by distributing these overhead expenses to the slots before the actual transmission begins. Besides, for the case where energy is harvested frequently and the harvested energy in each time slot is relatively low, LOKI greatly reduces the calculation complexity.

Further, several other optimal algorithms can be established through similar method. For example, a simple modification can be made to LOKI by replacing the constant transmit power with time-adaptive ones, as described in Algorithm 4.2. This algorithm increases the transmit power at each energy arrival after the transmission begins, while LOKI only utilizes the former accumulated energy once the transmitter starts to deliver the data. It is obvious that this algorithm is always *ahead* of LOKI and is thus optimal.

Algorithm 4.2: Another 2-competitive online algorithm for completion time minimization.

Input: D
Output: power level at each slot

$t \leftarrow 0$
$B \leftarrow 0$
repeat
    Harvest an energy packet of size e_t
    $B \leftarrow B + e_t$
    $t \leftarrow t + 1$
until $g(\frac{B}{t})t \geq D$
repeat
    if e_t amount of energy is harvested **then**
        $B \leftarrow B + e_t$
        P_0 equals the largest real number p satisfying $\lfloor \frac{B}{p} \rfloor \cdot g(p) \geq D$
    end if
    Transmit with power level P_0 at slot t
    $D \leftarrow D - g(P_0)$
    $B \leftarrow B - P_0$
    $t \leftarrow t + 1$
until transmission completed

However, more computation steps are required at the transmitter's side. Although computational cost is not explicitly considered here, the tradeoff between completion time and actual calculation costs exists among the optimal algorithms in practical systems, especially for low-cost sensor nodes or Internet of things (IoT) transceivers.

4.2.4 Delay-Optimal Energy Management

Former analysis mostly concentrates on the delay performance of a single task of packet, which does not take queuing delay into account. In the following, queuing model is adopted to characterize the overall system performance delivering multiple tasks or packets [6, 7].

4.2.4.1 Formulation

The system model is identical to Figure 4.1, in which an EH transmitter communicates with a receiver. The battery dynamic is expressed in (4.2) with energy causality constraint (4.1). To consider the queuing delay, multiple packets arrive into the data queue of the transmitter. Denote the queue length as Q_t and data arrival at slot t as D_t. Similarly, the rate function $g(p_t, \beta_t)$, in which p_t is the transmit power and β_t represents the channel condition at slot t, is assumed to be monotonically increasing and strictly concave. Assume that the second derivative of the rate function exists. Therefore, the queuing dynamics is expressed as

$$Q_t = [Q_{t-1} - g(p_{t-1}, \beta_{t-1})]^+ + D_{t-1}, \tag{4.30}$$

in which $[\cdot]^+ = \max(0, \cdot)$.

Note that for a fading channel with channel gain β_t, according to Shannon's formula, the maximum achievable rate at time slot t is $g(p_t, \beta_t) = \frac{1}{2}\log(1 + \beta_t p_t)$, which is still a special case of our discussion of the rate-power function previously.

Denoting the average queue length among the first T slots as $\bar{Q}_T \triangleq \frac{1}{T}\sum_{t=1}^{T} Q_t$, by Little's law, average queue-length minimization in the following also solves the delay minimization problem:

$$\begin{aligned} \min \quad & \bar{Q}_T \\ \text{s.t.} \quad & (4.1) \text{ and } (4.30), \quad \forall t \in \{1, \dots, T-1\} \end{aligned} \tag{4.31}$$

4.2.4.2 Offline Analysis

Since $g(p_t, \beta)$ is monotonically increasing with p_t, $Q_t > g(p_t, \beta_t)$ always holds under the optimal policy to avoid waste of energy. Therefore, Eq. (4.30) can be expressed as

$$\begin{aligned} Q_t &= Q_{t-1} - g(p_{t-1}, \beta_{t-1}) + D_{t-1} \\ &= Q_0 + \sum_{i=1}^{t-1}(D_i - g(p_i, \beta_i)), \end{aligned} \tag{4.32}$$

and the average queue length becomes

$$\bar{Q}_t = Q_0 + \frac{1}{T}\sum_{t=1}^{T-1}(T-t)(D_t - g(p_t, \beta_t)). \tag{4.33}$$

Similarly, the battery dynamic can be expressed as

$$B_t = B_0 + \sum_{i=1}^{t-1}(e_i - p_i).$$

Therefore, optimization problem (4.31) becomes

$$\begin{aligned}\min_{p_\tau} \quad & Q_0 + \frac{1}{T}\sum_{t=1}^{T-1}(T-t)(D_t - g(p_t, \beta_t)) \\ \text{s.t.} \quad & \sum_{\tau=1}^{t} p_\tau \le B_0 + \sum_{i=1}^{t-1} e_\tau, \quad \forall t \in \{0, 1, \dots, T-1\} \\ & \sum_{\tau=1}^{t} g(p_\tau, \beta_\tau) \le Q_0 + \sum_{\tau=1}^{t-1} D_\tau, \quad \forall t \in \{0, 1, \dots, T-1\}.\end{aligned} \tag{4.34}$$

Although problem (4.34) is not a convex problem due to its non-convex feasible region, it can be transformed into a convex problem by defining $r_t = g(p_t, \beta_t)$, which is the data rate given the channel condition and transmit power, and optimize over r_t instead of p_t. Define a mapping $h : \mathbb{R}_+^2 \mapsto \mathbb{R}_+$ by $r_t = g(h(r_t, \beta_t), \beta_t)$, and the new optimization problem is thus

$$\begin{aligned}\min_{r_\tau} \quad & Q_0 + \frac{1}{T}\sum_{t=1}^{T-1}(T-t)(D_t - r_t) \\ \text{s.t.} \quad & \sum_{\tau=1}^{t} h(r_\tau, \beta_\tau) \le B_0 + \sum_{\tau=1}^{t-1} e_\tau, \quad \forall t \in \{0, 1, \dots, T-1\} \\ & \sum_{\tau=1}^{t} r_\tau \le Q_0 + \sum_{i=1}^{t-1} D_\tau, \quad \forall t \in \{0, 1, \dots, T-1\}.\end{aligned} \tag{4.35}$$

Equation (4.35) is a convex optimization problem since the objective function and the left-hand sides of the constraints are convex functions. By convex optimization methods, the optimal power allocation can be efficiently obtained for the average delay minimization problem.

4.2.4.3 Online Analysis

With the statistic information about energy arrivals, e.g. assuming both energy arrival process and data arrival process are Markovian, delay minimization problem of an EH communication system can be modeled as an MDP and solved with policy iteration or value iteration [11].

For a discount factor $\xi \in (0, 1)$, an ξ-discounted delay minimization problem is to minimize $\frac{1}{1-\xi}\sum_{t=1}^{+\infty}\xi^t Q_t$. Let battery and data buffer be discretized and finite. Assuming static channel and that both energy arrivals e and data arrivals D at each time slot are independent and identically distributed (i.i.d.) discrete random variables, the problem becomes minimizing $Q_\xi = \mathbb{E}\left[\frac{1}{1-\xi}\sum_{t=1}^{+\infty}\xi^t Q_t\right]$. By Bellman's equation, the minimum of Q_ξ satisfies

$$Q_\xi^*(q, b) = \min_p \mathbb{E}_{D,e}\{q + \xi Q_\xi^*([q - g(p)]^+ + D, b - p + e)\}. \tag{4.36}$$

As proved in [7], the infimum of $Q_\xi(q, b)$ goes to the optimal average delay as ξ approaches to 1. Thus, with policy iteration, the optimal policy can be obtained. The comparison of the mean delay of an offline problem and an online problem with statistical energy arrival information is illustrated in Figure 4.3. We will further illustrate the

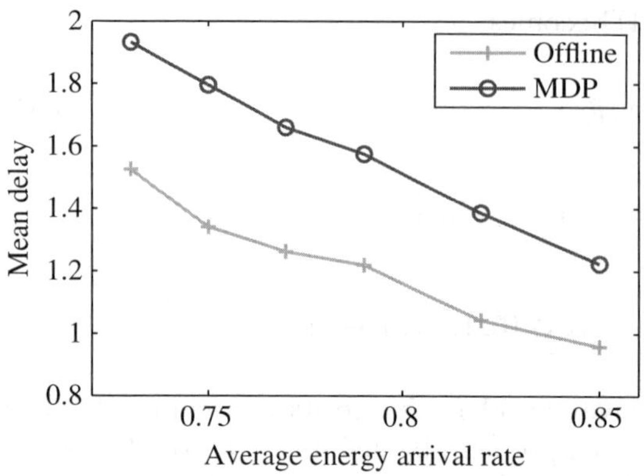

Figure 4.3 Mean delay under static channel with Poisson energy and data arrivals truncated at 5. Energy is scaled by factor 1 and data by factor 0.25. Mean data arrival rate is 0.5. The rate-power function is $g(p) = \log(1 + p)$.

usage of dynamic programming (DP) to design an online resource allocation policy in a cellular network with multiple BSs powered by renewable energy.

4.3 Traffic-Aware Base Station Sleeping in Renewable Energy-Powered Cellular Networks

Cellular networks have to provide guaranteed QoS to users. To this end, BS powered by the unstable renewable energy should also be able to access to the grid power. As illustrated in this section, the utilizing of the renewable energy should be jointly optimized with the wireless resource allocation and dynamic BS sleeping in order to minimize the grid power consumption for a given QoS requirement.

4.3.1 System Model of a Renewable Energy-Powered Cellular Network

Consider a wireless cellular network consisting of N_B BSs, denoted by a set $\{b|b = 1, 2, \dots, N_B\}$, each of which is powered jointly by a renewable energy source and the power grid. The system timeline is divided into T time periods (e.g. $T = 24$ and the length of each time period can be one hour, while in this chapter T can be adapted according to the system operation feature, and the length of each time period can also be adapted). More details of the model are provided as follows.

4.3.1.1 Power Consumption Model

In time period t, the average harvested renewable energy on BS b is denoted by $P^{(b)}_{\mathrm{H},t}T$, and the grid power input is $P^{(b)}_{\mathrm{G},t}$. Assume that each BS has a sufficiently large battery without the occurrence of battery overflow. For instance, a solar-powered BS is able to store all the harvested energy during the daytime. Assume that the total bandwidth W_0 is divided into N_c orthogonal subcarriers. The network will decide which BSs are powered on (in the *active* mode) and how many subcarriers on these BSs are activated. The BS energy consumption in the *active* mode is modeled as a constant power term plus a power term related to the radio-frequency (RF) module [23]:

$$P^{(b)}_{\mathrm{BS},t} = P_0 + \Delta_P P^{(b)}_{\mathrm{RF},t}, \tag{4.37}$$

where the constant power P_0 includes the baseband processor, the cooling system, etc. and Δ_P is the inverse of power amplifier efficiency factor and $P^{(b)}_{\mathrm{RF},t}$ is the RF transmit power that is a linear function of the number of active subcarriers, denoted by $n^{(b)}_t$, i.e.

$$P^{(b)}_{\mathrm{RF},t} = \frac{n^{(b)}_t}{N_c} P_T, \quad n^{(b)}_t \le N_c, \tag{4.38}$$

where P_T is the transmit power level when all the N_c subcarriers are activated. Assume a constant transmit power per subcarrier according to the LTE standard [24]. Substituting $P^{(b)}_{\mathrm{RF},t}$ in (4.37) into (4.38) leads to

$$P^{(b)}_{\mathrm{BS},t} = P_0 + \frac{n^{(b)}_t}{N_\mathrm{c}} \Delta_P P_\mathrm{T}. \tag{4.39}$$

Note that the power model does not reflect the power consumption of the broadcast, synchronization, and pilot channels, as they actually contribute to the total power consumption by a constant value.

In order to adapt to the possible traffic variations over time, the harvested energy can be reserved in the battery for future use by reducing the number of active subcarriers or by switching a BS to *sleep* mode, where two types of sleep modes are considered. The first one is the *deep sleep* mode, in which a BS is completely turned off for the whole time period. In this sleep mode, the BS power consumption is negligible, and the users in the sleeping cell are served by the neighboring BSs or wait for the wake-up. The second one is the *opportunistic sleep* mode. An active BS will turn to this mode for a $\varphi^{(b)}_t \in [0,1)$ fraction of the time period to save grid energy. It can be realized by time-domain BS sleep [25] where some subframes are turned off. In this mode, a user will be blocked due to the lack of sufficient transmission subframes. For example, if the traffic load is low and the renewable energy is insufficient, a larger value of $\varphi^{(b)}_t$ can be chosen to save energy without causing too many user blockings. Assume the power consumption in the opportunistic sleep mode is P_S. Denote $S^{(b)}_t$ as the state of BS b in time period t, which equals to 1 if it is in active mode (including opportunistic sleep mode) and equals to 0 otherwise. The BS power consumption model is summarized as

$$P^{(b)}_{\mathrm{BS},t} = \begin{cases} P_0 + \frac{n^{(b)}_t}{N_c} \Delta_P P_\mathrm{T} & \text{if } S^{(b)}_t = 1 \\ P_\mathrm{S} & \text{if } S^{(b)}_t = 1 \text{ with opportunistic sleep} \\ 0 & \text{if } S^{(b)}_t = 0. \end{cases} \tag{4.40}$$

Note that in reality, a BS in sleep mode still consumes a certain amount of power so that it can be reactivated. However, this amount of power is negligible compared with that in active mode. Hence, the sleep mode power consumption is approximated to zero.

4.3.1.2 Traffic Model

Assume there are K classes of users, each of which has a data rate requirement $R_k, k = 1, \dots, K$. Based on the distance between users and BSs, the network is divided into M disjoint areas with equal sizes, denoted by $A_m, m = 1, \dots, M$. In each area m at time period t, the users from class k are uniformly distributed, and they randomly arrive in a

Poisson process with arrival rate $\lambda_{mk,t}$. The average service rate is denoted by μ_{mk}. Hence, the traffic intensity of user class k in area m is given by

$$\rho_{mk,t} = \frac{\lambda_{mk,t}}{\mu_{mk}}. \tag{4.41}$$

Note that the partition of the M regions is not arbitrary, but based on the distance to the BSs to guarantee the approximation accuracy. Take a single-cell case (Figure 4.4), for example. When $M = 2$, the cell-edge users are grouped into one class, while the rests are in another. The example for the multi-cell case can be found in the simulation study. Further, increasing the value of M allows improving the approximation accuracy, while the computational complexity increases accordingly. Ideally, $M \to +\infty$ can yield the exact blocking probability, but the complexity is not affordable. The selection of M should balance the approximation accuracy and the computational complexity.

Users are served by the BS with the strongest signal strength, and thus the serving BS for users in an area is decided once the BSs' active/sleep states are given. The serving BS allocates bandwidth to users to meet their minimum rate requirement. Meanwhile, a newly arrived user will be blocked if no available subcarriers can be allocated to satisfy its rate requirement. The system can be viewed as a multi-server queue without buffer, while the server capacity varies according to the users' channel states. Intuitively, when the network traffic load is high, more BSs and subcarriers should be activated so that each BS takes care of a smaller area to guarantee the QoS. Otherwise, fewer BSs and subcarriers are required, and hence the power consumption can be reduced [26].

4.3.1.3 Channel Model

Small-scale fast fading and shadowing are assumed to be averaged out for sufficient channel realizations. Hence, only the distance-dependent path loss is considered, which facilitates the distance-based network division introduced in the traffic model. The SINR of user u in the coverage area of active BS-b is

$$\mathrm{SINR}_u = \frac{P_\mathrm{T}\Gamma(d_u^{(b)})^{-\kappa}}{\sigma^2 + \sum_{b':S^{(b')}=1,b'\neq b} \frac{n^{(b')}}{N_c} P_\mathrm{T}\Gamma(d_u^{(b')})^{-\kappa}}, \tag{4.42}$$

where Γ is the path-loss constant, κ is the path-loss exponent, $d_u^{(b)}$ is the distance between BS b and user u, and σ^2 is the noise power. Assume the $n^{(b)}$ subcarriers are

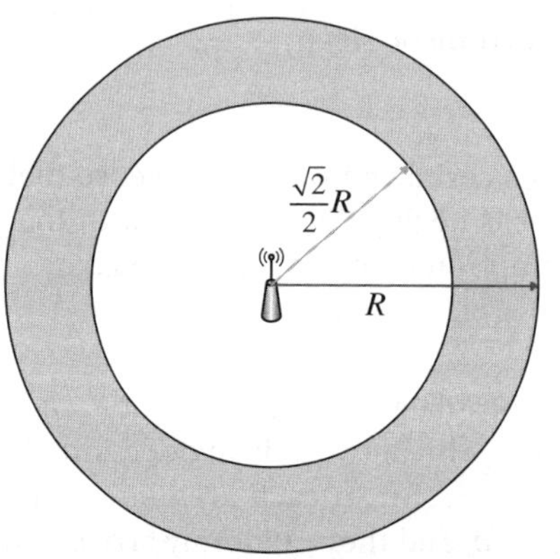

Figure 4.4 Single-cell Erlang's approximation settings for $M = 2$.

randomly chosen from the N_c subcarriers on each BS b and are opportunistically allocated to users in order to meet their data rate requirements. Due to the random subcarrier scheduling mechanism, the interference is considered to be averaged over the whole bandwidth, i.e. the perceived interference power is scaled by the ratio of active subcarriers $\frac{n^{(b')}}{N_c}$. Accordingly the maximum achievable transmission rate is

$$r_u = \frac{n^{(b)} W_0}{N_c} \log_2(1 + \text{SINR}_u). \tag{4.43}$$

Based on above model, the user blocking probability as the QoS metric and the relationship between the user blocking probability and power will be studied as follows.

4.3.2 Blocking Probability Analysis

The blocking probability is defined as the probability that a newly arrived user is blocked. In EH cellular systems, a blocking event may be caused by two factors. The first one is a high traffic load, which means that the required subcarriers are not available, namely, the *service blocking probability*. The second one is the BS's opportunistic sleep mode when the renewable energy is not sufficient. Such blocking probability is equal to the sleep ratio $\varphi_t^{(b)}$. The service blocking probability is analyzed first, and then the overall blocking probability that compromises both service blocking and opportunistic sleep is derived. The analysis is focusing on one time period, and thus the index t is ignored for brevity.

4.3.2.1 Service Blocking Probability

The service blocking probability is the probability that a newly arriving call is rejected by the BS when there are not sufficient subcarriers. It can be obtained by Erlang's formula [27], with the extension to the multi-class scenario. Specifically, denote the instantaneous set of users of class k in area m by $\mathcal{U}_{mk}$, which are uniformly distributed in area m, and the user number by $U_{mk} = |\mathcal{U}_{mk}|$. The bandwidth requirement of user u of class k in area m is given by

$$\Phi_{mk}(u) = \frac{R_k}{r_u}, \tag{4.44}$$

which varies according to rate requirement R_k for users of class k and location m. As each BS has a limited available bandwidth $\frac{n^{(b)}}{N_c} W_0$, the admission condition is that the total normalized bandwidth requirement should not exceed 1. Denote the total normalized bandwidth that is being used by

$$z_b = \sum_{m \in \mathcal{M}^{(b)}} \sum_{k=1}^{K} \sum_{u \in \mathcal{U}_{mk}} \Phi_{mk}(u), \tag{4.45}$$

where $\mathcal{M}^{(b)}$ is the set of regions served by BS b and BS b is assumed to be always active ($S^{(b)} = 1, \varphi^{(b)} = 0$) while deriving the service blocking probability. Hence, the service blocking probability can be expressed as

$$p_{\text{sv},mk} = \Pr(z_b < 1, z_b + \Phi_{mk}(u) \geq 1) \tag{4.46}$$

$$= \Pr(1 - \Phi_{mk}(u) \leq z_b < 1), \tag{4.47}$$

where (4.46) means that the total normalized bandwidth does not exceed 1 until the user u of class k arrives in area m. Calculation of the blocking probability (4.47) requires the integration of the probability over all the possible quantities and locations of users served by BS b, for which it is difficult to find analytical expressions. Therefore the extension to *Erlang's approximation* method proposed in [28] and to the multi-class multi-area scenario is adopted. The basic idea of Erlang's approximation method is to average the users' bandwidth requirements over all the possible positions. Assuming that all the users in the area have identical bandwidth requirements, the blocking probability can be calculated by Erlang's formula [27]. Specifically, the average normalized bandwidth requirement of class k users in area m is

$$\bar{\Phi}_{mk} = \int_{A_m} \frac{R_k}{r_{u(a)} A_m} da, \tag{4.48}$$

where $r_{u(a)}$ is the achievable data rate of user u at position a, which is expressed as (4.43). Hence, the admission condition (4.45) is changed to

$$\sum_{m \in \mathcal{M}^{(b)}} \sum_{k=1}^{K} U_{mk} \bar{\Phi}_{mk} < 1, \tag{4.49}$$

where U_{mk} is the number of active users of class k in area m. At the same time, the blocking probability of a user of class k in area m is modified to

$$p_{\text{sv},mk} = \Pr\left(1 - \bar{\Phi}_{mk} \leq \sum_{m' \in \mathcal{M}^{(b)}} \sum_{k'=1}^{K} U_{m'k'} \bar{\Phi}_{m'k'} < 1\right). \tag{4.50}$$

According to queueing theory (Section 3.7 [27] and Section 1.6 [29]), the stationary probability of user state $\boldsymbol{U}^{(b)} = \{U_{mk}\}_{m \in \mathcal{M}^{(b)}, k=1,\ldots,K}$ associated with BS b is

$$\pi^{(b)}(\boldsymbol{U}^{(b)}) = \prod_{m \in \mathcal{M}^{(b)}} \prod_{k=1}^{K} \frac{\rho_{mk}^{U_{mk}}}{U_{mk}!} \left(\sum_{\boldsymbol{U}^{(b)} \in \mathcal{U}^{(b)}} \prod_{m \in \mathcal{M}^{(b)}} \prod_{k=1}^{K} \frac{\rho_{mk}^{U_{mk}}}{U_{mk}!} \right)^{-1}, \tag{4.51}$$

where ρ_{mk} is defined as (4.41) and $\mathcal{U}^{(b)} = \{\boldsymbol{U}^{(b)} | \sum_{m \in \mathcal{M}^{(b)}} \sum_{k=1}^{K} U_{mk} \bar{\Phi}_{mk} < 1\}$ is the set of all possible active user states that satisfy the bandwidth constraint (4.49). As a consequence, the blocking probability can be calculated as

$$p_{\text{sv},mk} = \sum_{\boldsymbol{U}^{(b)} \in \bar{\mathcal{U}}_{mk}^{(b)}} \pi^{(b)}(\boldsymbol{U}^{(b)}), \quad m \in \mathcal{M}^{(b)}, \tag{4.52}$$

where $\bar{\mathcal{U}}_{mk}^{(b)} = \{\boldsymbol{U}^{(b)} : 1 - \Phi_{mk} \leq \sum_{m' \in \mathcal{M}^{(b)}} \sum_{k'=1}^{K} U_{m'k'} \bar{\Phi}_{m'k'} \leq 1\}$ is the set of user states where the newly arrived user of class k in area m is blocked. In addition, the probability of a newly arrived user in the coverage of BS b is

$$p_{\text{sv}}^{(b)} = \frac{\sum_{m \in \mathcal{M}^{(b)}} \sum_{k=1}^{K} p_{\text{sv},mk} \rho_{mk}}{\sum_{m \in \mathcal{M}^{(b)}} \sum_{k=1}^{K} \rho_{mk}}. \tag{4.53}$$

Notice that the service blocking probability can be tuned by adapting the BSs' working states $S^{(b)}, b = 1, \ldots, N_B$ and the number of active subcarriers in the active BSs $n^{(b)}, b \in \{b : S^{(b)} = 1\}$.

4.3.2.2 Relation Between $P_G^{(b)}$ and $\varphi^{(b)}$

Recall that the BS in the active mode can turn to opportunistic sleep mode with time ratio $\varphi^{(b)}$. Denote by $P_{\text{In}}^{(b)}$ the total input power to the BS. According to the balance between power input and consumption, the total input power satisfies

$$P_{\text{In}}^{(b)} = (1-\varphi^{(b)})P_{\text{BS}}^{(b)} + \varphi^{(b)}P_{\text{S}}, \quad P_{\text{In}}^{(b)} \le P_{\text{BS}}^{(b)}. \tag{4.54}$$

Then the relation between the opportunistic sleep time ratio and the input power is

$$\varphi^{(b)} = \frac{P_{\text{BS}}^{(b)} - P_{\text{In}}^{(b)}}{P_{\text{BS}}^{(b)} - P_{\text{S}}}, \quad P_{\text{In}}^{(b)} \le P_{\text{BS}}^{(b)}. \tag{4.55}$$

In what follows, we discuss the relationship between the opportunistic sleep time ratio and the grid power consumption, given the available harvested energy:

- *Case 1:* If the harvested energy is sufficient for the required energy, i.e. $E_{\text{C}}^{(b)} + P_{\text{H}}^{(b)}L \ge P_{\text{In}}^{(b)}L$, where $E_{\text{C}}^{(b)}$ is the battery energy budget and L is the length of a time period, then

$$P_{\text{G}}^{(b)} = 0, \quad \forall\, 0 \le \varphi^{(b)} \le 1, \tag{4.56}$$

  which means that the grid power is not needed. Note that the updating process of battery energy $E_{\text{C}}^{(b)}$ is described in Eq. (4.66).
- *Case 2:* On the other hand, for the case where $E_{\text{C}}^{(b)} + P_{\text{H}}^{(b)}L < P_{\text{In}}^{(b)}L$, then grid power is needed. The opportunistic sleep time ratio can be expressed as

$$\varphi^{(b)} = \frac{P_{\text{BS}}^{(b)} - (E_{\text{C}}^{(b)}/L + P_{\text{H}}^{(b)} + P_{\text{G}}^{(b)})}{P_{\text{BS}}^{(b)} - P_{\text{S}}}, \quad P_{\text{G}}^{(b)} \le P_{\text{BS}}^{(b)} - (E_{\text{C}}^{(b)}/L + P_{\text{H}}^{(b)}). \tag{4.57}$$

4.3.2.3 Overall Blocking Probability

In the considered system, if the BS is in the opportunistic sleep mode, a newly arrived user of class k in area m served by BS b will be blocked with probability 1. Otherwise, it will be blocked with probability $p_{\text{sv},mk}$. As a consequence, the overall blocking probability can be calculated as

$$\begin{aligned} p_{\text{blk},mk} &= \varphi^{(b)} + (1-\varphi^{(b)})p_{\text{sv},mk} \\ &= 1-(1-p_{\text{sv},mk})(1-\varphi^{(b)}). \end{aligned} \tag{4.58}$$

The blocking probability for a specific BS b satisfies

$$\begin{aligned} p_{\text{blk}}^{(b)} &= \frac{\sum_{m\in\mathcal{M}^{(b)}}\sum_{k=1}^{K} p_{\text{blk},mk}\rho_{mk}}{\sum_{m\in\mathcal{M}^{(b)}}\sum_{k=1}^{K}\rho_{mk}} \\ &= 1-(1-p_{\text{sv}}^{(b)})(1-\varphi^{(b)}), \end{aligned} \tag{4.59}$$

where $p_{\text{sv}}^{(b)}$ is expressed as (4.53). Notice that if BS b is in sleep mode ($S^{(b)} = 0$), the users in the sleeping cell must be associated with other active BSs. Hence, no blocking events are counted for this sleeping BS.

4.3.3 Power Grid Energy Minimization

Recall that the objective in this section is to reduce the carbon emissions in the cellular networks with mixed power supply by fully utilizing the renewable energy. The problem can be formulated as minimizing the grid power consumption under the users' QoS constraint. The following presents the mathematical formulation and the solutions of the problem in detail.

4.3.3.1 Problem Formulation

The traffic intensity in time period t for the K classes of users and M regions is denoted by an $M \times K$ matrix $\boldsymbol{\rho}_t = \{\rho_{mk,t}\}_{m=1,\dots,M,k=1,\dots,K}$. The EH power is denoted by a $1 \times N_{\mathrm{B}}$ vector $\boldsymbol{P}_{H,t} = [P_{\mathrm{H},t}^{(1)}, P_{\mathrm{H},t}^{(2)}, \dots, P_{\mathrm{H},t}^{(N_{\mathrm{B}})}]$. The traffic profile $\boldsymbol{\rho}_t$ and the renewable energy profile $\boldsymbol{P}_{H,t}$ are assumed to be noncausally known. By adjusting the BSs' on–off states $\boldsymbol{S}_t = [S_t^{(1)}, S_t^{(2)}, \dots, S_t^{(N_{\mathrm{B}})}]$, the number of active subcarriers of active BSs $\boldsymbol{n}_t = [n_t^{(1)}, n_t^{(2)}, \dots, n_t^{(N_{\mathrm{B}})}]$, and the opportunistic sleep time ratio $\varphi_t = [\varphi_t^{(1)}, \varphi_t^{(2)}, \dots, \varphi_t^{(N_{\mathrm{B}})}]$, the total power usage and thus the grid power input can be adapted in all the time periods $t = 1, 2, \dots, T$.

Denote the on–off states of BSs over T time periods $\boldsymbol{S} = \{\boldsymbol{S}_1, \boldsymbol{S}_2, \dots, \boldsymbol{S}_T\}$, number of subcarriers over T time periods $\boldsymbol{n} = \{\boldsymbol{n}_1, \boldsymbol{n}_2, \dots, \boldsymbol{n}_T\}$, and the opportunistic sleep time ratio over T time periods $\varphi = \{\varphi_1, \varphi_2, \dots, \varphi_T\}$. Then the following optimization problem is considered: given the traffic profile $\boldsymbol{\rho}_1, \boldsymbol{\rho}_2, \dots, \boldsymbol{\rho}_T$ and the renewable energy profile $\boldsymbol{P}_{H,1}, \boldsymbol{P}_{H,2}, \dots, \boldsymbol{P}_{H,T}$, adjust the BSs' working state $\boldsymbol{S}$, the resource allocation $\boldsymbol{n}$, and the sleep ratio φ to minimize the average grid power consumption while satisfying the weighted blocking probability. The problem can be formulated as

$$\min_{\boldsymbol{S},\boldsymbol{n},\varphi} \frac{\sum_{t=1}^{T} L_t \sum_{b=1}^{N_{\mathrm{B}}} P_{\mathrm{G},t}^{(b)}}{\sum_{t=1}^{T} L_t} \tag{4.60}$$

$$\text{s.t.} \quad \sum_{t=1}^{T} \sum_{b=1}^{N_{\mathrm{B}}} \omega_t^{(b)} p_{\mathrm{blk},t}^{(b)} \leq p_{\mathrm{target}}, \tag{4.61}$$

where L_t denotes the length of time period t, the blocking probability $p_{\mathrm{blk},t}^{(b)}$ is expressed as (4.59), and the weighting factor $\omega_t^{(b)}$, which satisfies $\sum_{t=1}^{T} \sum_{b=1}^{N_{\mathrm{B}}} \omega_t^{(b)} = 1$, reflects the system sensitivity to the blocking probability in each time period. The weighting factor allows for the case where users may require higher QoS at some particular time of the day. For instance, if higher QoS is required during peak load time (e.g. daytime) than low load time (e.g. nighttime), the weighting factor for the daytime can be set larger than that for the nighttime. The influence of the weighting factor settings will be studied in the simulations.

Note that the solution for problem (4.60) is not straightforward as the network traffic profile and the renewable energy profile in both time and space domains do not match with each other in general. One cannot just greedily use renewable energy as long as it is available, and then use grid energy if necessary. Storing some renewable energy for possible future use may reduce the overall grid power consumption. Here the time-averaged grid power consumption is considered. Nevertheless, this model can also be used to solve the problem where the cost of grid power varies with time. For example, by simply substituting the per-time period weight of grid power $L_t/(\sum_{t=1}^{T} L_t)$ with electricity price, one can minimize the charge for electricity.

4.3.3.2 Optimal DP Algorithm

The optimal solution to problem (4.60) with the constraint (4.61) can be found by exhaustive search through all possible policies. However, this approach is not practical due to its high complexity. First consider the following unconstrained optimization problem with a weighted combination of the power consumption and the blocking probability:

$$\min_{\boldsymbol{S},\boldsymbol{n},\varphi} \frac{\sum_{t=1}^{T} L_t \sum_{b=1}^{N_B} P_{G,t}^{(b)}}{\sum_{t=1}^{T} L_t} + \beta \sum_{t=1}^{T} \sum_{b=1}^{N_B} \omega_t^{(b)} p_{\text{blk},t}^{(b)}, \tag{4.62}$$

where the factor $\beta > 0$ plays the role of a Lagrangian multiplier and can be chosen to set the relative importance of the blocking probability with respect to the average grid power consumption. Denote the minimum objective value of problem (4.62) for a given β as $P_{\text{Gave},\beta}^{*} + \beta p_{\text{blk},\beta}^{*}$, where $P_{\text{Gave},\beta}^{*}$ and $p_{\text{blk},\beta}^{*}$ represent the average grid power and the weighted blocking probability, respectively. By solving problem (4.62) for different value of β, one can find a set of points $(P_{\text{Gave},\beta}^{*}, p_{\text{blk},\beta}^{*})$. Denote the function $P_{\text{Gave}}(p_{\text{blk},\beta}^{*}) = P_{\text{Gave},\beta}^{*}$. The following theorem shows the relationship between the unconstrained problem and the original constrained problem, i.e. the solution of problem (4.62) is also the solution of problem (4.60) with the constraint (4.61).

Theorem 4.2 Denote the optimal objective function of (4.60) by $P_{\text{Gave,min}}$. If $p_{\text{target}} = p_{\text{blk},\beta}^{*}$, then $P_{\text{Gave,min}} = P_{\text{Gave},\beta}^{*}$.

Proof: Firstly, the policy achieving $(P_{\text{Gave},\beta}^{*}, p_{\text{blk},\beta}^{*})$ of problem (4.62) is a feasible policy for problem (4.60) with the constraint (4.61). Hence,

$$P_{\text{Gave,min}} \leq P_{\text{Gave},\beta}^{*}. \tag{4.63}$$

Secondly, the equality in (4.63) always holds as otherwise:

$$P_{\text{Gave,min}} + \beta p_{\text{blk},\beta}^{*} < P_{\text{Gave},\beta}^{*} + \beta p_{\text{blk},\beta}^{*}, \tag{4.64}$$

which contradicts the optimality of $(P_{\text{Gave},\beta}^{*}, p_{\text{blk},\beta}^{*})$ for problem (4.62). □

Depending on Theorem 4.2, as the objective (4.62) is minimized, $P_{\text{Gave},\beta}^{*}$ must be the minimum average grid power to guarantee that the blocking probability is no more than $p_{\text{blk},\beta}^{*}$. Hence, the solution for (4.62) is also the solution for (4.60) where $p_{\text{target}} = p_{\text{blk},\beta}^{*}$. Hence, $P_{\text{Gave}}(p_{\text{blk}})$ indicates the minimum average grid power such that the blocking probability does not exceed p_{blk}. Notice that it is not guaranteed that all the values of $P_{\text{Gave}}(p_{\text{blk}})$ can be found by varying β. In this case, for a given target blocking probability p_{target}, if a corresponding point can be found by setting an appropriate value of β, the optimal solution for the original problem (4.60) with constraint (4.61) is found. Otherwise, a suboptimal result is obtained by adopting the policy related to the point with the largest blocking probability less than p_{target}, i.e.

$$\{\boldsymbol{S}, \boldsymbol{n}, \varphi\} = \arg\max\{p_{\text{blk},\beta}^{*} | p_{\text{blk},\beta}^{*} \leq p_{\text{target}}\}. \tag{4.65}$$

Problem (4.62) can be solved by the DP algorithm for deterministic systems (Chapter 2, [30]), which divides the whole problem into simple per-stage subproblems. The DP algorithm contains three key components: state, action, and cost function. In problem (4.62), the state is the amount of energy $\boldsymbol{E}_{\mathrm{C},t} = [E_{\mathrm{C},t}^{(1)}, E_{\mathrm{C},t}^{(2)}, \dots, E_{\mathrm{C},t}^{(B)}]$ in the battery at the beginning of time period t. For each BS b, $E_{\mathrm{C},t}^{(b)}$ evolves to time period $t+1$ as

$$E_{\mathrm{C},t+1}^{(b)} = E_{\mathrm{C},t}^{(b)} + L_t P_{\mathrm{H},t}^{(b)} - [(1-\varphi_t^{(b)}) L_t P_{\mathrm{BS},t}^{(b)} + \varphi_t^{(b)} L_t P_{\mathrm{S}}] \tag{4.66}$$

if the energy consumption in time period t does not exceed the available renewable energy, i.e.

$$[(1-\varphi_t^{(b)}) L_t P_{\mathrm{BS},t}^{(b)} + \varphi_t^{(b)} L_t P_{\mathrm{S}}] \leq E_{\mathrm{C},t}^{(b)} + L_t P_{\mathrm{H},t}^{(b)}. \tag{4.67}$$

If the harvested energy is not sufficient, i.e. (4.67) does not hold, the power grid will be plugged in to provide the required power. The grid power is chosen as

$$P_{\mathrm{G},t}^{(b)} \leq \max\left\{0, (1-\varphi_t^{(b)}) P_{\mathrm{BS},t}^{(b)} + \varphi_t^{(b)} P_{\mathrm{S}} - \frac{E_{\mathrm{C},t}^{(b)}}{L_t} - P_{\mathrm{H},t}^{(b)}\right\}. \tag{4.68}$$

The actions determine the BSs' working state $\boldsymbol{S}_t$, the number of active subcarriers $\boldsymbol{n}_t$, and the sleep ratio φ_t. Note that if $S_t^{(b)} = 0$, there are no active subcarriers ($n_t^{(b)} = 0$) and the BS remains sleep during t ($\varphi_t^{(b)} = 1$). The per-stage cost is a weighted combination of the grid power and the blocking probability, denoted by a function of the current action and state:

$$c_t(\boldsymbol{S}_t, \boldsymbol{n}_t, \varphi_t, \boldsymbol{E}_{\mathrm{C},t}) = \frac{L_t \sum_{b=1}^{N_{\mathrm{B}}} P_{\mathrm{G},t}^{(b)}}{\sum_{t=1}^{T} L_t} + \beta \sum_{b=1}^{N_{\mathrm{B}}} \omega_t^{(b)} p_{\mathrm{blk},t}^{(b)}. \tag{4.69}$$

The DP algorithm breaks the original problem down into subproblems with respect to the stage, where the objective is to minimize the cost of each time period plus that of the upcoming time periods. The per-time period subproblems are solved recursively. The *cost-to-go* function is defined recursively as

$$J_t(\boldsymbol{E}_{\mathrm{C},t}) = \begin{cases} \min\limits_{\boldsymbol{S}_t, \boldsymbol{n}_t, \varphi_t} c_t(\boldsymbol{S}_t, \boldsymbol{n}_t, \varphi_t, \boldsymbol{E}_{\mathrm{C},t}) & t = T \\ \min\limits_{\boldsymbol{S}_t, \boldsymbol{n}_t, \varphi_t} \{c_t(\boldsymbol{S}_t, \boldsymbol{n}_t, \varphi_t, \boldsymbol{E}_{\mathrm{C},t}) + J_{t+1}(\boldsymbol{E}_{\mathrm{C},t+1})\} & t < T. \end{cases} \tag{4.70}$$

which denotes the minimum cost of the subproblem with time period t as its initial stage. Performing a backward induction of the cost-to-go functions (4.70) from time period T to time period 1, the minimum cost equaling to $J_1(\mathbf{0})$ is obtained.

Assume the number of examined sleep ratios $\varphi_t^{(b)}$ is N_φ. Then, the cardinality of the action space for each cost-to-go function is $(N_c N_\varphi + 1)^{N_{\mathrm{B}}}$. Note that the number of BS actions $(S_t^{(b)}, n_t^{(b)}, \varphi_t^{(b)})$ is $(N_c N_\varphi + 1)$ instead of $2N_c N_\varphi$, as the BSs in sleep mode have only a single state. Hence, given the state in time period t, the cardinality of the state space in time period $(t+1)$ is no more than $(N_c N_\varphi + 1)^B$. That is, if the harvested energy of all BSs is enough for any resource allocation policy, each policy corresponds to a unique next-stage state. Otherwise, some policies will result in the same state, so the state space is less than $(N_c N_\varphi + 1)^B$.

Both the dimensions of the action space and the state space grow exponentially with the number of BSs N_B in the network, which, due to the *curse of dimensionality* [30], will result in an overwhelming computational complexity to find the optimal control policy if the network scale is large. As a consequence, the optimal DP algorithm is difficult to implement in practical systems, and therefore low-complexity solutions are favored. In the following, a two-stage near-optimal optimization algorithm is proposed to reduce the sizes of the state and action spaces.

4.3.3.3 Two-Stage DP Algorithm

The basic idea of the proposed two-stage optimization algorithm is to divide the action process into two stages. In the first stage, the number of active subcarriers in active BSs is assumed to be N_c, i.e. all the spectrum resources are used, and subcarrier allocation is not performed in this stage. In addition, the active BS sleep ratio $\varphi_t^{(b)}$ is assumed to be 0 for all $b = 1, \dots, N_B$, which means that the required power is always available. As a result, the actions at this stage only consist of the BSs' working states $\boldsymbol{S}$. The optimization problem can be written as

$$\min_{\boldsymbol{S}} \frac{\sum_{t=1}^{T} L_t \sum_{b=1}^{N_B} P_{G,t}^{(b)}}{\sum_{t=1}^{T} L_t} + \beta \sum_{t=1}^{T} \sum_{b=1}^{N_B} \omega_t^{(b)} p_{blk,t}^{(b)} \Big|_{\boldsymbol{n}_t=\boldsymbol{n}_0, \varphi_t=\varphi_0, \forall t}, \tag{4.71}$$

where $\boldsymbol{n}_0 = [n_0^{(1)}, \dots, n_0^{(N_B)}]$ and $\varphi_0 = [\varphi_0^{(1)}, \dots, \varphi_0^{(N_B)}]$ are the initial states. It can be set by *greedy initial state settings* expressed as $\boldsymbol{n}_0 = \boldsymbol{N}_c, \varphi_0 = \boldsymbol{0}$, i.e. for any $b = 1, 2, \dots, N_B$, if $S_t^{(b)} = 1$, the corresponding number of active subcarriers is $n_t^{(b)} = N_c$, and the sleep ratio is $\varphi_t^{(b)} = 0$, i.e. BS b activates all the subcarriers for the whole time period t. Other settings will be discussed later in simulations. The cost-to-go function is

$$J_t(\boldsymbol{E}_{C,t})|_{\boldsymbol{n}_t=\boldsymbol{n}_0, \varphi_t=\varphi_0} = \begin{cases} \min\limits_{\boldsymbol{S}_t} c_t(\boldsymbol{S}_t, \boldsymbol{n}_0, \varphi_0, \boldsymbol{E}_{C,t}) & t = T \\ \min\limits_{\boldsymbol{S}_t} \{c_t(\boldsymbol{S}_t, \boldsymbol{n}_0, \varphi_0, \boldsymbol{E}_{C,t}) + J_{t+1}(\boldsymbol{E}_{C,t+1})|_{\boldsymbol{n}_{t+1}=\boldsymbol{n}_0, \varphi_{t+1}=\varphi_0}\} & t < T. \end{cases} \tag{4.72}$$

Remark 4.1 The action space of each cost-to-go function in (4.72) is 2^{N_B}. Given the state in time period t, the maximum state space in time period $(t+1)$ is reduced from $(N_c N_\varphi + 1)^{N_B}$ to 2^{N_B}.

Problem (4.71) can be solved by the standard DP algorithm with a much lower complexity compared with the original DP problem (4.62). In the second stage, given the states of BSs $\boldsymbol{S}^* = \{\boldsymbol{S}_1^*, \boldsymbol{S}_2^*, \dots, \boldsymbol{S}_T^*\}$ obtained from the first stage, the number of active subcarriers and power allocation for each BS are adjusted separately. Since the subcarrier adaptation changes the interference profile, the per-BS resource allocation correlates with one another. An iterative resource allocation algorithm is accordingly proposed, which updates the per-BS resource allocation based on the allocation results of the other BSs and then iterates the process until the resource allocation solution remains unchanged for two consecutive iterations. The per-BS resource allocation optimization problem can be formulated as

$$\min_{n_t^{(b)},\varphi_t^{(b)}} \frac{\sum_{t=1}^{T} L_t \sum_{b=1}^{N_B} P_{G,t}^{(b)}}{\sum_{t=1}^{T} L_t} + \beta \sum_{t=1}^{T}\sum_{b=1}^{N_B} \omega_t^{(b)} p_{\text{blk},t}^{(b)}\Big|_{S^*,n_t^{(b')},\varphi_t^{(b')},\forall t,b'\neq b}. \tag{4.73}$$

The problem can be solved by the DP algorithm where the cost-to-go function is

$$J_t^{(b)}(E_{C,t}^{(b)})|_{S_t^*,n_t^{(b')},\varphi_t^{(b')},b'\neq b}$$
$$= \begin{cases} \min_{n_t^{(b)},\varphi_t^{(b)}} c_t(\boldsymbol{S}_t = \boldsymbol{S}_t^*, \boldsymbol{n}_t^{(b)}, \varphi_t^{(b)}, \boldsymbol{E}_{C,t}^{(b)})|_{n_t^{(b')},\varphi_t^{(b')},b'\neq b} & t = T \\ \min_{n_t^{(b)},\varphi_t^{(b)}} \{c_t(\boldsymbol{S}_t = \boldsymbol{S}_t^*, \boldsymbol{n}_t^{(b)}, \varphi_t^{(b)}, \boldsymbol{E}_{C,t}^{(b)})|_{n_t^{(b')},\varphi_t^{(b')},b'\neq b} & \\ \qquad + J_{t+1}(\boldsymbol{E}_{C,t+1})|_{S_t^*,n_{t+1}^{(b')},\varphi_{t+1}^{(b')},b'\neq b}\} & t < T. \end{cases} \tag{4.74}$$

Remark 4.2 The action space of each cost-to-go function in (4.74) is either $N_c N_\varphi$ ($S_t^{(b)*} = 1$) or 1 ($S_t^{(b)*} = 0$), and given the state in time period t, the maximum state space in time period $(t+1)$ is no more than $N_c N_\varphi$.

Remark 4.3 With the two-stage algorithm, the action space of each time period is reduced from $(N_c N_\varphi + 1)^{N_B}$ to $2^{N_B} N_B N_c N_\varphi$. Accordingly, given the state in time period t, the maximum state space in time period $(t+1)$ is reduced from $(N_c N_\varphi + 1)^{N_B}$ to $2^{N_B} N_B N_c N_\varphi$.

Algorithm 4.3: Two-stage DP optimization.

Input: Traffic load and energy arrival profiles over T time periods
Output: $\boldsymbol{S}^*$, $\boldsymbol{n}^*$, $\boldsymbol{\varphi}^*$

1: **The first stage**:
2: Solve the problem (4.71) to find $\boldsymbol{S}^*$
3: **The second stage**:
4: Set $\boldsymbol{n}_t = \boldsymbol{n}_0, \boldsymbol{\varphi}_t = \boldsymbol{\varphi}_0, \boldsymbol{n}'_t \neq \boldsymbol{n}_t, \boldsymbol{\varphi}'_t \neq \boldsymbol{\varphi}_t, t = 1, \dots, T$
5: **while** $\boldsymbol{n}_t \neq \boldsymbol{n}'_t$ or $\boldsymbol{\varphi}_t \neq \boldsymbol{\varphi}'_t$ for some $t = 1, \dots, T$ **do**
6: Set $\boldsymbol{n}'_t = \boldsymbol{n}_t, \boldsymbol{\varphi}'_t = \boldsymbol{\varphi}_t, t = 1, \dots, T$
7: **for** $b = 1$ to N_c **do**
8: Set $\boldsymbol{n}^{(b)} = \{n_1^{(b)}, n_2^{(b)}, \dots, n_T^{(b)}\}, \boldsymbol{\varphi}^{(b)} = \{\varphi_1^{(b)}, \varphi_2^{(b)}, \dots, \varphi_T^{(b)}\}$
9: Find $\boldsymbol{n}^{(b)*}, \boldsymbol{\varphi}^{(b)*}$ which solve the problem (4.73) by fixing $\boldsymbol{n}^{(b')}$ and $\boldsymbol{\varphi}^{(b')}, b' \neq b$
10: Update $\boldsymbol{n}_t, \boldsymbol{\varphi}_t, t = 1, \dots, T$ by setting $\boldsymbol{n}^{(b)} = \boldsymbol{n}^{(b)*}, \boldsymbol{\varphi}^{(b)} = \boldsymbol{\varphi}^{(b)*}$
11: **end for**
12: **end while**

The algorithm is summarized as Algorithm 4.3. Due to the heuristic nature, the proposed two-stage DP algorithm may not achieve the optimal solution. However, its optimality can be guaranteed when the traffic is uniformly distributed in space domain and the energy arrival process in each BS is the same. In this case, to match the resource allocation with traffic and efficiently utilize energy, each active BS activates all its subcarriers. As a result, the two-stage DP algorithm performs the same with the optimal DP algorithm. Both of them just need to optimize optimal BS sleeping policy as in the first-stage DP algorithm. The result is summarized as the following proposition.

Proposition 4.1 *If for any t, $\rho_{mk,t} = \tilde{\rho}_t, \forall m = 1, \dots, M, k = 1, \dots, K, P^{(1)}_{\mathrm{H},t} = \cdots = P^{(N_\mathrm{B})}_{\mathrm{H},t}$, the performance of two-stage DP algorithm is the same with that of optimal DP algorithm.*

The intuition behind the proposition is as follows. Firstly, it is easy to prove that the blocking probability $p^{(b)}_{\mathrm{blk},t}$ is a convex function of the traffic handled by BS b. Hence, the average blocking probability $\sum_{b=1}^{N_\mathrm{B}} \omega^{(b)}_t p^{(b)}_{\mathrm{blk},t}$ is minimized when the active BSs evenly share the network traffic. Based on the distance-based association mechanism, uniform BS sleeping can guarantee even load sharing. Secondly, for a given grid power consumption, the total number of active subcarriers is fixed. In this case, the network capacity is maximized when $n^{(b)}_t = \tilde{n}, \forall b : S^{(b)}_t = 1$ according to (4.42) and (4.43) due to the concavity of the rate function, and the average blocking probability is then minimized. Finally, in this case $\tilde{n} = N_\mathrm{c}$, the ratio between transmit power and fixed power of all the BSs is maximized. Hence, the energy is most efficiently used to achieve the minimum blocking probability.

The proposition is further validated by simulations. The simulations also show that in some other cases, for example, an asymmetric traffic distribution, the performance of the two-stage DP algorithm may not achieve the optimal result.

4.3.3.4 Heuristic Algorithms

Motivated by the two-stage DP algorithm where the BSs' on–off states are determined in the first stage and the per-BS resource allocation is determined in the second stage, some low-complexity heuristic algorithms are further proposed for comparisons, and they also operate in a two-stage manner. Specifically, the BSs' on–off states can be adjusted by the following algorithms:

- *Non-sleep policy*: In this policy, all the BSs are active in each time period. It is the policy used in the conventional cellular network, which can be viewed as the baseline.
- *Threshold-based sleep policy*: In this policy, the number of active BSs is decided by the network traffic intensity. Assume the maximum supportable traffic is $\theta_{\max}$. Set Q as BSs' on–off patterns, in each of which $B_{\min} + iB_q$ uniformly chosen BSs are active, where $B_{\min}$ is the minimum number of active BSs required to guarantee the network coverage and B_q is an integer satisfying $B_{\min} + QB_q \leq N_\mathrm{B}$. Then define a set of thresholds $0 = \theta_0 < \theta_1 < \cdots < \theta_Q = \theta_{\max}$. If the integrated network traffic intensity satisfies $\theta_{i-1} < \sum_m \sum_k \rho_{mk} \leq \theta_i$, the ith pattern is chosen.

Once the BSs' on–off states are decided, the number of active subcarriers and the opportunistic sleep ratio are tuned in each BS distributively based on the algorithms listed below:

- *Maximum resource block utilization*: In this policy, all the blocks are activated for transmission, i.e. $n_t^{(b)} = N_c$ for all t. It can be considered as the baseline for resource allocation.
- *Traffic-aware resource block utilization*: Based on the intuition that higher traffic intensity requires more wireless resources, a heuristic is to set the number of activated subcarriers proportional to the traffic intensity with the scaling factor η_1, i.e.

$$n_t^{(b)} = \min\{N_c, \lceil \eta_1 \rho_t^{(b)} N_c \rceil\}, \quad \eta_1 > 0, \tag{4.75}$$

where $\lceil x \rceil$ is the minimum integer no smaller than x and $\rho_t^{(b)}$ is the aggregated traffic intensity of BS b.
- *Joint traffic–energy-aware resource block utilization*: As the outages are caused not only by the lack of wireless resources but also by the lack of power, the power budget should be taken into consideration. In this case, the number of active subcarriers is also proportional to the available power besides the traffic intensity with scaling factor η_2:

$$n_t^{(b)} = \min\left\{N_c, \left\lceil \eta_2 \rho_t^{(b)} \frac{E_{B,t}^{(b)} + E_{G,t}^{(b)} + L_t P_{H,t}^{(b)}}{\sum_{k=t}^{T} L_k (P_0 + \Delta_P P_T)} N_c \right\rceil\right\}, \tag{4.76}$$

where $\eta_2 > 0$. As the grid energy can be used flexibly, given the average grid power $P_{Gave}^{(b)}$, one can view the grid energy as a virtual battery with initial state $E_{G,1}^{(b)} = \sum_{t=1}^{T} L_t P_{Gave}^{(b)}$, which evolves as $E_{G,t+1}^{(b)} = \max\{0, E_{G,t}^{(b)} - L_t P_{G,t}^{(b)}\}$. Note that $P_0 + \Delta_P P_T$ in the denominator is for normalization.

The sleep ratio $\varphi_t^{(b)}$ in all these policies is set to zero as long as the required power is available either from EH or from power grid. Hence, the grid power is calculated as

$$P_{G,t}^{(b)} = \min\left\{\frac{E_{G,t}^{(b)}}{L_t}, \max\left\{0, P_0 + \frac{n_t^{(b)}}{N_c}\Delta_P P_T - \frac{E_{B,t}^{(b)}}{L_t} - P_{H,t}^{(b)}\right\}\right\}, \tag{4.77}$$

and $\varphi_t^{(b)} > 0$ only when the total energy is not sufficient.

Notice that given the parameters θ_i, η_1, and η_2, the heuristic algorithms only depend on the traffic and the energy conditions of the current time period, i.e. they are myopic. The complexity is much lower than the DP algorithm. However, the QoS is not guaranteed, which is demonstrated in the simulation results.

4.3.4 Numerical Simulations

In the simulations, the energy consumption model of the BSs is from the EARTH project [23], and the channel model is from 3GPP LTE [24]. In the macro-cell scenario, set $P_0 = 712.2$ W, $\Delta_P = 15.96$, the maximum transmit power $P_{max} = 40$ W, and the cell radius $R = 1000$ m. The opportunistic sleep mode power is $P_S = 50$ W. The bandwidth is set to $W_0 = 10$ MHz, and the number of subcarriers is set to $N_c = 600$. The path loss is $PL^{dB} = 34.5 + 35 \log_{10}(l)$, and the noise power density is -174 dBm/Hz. The relation between the QoS and the resource allocation in single-cell scenario is shown first. Then the simulation is extended to the sectorized multi-cell scenario.

4.3.4.1 Single-Cell Case

For the single-cell case, the superscript b is removed as there is only one BS. The number of user classes is also set as $K = 1$. The circular cell area is divided into $M = 2$ regions with equal areas, as shown in Figure 4.4, and it is easy to find that the inner circular region is of radius $\frac{\sqrt{2}}{2}R$. Accordingly, the user data requirement is set to $r_K = r_0 = 2$ Mbps, the user service rate is set to $\mu_{1K} = \mu_{2K} = \mu = 1$ s^{-1}, and the arrival rate $\lambda_{1K} = \lambda_{2K} = \frac{\lambda}{2}$. The total traffic intensity is denoted by $\rho = \lambda/\mu$, and the total input power is denoted by P_{In}, which includes harvested power, grid power, and battery power. The relation between the number of active subcarriers and the blocking probability is depicted in Figure 4.5. Take $\rho = 5, P_{\text{In}} = 1$ kW as an example. When the number of active subcarriers is less than 300, the blocking is mainly caused by the limited available subcarriers. Hence, $n < 300$ is called the *bandwidth-limited* region. On the contrary, if $n \geq 300$, the available power is insufficient to activate all the subcarriers, which means $\varphi > 0$. In this case, then the blocking is also caused by opportunistic sleep. Correspondingly, $n \geq 300$ is called the *energy-limited* region. As a result, there is a minimum outage probability as shown by the star on each curve, corresponding to the optimal number of active subcarriers. In addition, if a certain blocking probability can be achieved in both the bandwidth-limited region and the energy-limited region, the policy in the

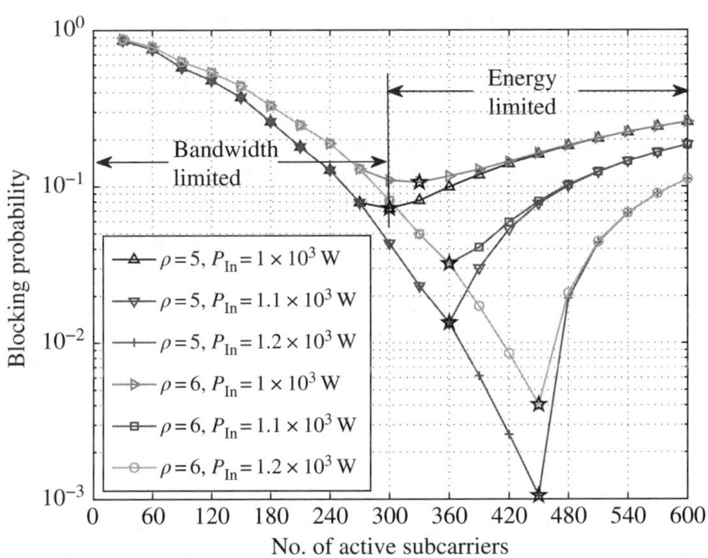

Figure 4.5 Relationship between the blocking probability and the number of active subcarriers, where ρ is the traffic intensity and P_{In} is the total available input power. The "star" is the minimum blocking probability for each parameter settings.

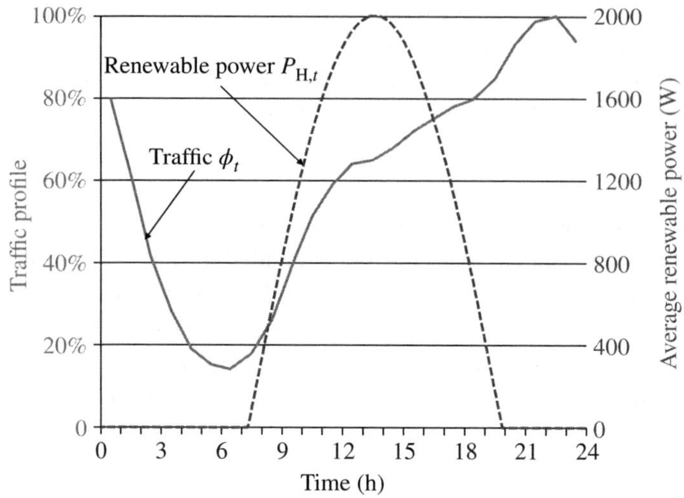

Figure 4.6 Daily traffic (solid line). Source: Auer et al. 2011 [23]. Reproduced with permission of IEEE. Renewable energy profile (dashed line). Source: Gorlatova et al. 2013 [9]. Reproduced with permission of IEEE.

bandwidth-limited region consumes less power than P_{In}, while in the energy-limited region, the policy consumes all the available power P_{In}.

Then the performance of the DP algorithm is evaluated with a given traffic profile (from [23]) and energy arrival statistics (from [9]) for one day as illustrated in Figure 4.6. The standard DP algorithm (4.70) is run for the single-cell case as the computational complexity is affordable. The total number of time periods is set $T = 24$, and the length of each time period is $L_t = 1$ hour. The traffic profile is $\lambda_t = \phi_t \lambda_{max}$, where the maximum traffic intensity $\lambda_{max} = 10$ s^{-1} and the time-dependent scaling factor $0 < \phi_t \leq 1$.

The tradeoff between the average blocking probability ($\omega_t = 1/T$) and the grid energy consumption for different policies is depicted in Figure 4.7. The joint traffic- and energy-aware policy performs better than the traffic-aware policy in almost all conditions by choosing a proper value of the scaling factor of the heuristic algorithm η_2 ($\eta_2 = 0.26, 0.3$ for $\eta_1 = 0.18, 0.22$, respectively). In addition, by adjusting the values of η_1 and η_2, different curves are obtained. For instance, the traffic-aware heuristic algorithm with a smaller value of η_1 (0.18) performs closer to the optimal than that with larger value (0.22) for the low grid power input regime (< 610 W), and that with larger η_1 (0.22) is near optimal for the high grid energy input regime (> 610 W). Notice that there is a sudden change on the traffic-aware algorithm curve. The left side of the change point is not achievable, because the number of active subcarriers is fixed for the given traffic load in this algorithm, even though there is additional energy.

4.3.4.2 3-Sector Case

Consider the sectorized multi-cell setup as shown in Figure 4.8a, where each site has three colocated BSs. In this setup, the dominant interference for a user in cell 1, 2, or 3 is from the other two cells. The interference from BSs at further locations can be considered as a low-power background noise. The renewable energy profile as in Figure 4.6 is the same for three BSs.

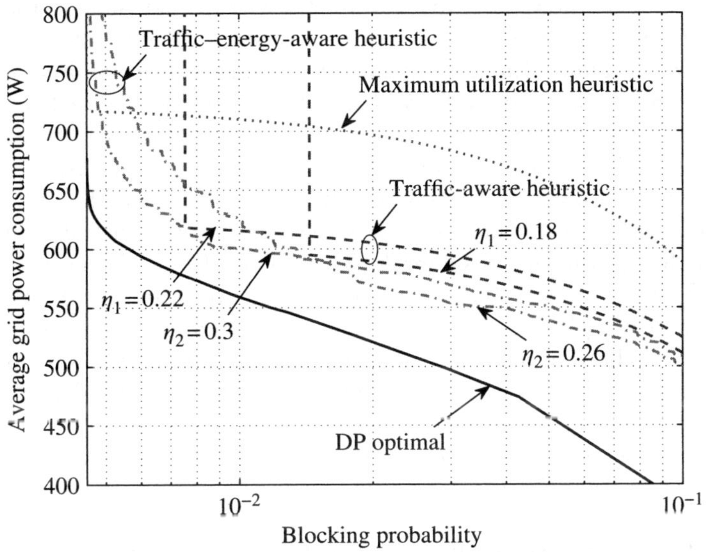

Figure 4.7 Tradeoff curves between outage probability and grid energy consumption with different policies.

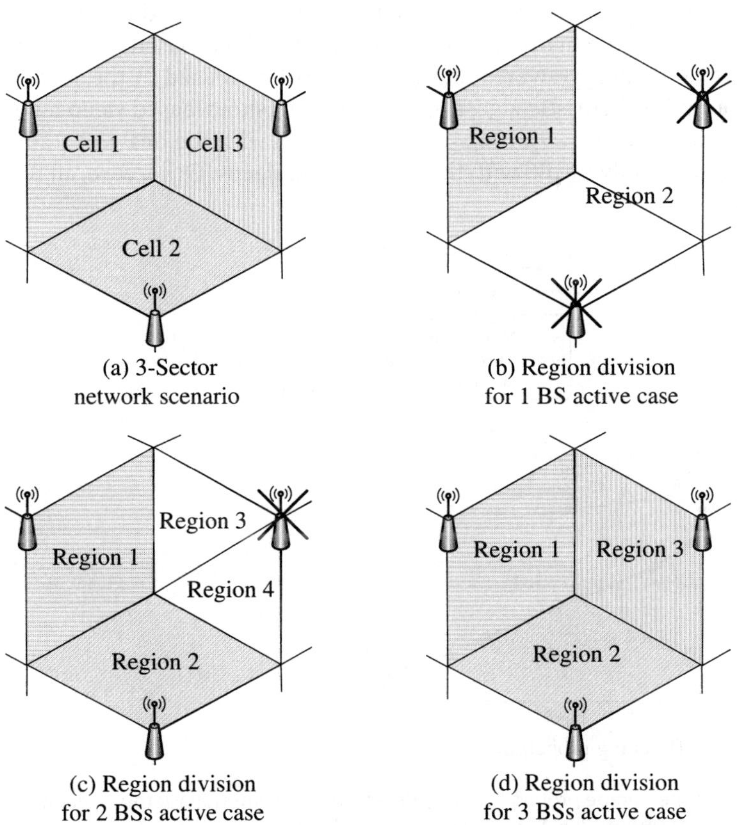

Figure 4.8 Sectorized multi-cell Erlang's approximation settings.

The parameter settings are as follows. The division for the regions depends on the BSs' on–off states. Specifically, if only one BS is active, as shown in Figure 4.8b, the cluster is divided into $M = 2$ regions. The first region is the original coverage area of the active BS, and the second is that of the others. If two BSs are active (Figure 4.8c), $M = 4$. The coverage area of the sleep BS is divided into two regions, which are served by the two active BSs, respectively. Finally, if all BSs are active (Figure 4.8d), $M = 3$, and each region is covered by its own BS. The service rate is assumed to be the same in all cells $\mu_K = 1\ \text{s}^{-1}$. The traffic in the studied area follows the profile illustrated in Figure 4.6. Assume that the arrival rate in sector b in time period t is $\lambda_t^{(b)} = \psi^{(b)}\lambda_t$, where $0 \leq \psi^{(b)} \leq 1$ and $\sum_b \psi^{(b)} = 1$. The simulations are run for two setups: an asymmetric traffic distribution $\psi^{(1)} : \psi^{(2)} : \psi^{(3)} = 1 : 2 : 3$ and a symmetric distribution $\psi^{(1)} : \psi^{(2)} : \psi^{(3)} = 1 : 1 : 1$. The renewable energy profiles of three BSs are assumed identical as in reality the renewable energy (for example the solar power) intensity will be highly correlated in the nearby area and the size of the solar panel of the BSs is assumed the same. The same energy profile depicted in Figure 4.6 is adopted for all the BSs. For the proposed two-stage DP algorithm, the greedy initial state settings are adopted.

Figure 4.9 depicts the tradeoff curves of different algorithms with symmetric traffic distribution. There is one class of users, i.e. $K = 1$. Notice that the *BS on–off* algorithm only optimizes the BS on–off state assuming all subcarriers are active ($n_t^{(b)} = N_c$), so there is no opportunistic sleep ($\varphi_t^{(b)} = 0$). It is actually the first stage in the two-stage algorithm. Also notice that for the heuristic non-sleep and threshold-based sleep algorithms, the joint traffic- and energy-aware adaptation algorithm is used in the second stage as it is better than other heuristic algorithms. In the threshold-based sleep algorithm, two thresholds are set as $\theta_1 < \theta_2$. If $\lambda_t \leq \theta_1$, only the BS with the heaviest traffic load is active. If $\theta_1 < \lambda_t < \theta_2$, only the BS with the lightest load sleeps. Otherwise, all BSs

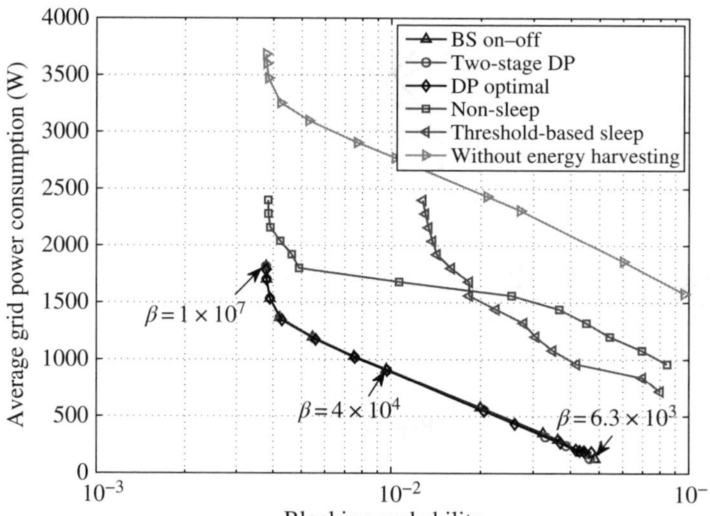

Figure 4.9 Tradeoff between the outage probability and grid energy consumption for the 3-sector case with symmetric traffic distribution. $K = 1$, $\psi^{(1)} : \psi^{(2)} : \psi^{(3)} = 1 : 1 : 1$.

are active. In this figure, set $\lambda_{max} = 7.5, \theta_1 = 3, \theta_2 = 6$. It can be seen that the two-stage DP algorithm performs almost the same as the optimal DP, and the optimality can be achieved by purely adjusting the BS sleeping states, which is consistent with the findings in Proposition 4.1. The threshold-based heuristic sleep algorithm performs better than the non-sleep algorithm when the grid power is less than 1550 W. In addition, the figure shows that using renewable energy can greatly reduce the power usage from grid, as compared with the scenario without EH.

The tradeoff between the blocking probability and grid energy consumption under an asymmetric traffic distribution for a single-user class ($K = 1$) is shown in Figure 4.10. Unlike the symmetric traffic case, the proposed two-stage DP algorithm performs close to the optimal DP algorithm and is notably better than the BS on–off algorithm. In other words, due to the asymmetric traffic distribution, the adaptation of number of active subcarriers and opportunistic sleep ratio further improves the performance in addition to the BS on–off optimization. The reason is that reducing the number of active subcarriers reduces the available wireless radio resources (bringing more blockings in the own cell) on one hand but on the other hand reduces the interference to the neighboring cells (reducing blockings in neighbor cells). In the asymmetric traffic distribution scenario, if the number of active subcarriers of a low traffic load BS is reduced, it is more effective to reduce the interference to other cells as the blocking probability of its own is already low. Therefore, there can be benefits from adapting the number of active subcarriers. On the contrary, there is no improvement by reducing active subcarriers in the symmetric traffic distribution scenario.

For the case of multiple user classes, Figures 4.11 and 4.12 illustrate the tradeoff with $K = 2$ user classes (rate requirements are $R_1 = 2$ Mbps and $R_2 = 0.5$ Mbps) for the symmetric and the asymmetric traffic distribution, respectively. Assume $\lambda_{max} = 12, \theta_1 = 3, \theta_2 = 10$, and each user class occupies half of the traffic. It can be

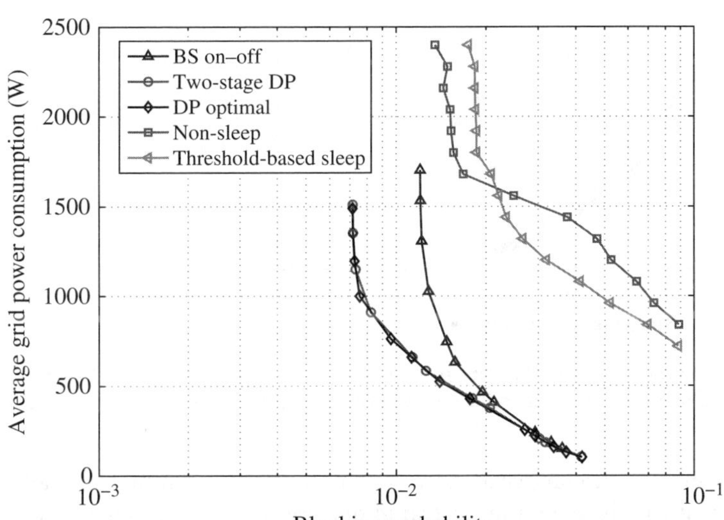

Figure 4.10 Tradeoff between the outage probability and grid energy consumption for the 3-sector case with asymmetric traffic distribution. $K = 1, R_1 = 2$ Mbps, $\psi^{(1)} : \psi^{(2)} : \psi^{(3)} = 1 : 2 : 3$.

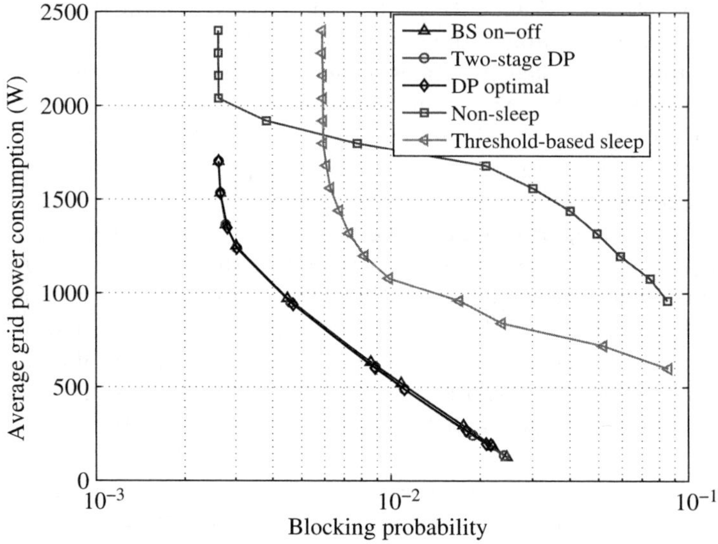

Figure 4.11 Tradeoff between the outage probability and grid energy consumption for the 3-sector case with multiple user classes and symmetric traffic distribution. $K = 2, \psi^{(1)} : \psi^{(2)} : \psi^{(3)} = 1 : 1 : 1$.

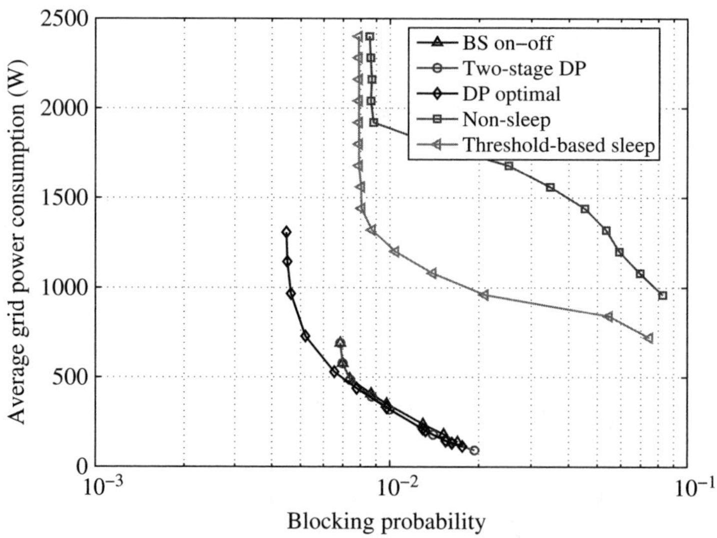

Figure 4.12 Tradeoff between the outage probability and grid energy consumption for the 3-sector case with multiple user classes and asymmetric traffic distribution. $K = 2, \psi^{(1)} : \psi^{(2)} : \psi^{(3)} = 1 : 2 : 3$.

seen that the performance is similar with the $K = 1$ case for the symmetric traffic distribution, but different for the asymmetric traffic distribution. In particular, when a low blocking probability is targeted, the two-stage DP algorithm is not close to the optimal solution anymore. The results also reveal that no matter how large β is set, the BS on–off algorithm cannot activate all the three BSs. This is because the algorithm assumes that all subcarriers are active, which causes very high interference when all the three BSs are active. However, the optimal DP algorithm can activate all three BSs via joint optimization of the number of active subcarriers in order to fine-tune the inter-cell interference.

4.4 Summary

In this chapter, we have illustrated the cross-layer design for EH wireless systems for a single link and a cellular network with multiple BSs, respectively. For a single wireless link powered by renewable energy, delay-related performance metric is optimized, whereas the optimal competitive ratio of completion time minimization for a single packet is characterized with several achieving power control algorithms. The average queueing delay of multiple packets is further studied with the comparison of offline and online power control algorithms. For a cellular network with multiple BSs powered with both renewable energy and grid power, bandwidth allocation and BS sleeping are jointly optimized with the utilization of two types of energy, via a low-complexity DP-based algorithm, to minimize the grid power consumption with a given outage probability constraint. Tradeoff between grid power consumption and outage probability is investigated through extensive simulations.

References

1 Yang, J. and Ulukus, S. (2012). Optimal packet scheduling in an energy harvesting communication system. *IEEE Transactions on Communications* 60 (1): 220–230. https://doi.org/10.1109/TCOMM.2011.112811.100349.
2 Tutuncuoglu, K. and Yener, A. (2012). Optimum transmission policies for battery limited energy harvesting nodes. *IEEE Transactions on Wireless Communications* 11 (3): 1180–1189. https://doi.org/10.1109/TWC.2012.012412.110805.
3 Ozel, O., Tutuncuoglu, K., Yang, J. et al. (2011). Transmission with energy harvesting nodes in fading wireless channels: optimal policies. *IEEE Journal on Selected Areas in Communications* 29 (8): 1732–1743. https://doi.org/10.1109/JSAC.2011.110921.
4 Vaze, R. (2013). Competitive ratio analysis of online algorithms to minimize packet transmission time in energy harvesting communication system. In: *2013 Proceedings IEEE INFOCOM*, 115–1123. Turin, Italy: IEEE. https://doi.org/10.1109/INFCOM.2013.6566902.
5 Zheng, X., Zhou, S., and Niu, Z. (2016). On the online minimization of completion time in an energy harvesting system. In: *2016 14th International Symposium on Modeling and Optimization in Mobile, Ad Hoc, and Wireless Networks (WiOpt)*, 1–8. Tempe, AZ, USA: IEEE. https://doi.org/10.1109/WIOPT.2016.7492927.
6 Cong, Y. and Zhou, X. (2016). Offline delay-optimal transmission for energy harvesting nodes. In: *2016 IEEE Global Communications Conference (GLOBECOM)*, 1–6. Washington, DC, USA: IEEE. https://doi.org/10.1109/GLOCOM.2016.7842035.
7 Sharma, V., Mukherji, U., Joseph, V., and Gupta, S. (2010). Optimal energy management policies for energy harvesting sensor nodes. *IEEE Transactions on Wireless Communications* 9 (4): 1326–1336. https://doi.org/10.1109/TWC.2010.04.080749.

8 Cui, Y., Lau, V.K.N., and Zhang, F. (2015). Grid power-delay tradeoff for energy harvesting wireless communication systems with finite renewable energy storage. *IEEE Journal on Selected Areas in Communications* 33 (8): 1651–1666. https://doi.org/10.1109/JSAC.2015.2391836.
9 Gorlatova, M., Wallwater, A., and Zussman, G. (2013). Networking low-power energy harvesting devices: measurements and algorithms. *IEEE Transactions on Mobile Computing* 12 (9): 1853–1865. https://doi.org/10.1109/TMC.2012.154.
10 Gatzianas, M., Georgiadis, L., and Tassiulas, L. (2010). Control of wireless networks with rechargeable batteries. *IEEE Transactions on Wireless Communications* 9 (2): 581–593. https://doi.org/10.1109/TWC.2010.080903.
11 Bertsekas, D.P. (2000). *Dynamic Programming and Optimal Control*. Belmont, MA: Athena Scientific.
12 Niyato, D., Hossain, E., and Fallahi, A. (2007). Sleep and wakeup strategies in solar-powered wireless sensor/mesh networks: performance analysis and optimization. *IEEE Transactions on Mobile Computing* 6 (2): 221–236. https://doi.org/10.1109/TMC.2007.30.
13 Medepally, B., Mehta, N.B., and Murthy, C.R. (2009). Implications of energy profile and storage on energy harvesting sensor link performance. In: *2009 IEEE Global Telecommunications Conference*, 1–6. Honolulu, HI, USA: IEEE. https://doi.org/10.1109/GLOCOM.2009.5425655.
14 Chen, S., Sinha, P., Shroff, N.B., and Joo, C. (2011). Finite-horizon energy allocation and routing scheme in rechargeable sensor networks. In: *2011 Proceedings IEEE INFOCOM*, 2273–2281. IEEE. https://doi.org/10.1109/INFCOM.2011.5935044.
15 Pappas, N., Jeon, J., Ephremides, A., and Traganitis, A. (2012). Optimal utilization of a cognitive shared channel with a rechargeable primary source node. *Journal of Communications and Networks* 14 (2): 162–168. https://doi.org/10.1109/JCN.2012.6253064.
16 Krikidis, I., Charalambous, T., and Thompson, J.S. (2012). Stability analysis and power optimization for energy harvesting cooperative networks. *IEEE Signal Processing Letters* 19 (1): 20–23. https://doi.org/10.1109/LSP.2011.2175382.
17 Bacinoglu, B.T. and Uysal-Biyikoglu, E. (2014). Finite horizon online lazy scheduling with energy harvesting transmitters over fading channels. In: *2014 IEEE International Symposium on Information Theory*, 1176–1180. Honolulu, HI, USA: IEEE. https://doi.org/10.1109/ISIT.2014.6875018.
18 Bacinoglu, B.T. and Uysal-Biyikoglu, E. (2014). Finite-horizon online transmission scheduling on an energy harvesting communication link with a discrete set of rates. *Journal of Communications and Networks* 16 (3): 300–393. https://doi.org/10.1109/JCN.2014.000049.
19 Han, T. and Ansari, N. (2013). On optimizing green energy utilization for cellular networks with hybrid energy supplies. *IEEE Transactions on Wireless Communications* 12 (8): 3872–3882. https://doi.org/10.1109/TCOMM.2013.051313.121249.
20 Gong, J., Thompson, J.S., Zhou, S., and Niu, Z. (2014). Base station sleeping and resource allocation in renewable energy powered cellular networks. *IEEE Transactions on Communications* 62 (11): 3801–3813. https://doi.org/10.1109/TCOMM.2014.2359883.
21 Borodin, A. and El-Yaniv, R. (1998). *Online Computation and Competitive Analysis*. New York: Cambridge University Press.

22 Basar, T. and Olsder, G. (1999). *Dynamic Noncooperative Game Theory*, 2e. Society for Industrial and Applied Mathematics.

23 Auer, G., Giannini, V., Godor, I. et al. (2011). Cellular energy efficiency evaluation framework. In: *2011 IEEE Vehicular Technology Conference (VTC Spring)*, 1–6. Budapest, Hungary: IEEE. https://doi.org/10.1109/VETECS.2011.5956750.

24 Sesia, S., Baker, M., and Toufik, I. (2009). *LTE - The UMTS Long Term Evolution: From Theory to Practice*. Wiley.

25 Wang, R., Thompson, J.S., Haas, H., and Grant, P.M. (2011). Sleep mode design for green base stations. *IET Communications* 5 (19): 2606–2616. https://doi.org/10.1049/iet-com.2011.0104.

26 Niu, Z., Wu, Y., Gong, J., and Yang, Z. (2010). Cell zooming for cost-efficient green cellular networks. *IEEE Communications Magazine* 48 (11): 74–79. https://doi.org/10.1109/MCOM.2010.5621970.

27 Kleinrock, L. (1976). *Queueing Systems*, vol. 1. Wiley.

28 Karray, M.K. (2010). Analytical evaluation of QoS in the downlink of OFDMA wireless cellular networks serving streaming and elastic traffic. *IEEE Transactions on Wireless Communications* 9 (5): 1799–1807. https://doi.org/10.1109/TWC.2010.05.091501.

29 Kelly, F. (1979). *Reversibility and Stochastic Networks*. Wiley: New York.

30 Bertsekas, D.P. (2005). *Dynamic Programming and Optimal Control*, 3e. Belmont, MA: Athena Scientific.

Part II

Energy Harvesting Network Optimization

5

Energy Harvesting *Ad Hoc* Networks

In this chapter, we focus on the *ad hoc* networks, where multiple energy harvesting (EH) nodes are deployed in a given area and try to utilize the shared wireless channels via opportunistic access. Thus, a smart scheduling scheme is required to coordinate the transmissions among the multiple nodes without central control. In the EH communication scenario, the scheduling scheme needs to consider not only the channel state information (CSI) at each transmission node, as what a conventional scheduler does, but also the energy state information (ESI) to optimize the design metric, i.e. throughput. In particular, random energy arrivals at individual EH nodes introduce another dimension of resource dynamics other than the channel and make the scheduling problem for *ad hoc* networks more complicated. However, other than the multiuser gain coming from the diversified channel power gains across different transmitter–receiver links, a new class of diversity, termed as *energy diversity*, can be exploited by utilizing the ESI at these transmitters powered by different energy harvesters, where the harvested energy randomly varies across different nodes.

In this chapter, we first propose a distributed opportunistic scheduling (DOS) framework with the two-stage probing and the save-then-transmit energy utilization schemes to fully utilize both the ESI and CSI at each node and employ an optimal stopping framework to solve the expected throughput maximization problem for the considered network. Then, we study the multiuser gain with the emphasis on energy diversity, where the scaling law of the expected throughput over the number of users is investigated.

5.1 Distributed Opportunistic Scheduling

5.1.1 System Model

We consider a heterogeneous single-hop *ad hoc* network, where all the I transmitter–receiver pairs have independent but not necessarily identical statistical information of CSI and ESI. All pairs contend for one shared channel by random access. For each link, the transmitter is powered by a renewable energy source and utilizes a small rechargeable battery to temporally store the harvested energy. Note that the transmitter could keep harvesting energy until it initiates a data transmission. In addition, we do not consider the effect of inefficiency in energy storage and retrieval, nor the energy consumed other than data transmission, which can be approximately neglected by properly

Energy Harvesting Wireless Communications, First Edition.
Chuan Huang, Sheng Zhou, Jie Xu, Zhisheng Niu, Rui Zhang and Shuguang Cui.

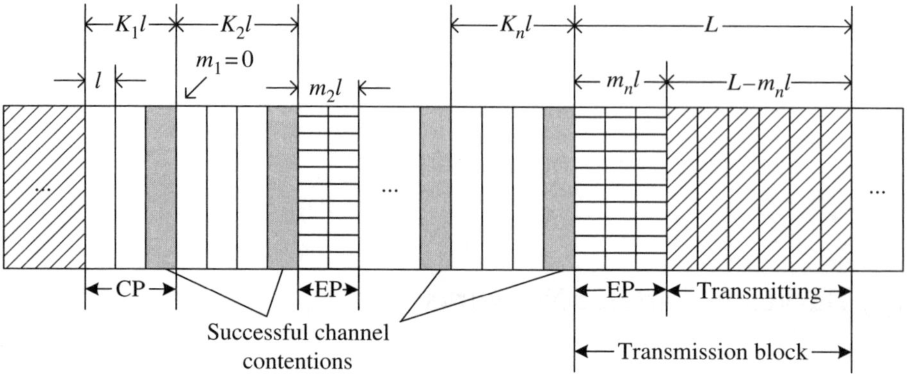

Figure 5.1 One realization for the DOS with two-stage probing.

adjusting the energy model [1–4]. Denote the duration of one channel contention as $l > 0$ and the length of one transmission block as L, which is an integer multiple of l.

As illustrated in Figure 5.1, the DOS procedure of the whole network takes place in two stages: first, each transmitter probes the channel via random access and harvests energy at the same time, and then the successful transmitter may start the energy probing (EP) (to potentially increase the average transmission rate over the transmission block[1]) before the data transmission process.

Channel probing: In the first stage, a successful channel contention is defined as follows: all transmitters first independently contend for the channel until there is only one contending in a particular time slot. Furthermore, one round of channel probing (CP) is defined as the process to achieve one successful channel contention. Denote the probability that transmitter i contends for the channel as q_i, $1 \leq i \leq I$, with $0 \leq q_i \leq 1$. As such, the probability that the ith transmitter successfully occupies the channel is given by $Q_i = q_i \prod_{j \neq i}(1 - q_j)$. Then, the probability to achieve one successful channel contention at each time slot is given by $Q = \sum_{i=1}^{I} Q_i$, and it is easy to check that $Q \leq 1$ [5]. Accordingly, for the nth round of CP, $n \geq 1$, we use K_n to denote the number of time slots needed to achieve a successful channel contention, which is a random variable and satisfies the geometric distribution with parameter Q [6–8]. In this way, the expected duration of one round of CP is given as l/Q. Denote the transmitted signal at transmitter i as x^i, and the received signal y^i is thus given by $y^i = h^i x^i + z^i$, where h^i is the complex channel gain and z^i is the circularly symmetric complex Gaussian (CSCG) noise with zero mean and variance σ^2 at the receiver. Across different links, $\{h^i\}_{1 \leq i \leq I}$ are independent with finite mean and variance, while not necessarily identically distributed. After one round of CP, the successful transmitter can perfectly estimate the corresponding channel gain via certain feedback mechanisms, and thus h^i is assumed a known constant during the whole transmission block. After CP, the successful transmitter chooses one of the following actions based on its local CSI and ESI:

(a) Releases the channel (if the CSI and ESI indicate that the transmission rate is lower than a threshold) and let all links re-contend.

1 If the successful transmitter experiences a bad channel condition and a low energy level, it may skip the transmission.

(b) Directly transmits until the end of the transmission block.
(c) Holds the channel, starts EP.

Note that to complete one data transmission, it may take n rounds of CPs as depicted in Figure 5.1. It is worth noting that each transmitter keeps harvesting energy until it starts a transmission, and after each round of CP, only the successful transmitter makes a choice among three actions as listed above.

Energy probing: When the successful transmitter decides not to take action (a) or (b) defined above, it starts the second stage EP, i.e. action (c), to obtain more energy. During this stage, the transmitter chooses to continue harvesting energy slot by slot and then ends EP by action (a) or (b), i.e. either releasing the channel or transmitting over the rest of the transmission block. As it is depicted in Figure 5.1, one transmission is fulfilled with n rounds of CPs and m_n extra slots of EP.

For transmitter i, let $B^i_{n,m} \in \mathbf{\Delta}$ denote the energy level of the battery after the nth round of CP and m additional time slots for EP, where $\mathbf{\Delta} = \{0, \delta, 2\delta, \dots, B_{\max}\delta\}$ is the set of all possible energy states, with δ being the minimum energy unit and $B_{\max}\delta$ the capacity of the battery. We use E^i_t to denote the EH rate of transmitter i at time t. As noted in the previous section, we consider the following two types of scenarios:

(1) *Constant EH rate model*: $\{E^i_t\}_{t\geq 1}$ are constants for each i, i.e. $E^i_t = E^i \in \mathbf{\Delta}$ for all $t \geq 1$, and $\{E^i\}$ can thus be learned and assumed noncausally known before transmissions.
(2) *Independent and identically distributed (i.i.d.) EH rate model*: The EH rates among different transmitters are independent. For transmitter i, $\{E^i_t\}_{t\geq 1}$ are i.i.d. across t, with finite mean, and the probability density function (PDF) $\Pr\{E^i_t = e\delta\} = F^i(e)$, where $e \in \{0, 1, 2, \dots\}$.

Under the save-then-transmit scheme, the energy level will keep non-decreasing and drop to zero after the transmission, which forms a Markov chain (as described in Section 5.1.3 later). Thus, the energy level $B^i_{n,m}$ can be written as

$$B^i_{n,m} = \min\left\{ B^i_{n,0} + l\sum_{k=0}^{m} E^i_k, B_{\max}\delta \right\}, \tag{5.1}$$

where $n \geq 1$, $0 \leq m \leq L/l$, and $\min\{x, y\}$ denotes the smaller value between two real numbers x and y. Note that $B^i_{n,0}$ indicates the energy level after the successful contention round before taking any action. If $m = 0$, i.e. transmitter i does not do EP, we let $\sum_{k=0}^{m} E^i_k = E^i_0 = 0$.

5.1.2 Transmission Scheduling

In this section, we target to derive the optimal scheduling policy that maximizes the average throughput for the considered network with the proposed two-stage access strategy, conditioned on the given battery state distribution. We point out that the results obtained in this section are based on the assumption that the energy level at transmitter i is stationary with a given distribution Π^i, for $1 \leq i \leq I$, which will be validated in Section 5.1.3.

5.1.2.1 Problem Formulation

After the nth round of CP and m additional time slots, the CSI and the ESI at the successful transmitter are given as $\mathcal{F}^i_{n,m} = \{h^i_n, B^i_{n,m}\}$. Note that the channel gain h^i_n is now indexed by n, which is determined at the end of the nth round of CP and assumed fixed during the whole data transmission block. In particular, $\mathcal{F}^i_{n,0} = \{h^i_n, B^i_{n,0}\}$ denotes the initial information right after the nth round of CP. For convenience, we omit the index i for either the CSI or the ESI in the sequel and retrieve it when necessary.

By adopting the save-then-transmit scheme at the transmitters to fully take advantage of each channel use, the transmission rate over L/l time slots with state $\mathcal{F}_{n,m}$ is defined as

$$R_n(m) = \left(1 - \frac{ml}{L}\right) \log\left(1 + |h_n|^2 \frac{B_{n,m}}{(L - ml)\sigma^2}\right). \tag{5.2}$$

When $ml = L$, we set $R_n(m) = 0$ since there is no transmission in this case. Note that $\mathbb{E}[R_n(m)] < \infty$ and $\mathbb{E}[(R_n(m))^2] < \infty$, which results from the fact that h_n has finite mean and variance and the energy level $B_{n,m}$ is also finite.

Let N be the stopping rule for CP and M_n be the stopping rule for EP associated with the nth CP for $1 \le n \le N$, which together tell the transmitter when to start the data transmission. Then, under these stopping rules, the transmission rate would be $R_N(M_N)$, and we let T_N be the total time duration for completing one data transmission. Here, T_N contains the duration of $N-1$ rounds of CP, which is given by $l\sum_{n=1}^{N-1} K_n$ and $l\sum_{n=1}^{N-1} M_n$ time slots in which the transmitter probes the energy but gives up the channel after EP. Also, after the Nth round of CP with the time $K_N l$, the transmitter may use M_N slots for the EP and transmit within the duration $L - M_N l$ afterward. Accordingly, we obtain

$$T_N = l\sum_{n=1}^{N-1} M_n + l\sum_{n=1}^{N} K_n + L. \tag{5.3}$$

If such a process is executed J times with $R_{N_j}(M_{N_j})L$ bits transmitted at each transmission, $1 \le j \le J$, we obtain the average throughput λ per transmission of the network:

$$\frac{\lim_{J\to\infty} \frac{1}{J} L \sum_{j=1}^{J} R_{N_j}(M_{N_j})}{\lim_{J\to\infty} \frac{1}{J} \sum_{j=1}^{J} T_{N_j}} = \frac{L\mathbb{E}[R_N(M_N)]}{\mathbb{E}[T_N]} = \lambda.$$

Again, we point out that the energy level is stationary at the N_jth round of CP for $j \ge 1$, as we assumed.

Our target is to maximize λ by adjusting the stopping rule N and $\{M_n\}_{1\le n\le N}$. It is easy to see that maximizing λ is in fact a "rate-of-return" stopping problem [9, 10] (for which the specific definition is given later). Instead of directly solving this problem, we examine the "net reward" of the considered network, which is given as

$$\begin{aligned} r_N(\lambda) &= R_N(M_N)L - \lambda T_N \\ &= (R_N(M_N) - \lambda)L - \lambda l \left[K_N + \sum_{n=1}^{N-1}(K_n + M_n)\right], \end{aligned} \tag{5.4}$$

for some $\lambda > 0$. The term $(R_N(M_N) - \lambda)L$ can be interpreted as the reward of transmission, $\lambda l K_n$ as the cost of CP, and $\lambda l M_n$ as the cost of failed EP for $1 \le n \le N-1$. We set $r_{-\infty}(\lambda) = -\infty$ since it is irrational that the system does not send any data forever. Then, we define the maximum value of the expected net reward with $\lambda > 0$ as

$$S^*(\lambda) = \sup_{N\in\mathcal{N},\{M_n\}_{1\le n\le N}} \mathbb{E}[r_N(\lambda)], \tag{5.5}$$

where sup(·) denotes the least upper bound for a set of real numbers and

$$\mathcal{N} \triangleq \{N : N \geq 1, \quad \mathbb{E}[T_N] < \infty, \ \text{for} \ M_n \in [0, L/l] \ \text{with} \ 1 \leq n \leq N\}. \tag{5.6}$$

Remark 5.1 Problem (5.5) is approximately *time invariant* when the number of pairs is large. We observe that before the system starts the Nth round of CP, the accumulated cost $\lambda l \sum_{n=1}^{N-1}(K_n + M_n)$ over the past $N-1$ rounds of CP has already been finalized, with no need to be further considered in the remaining decision process. At the beginning of the Nth round of CP, the expected optimal reward is approximately the same since the successful transmitter has no information of other pairs and the probability that the current successful transmitter occupies the channel again is small. For example, in an *ad hoc* network with K pairs where each pair fairly competes for the channel use with probability $1/K$, such a probability is $\frac{1}{K}(1-1/K)^{(K-1)}$ [5], which is as small as 0.067 even when $K = 6$. In other words, the system can obtain the expected optimal reward about $S^*(\lambda)$ whenever a new round of CP is about to start. Therefore, we conclude that problem (5.5) is approximately time invariant.

Recall from Section 5.1.1 that after each round of CP, the successful transmitter will choose one of three actions (i.e. transmitting, giving up the channel, or starting EP) according to the stopping rule of CP, which needs the expected reward of EP depending on the stopping rule of EP. Thus, we will first introduce the formulation and the optimal stopping rule for EP and then for CP.

Formulation for EP: When the successful transmitter starts EP after the nth round of CP, where $1 \leq n \leq N$, it will end up with one of the two actions: transmitting or giving up the channel without transmission. Specifically, we define the expected optimal reward at the kth slot of EP, $0 \leq k \leq L/l$, as

$$U_k(\mathcal{F}_{n,k}) = \max_{k \leq M_n \leq L/l} \mathbb{E}[\max\{(R_n(M_n) - \lambda)L, -\lambda l M_n + S^*(\lambda)\} \mid \mathcal{F}_{n,k}], \tag{5.7}$$

where $-\lambda l M_n + S^*(\lambda)$ is the expected value of giving up the channel after M_n slots of EP. If $k = 0$, $U_0(\mathcal{F}_{n,0})$ denotes the maximum of the expected net reward right after the nth round of CP. In other words, we want to find the optimal stopping rule M_n^* of EP that attains

$$U_0(\mathcal{F}_{n,0}) = \max_{0 \leq M_n \leq L/l} \mathbb{E}[\max\{(R_n(M_n) - \lambda)L, -\lambda l M_n + S^*(\lambda)\} \mid \mathcal{F}_{n,0}]. \tag{5.8}$$

Note that M_n^* exists since problem (5.8) is an optimal stopping problem over a finite time horizon [9–11].

Formulation for CP: By choosing $\{M_n^*\}_{1 \leq n \leq N}$, we define

$$\lambda^* = \sup_{N \in \mathcal{N}} \frac{L\mathbb{E}[R_N(M_N^*)]}{\mathbb{E}[T_N]}, \quad N^* = \arg\sup_{N \in \mathcal{N}} \frac{L\mathbb{E}[R_N(M_N^*)]}{\mathbb{E}[T_N]}. \tag{5.9}$$

Note that if the optimal stopping rule $N^* \notin \mathcal{N}$, we would claim that N^* does not exist. Thus, λ^* is the optimal average throughput of the original rate-of-return problem.

The connection between the transformed problem (5.5) and the original problem (5.9) is introduced in the following lemma. It is worth noticing that with the optimal stopping rule $\{M_n^*\}_{1\leq n\leq N}$ for EP, problem (5.5) boils down to a one-level stopping problem with stopping rule N.

Lemma 5.1

(i) If there exists λ^* such that $S^*(\lambda^*) = 0$, this λ^* is the optimal throughput defined in (5.9). Moreover, if $S^*(\lambda^*) = 0$ is attained at $N^*(\lambda^*)$, the stopping rule N^* defined in (5.9) is the same as $N^*(\lambda^*)$, i.e. $N^* = N^*(\lambda^*)$.
(ii) Conversely, if (5.9) is true, there is $S^*(\lambda^*) = 0$, which is attained at N^* given by (5.9).

This lemma directly follows Theorem 1 in Chapter 6 of [9].
The next proposition secures the existence of the optimal stopping rule for CP.

Proposition 5.1 *With the EP stopping rule $\{M_n^*\}_{0\leq n\leq N}$, the optimal stopping rule $N^*(\lambda)$ for problem (5.5) exists. Moreover, for $N \geq 1$, the following equation holds*

$$S^*(\lambda) = U_0(\mathcal{F}_{N,0}) - \lambda l K_N. \tag{5.10}$$

Proof: For the first part of Proposition 5.1, it follows by Theorem 1 in Chapter 3 of [9] that $N^*(\lambda)$ exists and $S^*(\lambda)$ is attained by this $N^*(\lambda)$ if the following two conditions are satisfied:

(C1) $\limsup_{N\to\infty} r_N(\lambda) \leq r_{-\infty}(\lambda)$ a.s.,
(C2) $\mathbb{E}[\sup_{N\geq 1} r_N(\lambda)] < \infty$,

where $r_N(\lambda)$ is given by (5.4). As we pointed out in Section 5.1.1, the energy level $B_{N,0}$ is stationary for $N \geq 1$. Although $\{R_N(M_N^*)\}_{N\geq 1}$ are independent, it may not be identically distributed with respect to h_N and $B_{N,0}$. However, it is not too difficult to show that (C1) and (C2) hold. The idea is that we first consider that every transmitter has the same statistics and then we apply the channel contention probability as the summation coefficients over all transmitters.

For (C1), if we assume that all transmitters have the same statistics as transmitter i, then $\{R_N^i(M_N^*)\}_{N\geq 1}$ become i.i.d. Since $\mathbb{E}[R_N^i(M_N^*)] < \infty$ according to Section 5.1.2 and the accumulated cost $\lambda T_N = \lambda l\left(K_N + \sum_{n=1}^{N-1}(K_n + M_n^*)\right) \to \infty$ as $N \to \infty$ a.s., we obtain that $\mathbb{P}\{\limsup_{N\to\infty} r_N^i(\lambda) = -\infty\} = 1$. Recalling from Section 5.1.1 that the channel is occupied by transmitter i with probability Q_i and $\sum_{i=1}^{I} \frac{Q_i}{Q} = 1$, we obtain that

$$\begin{aligned} 1 &= \sum_{i=1}^{I} \frac{Q_i}{Q} \mathbb{P}\left\{\limsup_{N\to\infty} r_N^i(\lambda) = -\infty\right\} \\ &= \mathbb{P}\left\{\limsup_{N\to\infty} r_N(\lambda) = -\infty\right\}, \end{aligned}$$

which proves that (C1) holds.

For (C2), it can be shown that

$$\begin{aligned}\mathbb{E}\left[\sup_{N\geq 1} r^i_N(\lambda)\right] &= \mathbb{E}\left[\sup_{N\geq 1}((R^i_N(M^*_N) - \lambda)L - \lambda T_N)\right] \\ &\leq \mathbb{E}\left[\sup_{N\geq 1}(R^i_N(M^*_N) - \lambda(lN + L))\right], \end{aligned} \tag{5.11}$$

due to the fact that $K_n \geq 1$ and $M^*_n \geq 0$ for $1 \leq n \leq N$. Since $\mathbb{E}[(R^i_N(M^*_N))^2] < \infty$, it follows that the right-hand side of (5.11) is finite by Theorem 1 in Chapter 4 of [9]. Similar to the technique in the proof of (C1), we have

$$\mathbb{E}\left[\sup_{N\geq 1} r_N(\lambda)\right] = \sum_{i=1}^{I} \frac{Q_i}{Q}\mathbb{E}\left[\sup_{N\geq 1} r^i_N(\lambda)\right] < \infty,$$

which shows that (C2) also holds.

For the second part, we know that with the cost $\lambda l K_N$ at the Nth CP for any $N \geq 1$, the successful transmitter could choose one of three actions: transmits immediately with reward $(R_N(0) - \lambda)L$, gives up the channel immediately and obtains the optimal expected net reward $S^*(\lambda)$ based on the property of time invariance described in Section 5.1.2.1, or starts EP and obtains the expected net reward $\mathbb{E}[\mathcal{U}_1(\mathcal{F}_{N,1}) \mid \mathcal{F}_{N,0}]$. Thus, by the optimal stopping theory [9, 10], $S^*(\lambda)$ satisfies the *optimality equation* under (C2) as

$$S^*(\lambda) = -\lambda l K_N + \max\{S^*(\lambda), (R_N(0) - \lambda)L, \mathbb{E}[\mathcal{U}_1(\mathcal{F}_{N,1}) \mid \mathcal{F}_{N,0}]\},$$

which is equivalent to (5.10). □

Remark 5.2 Equation (5.10) is obtained from the optimality equation of the CP. The calculation of the optimal throughput relies on this equation, which will be shown in Section 5.1.4.

Now, we are ready to derive the optimal stopping rules N^* and $\{M^*_n\}$ that jointly maximize the expected value of $r_N(\lambda)$ for the two different EH models. As we mentioned above, the stopping rule N for CP relies on the form of M_N (the stopping rule for EP). We will find the optimal stopping rule M^*_N before N^*. After obtaining the forms of the optimal stopping rules, the calculation for the optimal throughput will be discussed.

5.1.2.2 Optimal Stopping Rule for Constant EH Model

For notation simplicity, we omit the index N of CP when we derive the stopping rule M in this subsection. Then, we will derive the stopping rule N based on the results of EP.

When the EH rate is constant, the transmission rate $R(M)$ is deterministic for a given $\mathcal{F}_0$ over the transmission block. Then, we obtain a simplified version of $\mathcal{U}_0(\mathcal{F}_0)$ (5.8) as

$$\mathcal{U}_0(\mathcal{F}_0) = \max_{0\leq M\leq L/l} \max\{(R(M) - \lambda)L, -\lambda l M + S^*(\lambda)\}.$$

The value of $\mathcal{U}_0(\mathcal{F}_0)$ can be obtained simply by comparing $-\lambda l M + S^*(\lambda)$ and $(R(M) - \lambda)L$, whose values can be computed individually. Clearly, the first one achieves its maximum $S^*(\lambda)$ at $M = 0$. For the second term, only $R(M)$ is changing over M with a given $\mathcal{F}_0$. Therefore, we settle down to the following auxiliary problem:

$$V^* = \arg \max_{0 \le V \le L/l} R(V). \tag{5.12}$$

Then, we could use the optimal V^* to find M^* without difficulty. Note that when $Vl = L$, it follows that $R(V) = 0$ according to our definition in Section 5.1.1, which implies that $V = L/l$ cannot be optimal, and thus we take $0 \le V \le L/l - 1$. We first consider a related continuous version of $R(V)$ by relaxing Vl/L as ρ, $0 \le \rho < 1$:

$$\max_{0 \le \rho < 1} R(\rho) = \max_{0 \le \rho < 1} (1-\rho) \cdot \log \left(1 + |h|^2 \frac{\min\{B_0 + \rho LE, B_{\max}\delta\}}{(1-\rho)L\sigma^2} \right). \tag{5.13}$$

After solving (5.13), we will show how to obtain the optimal solution of problem (5.12).

First, we establish some properties for the objective function of problem (5.13).

Proposition 5.2 *For arbitrary $a, b \ge 0$, we have that:*

(1) *The function $y(x) = (1-x)\log\left(1 + \frac{a+bx}{1-x}\right)$ is concave over $[0, 1)$, and $\lim_{x\to 1^-} y'(x) < 0$.*
(2) *The function $g(x) = (1-x)\log\left(1 + \frac{a}{1-x}\right)$ is concave and nonincreasing over $[0, 1)$.*

Proof: For (1), we show the concavity of function $y(x)$ by checking its second-order derivative over $[0, 1)$, which is given by

$$y''(x) = -\frac{(a+b)^2}{(1-x)[a+1+(b-1)x]^2} \le 0.$$

Therefore, $y(x)$ is concave over $[0, 1)$ [12]. To prove the second part of (1), we check the first-order derivative of $y(x)$, which is given by

$$y'(x) = -\log\left(1 + \frac{a+bx}{1-x}\right) + \frac{a+b}{1-x+a+bx}. \tag{5.14}$$

It is easy to see that as $x \to 1^-$, the first term of the right-hand side of (5.14) goes to negative infinity, while the second term is bounded. Hence, $y'(x)$ is strictly negative as $x \to 1^-$. Therefore, part (1) is proved.

Next, we prove (2). By checking the second-order derivative of $g(x)$, we obtain

$$g''(x) = -\frac{a^2}{(1-x)(a+1-x)^2} \le 0,$$

which implies that $g(x)$ is concave. For the second part of (2), we consider the first-order derivative of $g(x)$, which is given by

$$g'(x) = -\log\left(1 + \frac{a}{1-x}\right) + \frac{a}{1-x+a}. \tag{5.15}$$

Since $g''(x) \le 0$, it follows that

$$\max_{0 \le x < 1} g'(x) = g'(0) = -\log(1+a) + \frac{a}{1+a}.$$

Moreover, due to the fact that $\frac{d}{da}\left(-\log(1+a)+\frac{a}{1+a}\right)=-\frac{a}{(1+a)^2}\leq 0$ for arbitrary $a \geq 0$, we obtain

$$\max_{0\leq x<1} g'(x) = g'(0) \leq \left(-\log(1+a)+\frac{a}{1+a}\right)\Big|_{a=0} = 0,$$

which proves the second part of (2). □

Since $\rho \in [0,1)$, when $\frac{B_{\max}\delta - B_0}{LE} \geq 1$, $R(\rho)$ is simply concave over ρ on $[0,1)$ according to part (1) of Proposition 5.2. When $\frac{B_{\max}\delta - B_0}{LE} < 1$, according to Proposition 5.2, $R_N(\rho)$ is concave over $\left[0, \frac{B_{\max}\delta-B_0}{LE}\right]$ and is nonincreasing on $\left[\frac{B_{\max}\delta-B_0}{LE}, 1\right)$. Thus, $R(\rho)$ cannot achieve its maximum on $\left(\frac{B_{\max}\delta-B_0}{LE}, 1\right)$. Therefore, we treat this fact as a new constraint over ρ and rewrite problem (5.13) as

$$\max G(\rho) = \max(1-\rho)\log\left(1+|h|^2\frac{B_0+\rho LE}{(1-\rho)L\sigma^2}\right)$$
$$\text{s.t.}\quad B_0+\rho LE \leq B_{\max}\delta, \quad 0\leq \rho < 1. \tag{5.16}$$

Next, we establish the following proposition to solve problem (5.16), where the obtained solution is optimal for problem (5.13) as well.

Proposition 5.3 *The optimal solution ρ^* for problem (5.16) is given by*

$$\rho^* = \begin{cases} \min\left\{\rho_0, \frac{B_{\max}\delta-B_0}{LE}\right\}, & \text{when } \frac{C+D}{1+C} \geq \log(1+C);\\ 0, & \text{otherwise,}\end{cases}$$

where $C=\frac{|h|^2B_0}{L\sigma^2}$, $D=\frac{|h|^2E}{\sigma^2}$, and ρ_0 is the unique solution for the equation $\log\left(1+\frac{C+D\rho}{1-\rho}\right)=\frac{C+D}{1-\rho+C+D\rho}$ when $\frac{C+D}{1+C}\geq \log(1+C)$.

Proof: According to part (1) of Proposition 5.2, we obtain that $G(\rho)$ is concave over $\rho \in [0,1)$, which means that $G'(\rho)=\frac{dG(\rho)}{d\rho}$ is decreasing over $[0,1)$ and attains its maximum at $\rho=0$. Then, finding the maximum of $G(\rho)$ boils down to two cases:

(1) $G'(\rho)|_{\rho=0}<0$: It follows that $G(\rho)$ is decreasing over $[0,1)$ and $\rho^*=0$ is the optimum.
(2) $G'(\rho)|_{\rho=0}\geq 0$: The point ρ_0, satisfying $G'(\rho)|_{\rho=\rho_0}=0$, lies on the right-hand side of $\rho=0$. By part (1) of Proposition 5.2, $G'(\rho)<0$ as $\rho\to 1^-$, which implies that $\rho_0\in[0,1)$. Since the optimal point $\rho^* \leq \frac{B_{\max}\delta-B_0}{LE}$ due to (5.16), it follows that $\rho^*=\min\left\{\rho_0, \frac{B_{\max}\delta-B_0}{LE}\right\}$.

Note that $G'(\rho)|_{\rho=0}\geq 0$ is equivalent to $\frac{C+D}{1+C}\geq \log(1+C)$, where $C=\frac{|h|^2B_0}{L\sigma^2}\geq 0$, $D=\frac{|h|^2E}{\sigma^2}\geq 0$, and $G'(\rho)|_{\rho=\rho_0}=0$ is equivalent to

$$\log\left(1+\frac{C+D\rho_0}{1-\rho_0}\right)=\frac{C+D}{1-\rho_0+C+D\rho_0}. \tag{5.17}$$

Next, we show that when $\frac{C+D}{1+C}\geq\log(1+C)$, (5.17) has a unique solution. For $\rho\in[0,1)$, the left-hand side of (5.17) is increasing over ρ from $\log(1+C)$ to $+\infty$. For its right-hand side, we have the following two cases:

(1) $D \geq 1$: The right-hand side of (5.17) decreases from $\frac{C+D}{1+C}$ to 1. Since $\frac{C+D}{1+C} \geq \log(1+C)$, there exists a unique solution ρ_0 for (5.17).
(2) $0 \leq D < 1$: The right-hand side of (5.17) increases from $\frac{C+D}{1+C}$ to 1. If the first-order derivative of the left-hand side of (5.17) is always greater than that of the right-hand side, there must be only one solution for (5.17) when $\frac{C+D}{1+C} \geq \log(1+C)$. Thus, we check their first-order derivatives: for the left-hand side of (5.17), we obtain

$$\frac{\mathrm{d}}{\mathrm{d}\rho}\log\left(1+\frac{C+D\rho}{1-\rho}\right) = \frac{C+D}{(1-\rho)(1+C+(D-1)\rho)}; \tag{5.18}$$

for the right-hand side, we have

$$\frac{\mathrm{d}}{\mathrm{d}\rho}\left(\frac{C+D}{1-\rho+C+D\rho}\right) = \frac{(C+D)(1-D)}{(1+C+(D-1)\rho)^2}. \tag{5.19}$$

Thus, by calculating the difference between (5.18) and (5.19), we arrive at

$$\frac{C+D}{(1-\rho)(1+C+(D-1)\rho)} - \frac{(C+D)(1-D)}{(1+C+(D-1)\rho)^2}$$
$$= \frac{(C+D)^2}{(1-\rho)(1+C+(D-1)\rho)^2} \geq 0. \tag{5.20}$$

Therefore, there exists a unique solution ρ_0 satisfying (5.17).

In conclusion, the proposition is proved.

Remark: Since it is proved that ρ_0 is unique in (5.17), ρ_0 can be found just by adopting a simple one-dimensional searching method, e.g. bisection search. □

Based on the optimal solution ρ^*, the optimal V^* for $R(V)$ in (5.12) can be obtained easily: we only need to compare $R(\lfloor \rho^* L/l \rfloor)$ against $R(\lceil \rho^* L/l \rceil)$, and V^* should attain the larger value. Specifically, we have the following result.

Proposition 5.4 *The optimal V^* of problem (5.12) is given by*

$$V^* = \begin{cases} \lfloor \rho^* L/l \rfloor & \text{if } R(\rho^* L/l \rfloor) \geq R(\lceil \rho^* L/l \rceil) \\ \lceil \rho^* L/l \rceil & \text{if } R(\lceil \rho^* L/l \rceil) > R(\lfloor \rho^* L/l \rfloor) \\ 0 & \text{otherwise} \end{cases} \tag{5.21}$$

where ρ^* is obtained by Proposition 5.3. Thus, the optimal stopping rule M^* is given by

$$M^* = \begin{cases} 0 & \text{if } (R(V^*) - \lambda)L < S^*(\lambda) \\ V^* & \text{otherwise.} \end{cases} \tag{5.22}$$

The optimal reward $U_0(\mathcal{F}_0)$ with constant EH rate model is

$$U_0(\mathcal{F}_0) = \max\{(R(V^*) - \lambda)L, S^*(\lambda)\}. \tag{5.23}$$

Next, the following proposition formally quantifies the optimal stopping rule N^* and the equation to compute the optimal throughput λ^*.

Proposition 5.5 *The optimal stopping rule to solve problem (5.5) is given by*

$$N^* = \min\{n \geq 1 : R_n(V^*) \geq \lambda^*\}, \tag{5.24}$$

with V^* given in Proposition 5.4. Moreover, λ^* satisfies the following equation:

$$\sum_{i=1}^{I} Q_i \mathbb{E}[(R^i(V^*) - \lambda^*)^+] = \frac{\lambda^* l}{L}, \tag{5.25}$$

where the function $(x)^+$ means $\max\{x, 0\}$ for some real number x and Q_i is the probability of a successful channel contention at transmitter i, defined in Section 5.1.1. The index n for $R^i(V^*)$ in (5.25) is removed since $\{R_n(V^*)\}_{n\geq 1}$ are ergodic for $1 \leq i \leq I$.

Proof: Following (5.23) in Proposition 5.4, the stopping rule N^* has the form

$$N^* = \min\{n \geq 1 : (R_n(V^*) - \lambda^*)L \geq S^*(\lambda^*)\}. \tag{5.26}$$

Thus, we can obtain N^* by plugging $S^*(\lambda^*) = 0$ into (5.26), which results in (5.24). Finally, Eq. (5.25) can be obtained by plugging $S^*(\lambda^*) = 0$ into (5.10) and taking the expectation on both sides. □

Remark 5.3 Note that the stopping rule (5.26) implies that each transmitter has the same threshold that is globally determined even when all transmitters have different statistics of the CSI and ESI. The intuition is similar to that in [7]: in order to guarantee the overall system performance, the transmitter with a bad channel condition and a low energy level should "sacrifice" its own reward, while the one with good conditions should transmit more data.

Directly following Propositions 5.4 and 5.5, the next proposition gives the DOS under the constant EH model.

Proposition 5.6 *After the nth round of CP, it is optimal for the successful transmitter to take one of the following two options:*

(1) Release the channel immediately if $R_n(V^*) < \lambda^*$ (which is equivalent to $M^* = 0$), and let all transmitters perform the next round of CP.
(2) Otherwise, transmit after V^* slots for EH, where V^* is given by Proposition 5.4.

5.1.2.3 Optimal Stopping Rule for i.i.d. EH Model

Similarly as in the previous subsection, we first consider problem (5.8) to find the optimal stopping rule M^* and then the optimal stopping rule N^* afterward.

Under the i.i.d. EH model, $U_0(\mathcal{F}_0)$ has the form in (5.8). As we mentioned in Section 5.1.2.1, it is a finite-horizon stopping problem [9, 11], and the solution of problem (5.8) could be directly generalized in the next proposition.

Proposition 5.7 *For $0 \leq k \leq L/l$ and some $\lambda > 0$, the* optimality equation *for problem (5.8) is given by*

$$U_k(\mathcal{F}_k) = \max\{(R(k) - \lambda)L, -\lambda kl + S^*(\lambda), \mathbb{E}[U_{k+1}(\mathcal{F}_{k+1}) \mid \mathcal{F}_k]\}, \tag{5.27}$$

and the optimal stopping rule has the following form:

$$M^* = \min\{0 \leq k \leq L/l : U_k(\mathcal{F}_k) = \max\{(R(k) - \lambda)L, -\lambda kl + S^*(\lambda)\}\}. \tag{5.28}$$

The stopping rule M^* given in (5.28) suggests that the EP would stop at M^* by either transmitting or giving up the channel, which also indicates the final decision for the current round of CP. Thus, the optimal stopping rule N^* could be obtained by reorganizing (5.28).

Proposition 5.8 *The optimal stopping rule of CP under the i.i.d. EH model has the form*

$$N^* = \min\{n \geq 1 : U_{M^*}(\mathcal{F}_{n,M^*}) = (R_n(M^*) - \lambda^*)L\}, \tag{5.29}$$

where M^* is the optimal stopping rule of EP given in Proposition 5.7. The optimal throughput λ^* satisfies the following equation:

$$\sum_{i=1}^{I} Q_i \mathbb{E}[\mathbb{E}[\max\{R^i(M^*) - \lambda^*, -\lambda^* M^* l/L\} \mid \mathcal{F}_0]^+] = \frac{\lambda^* l}{L}. \tag{5.30}$$

The proof is analogous to the constant EH rate case, which is omitted here.

The next proposition, which directly follows Propositions 5.7 and 5.8, concludes the overall DOS under i.i.d. EH model.

Proposition 5.9 *After the nth round of CP, it is optimal for the successful transmitter to take one of the following two options:*

(1) If $\max\{(R_n(0) - \lambda^*)L, \mathbb{E}[U_1(\mathcal{F}_{n,1}) \mid \mathcal{F}_{n,0}]\} < 0$, release the channel immediately, and let all transmitters start the next round of CP.
(2) Otherwise, start EP following the optimal stopping rule M_N^* given in Proposition 5.7.

Remark 5.4 Propositions 5.6 and 5.9 summarize the DOS under the constant and i.i.d. EH models, respectively. We observe that under the constant EH model, the EP could be "forecasted" by finding the optimal V^*; then the decision of transmission would be made before starting EP. On the contrary, when the EH rates are i.i.d., such decision can only be made step by step during the EP.

5.1.3 Battery Dynamics

In this section, we validate the assumption made in Section 5.1.2 that the energy level at each transmitter is stationary with some distribution. Firstly, we show that under the constant EH model, the energy level stored at each transmitter forms a Markov chain over time, while the state transition probabilities for different transmitters are coupled together. However, we propose an iterative algorithm to compute the corresponding steady-state distribution, which is shown converging to the global optimal point. Then, we extend our analysis to the case with i.i.d. EH rate model.

5.1.3.1 Battery with Constant EH Model

Note that after CP, if the successful transmitter releases the channel immediately, then the next round of CP starts, and the battery continues to be charged. If the transmitter starts the transmission, its energy level will become zero at the end of the transmission block according to Section 5.1.1. During this time, all other transmitters will keep harvesting energy within this period. Thus, the energy level transition over the transmission

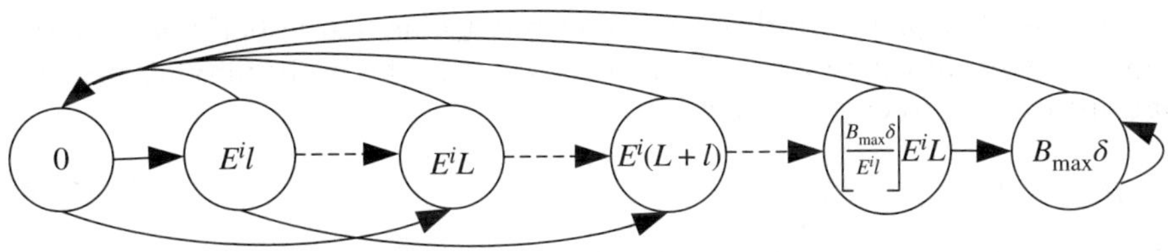

Figure 5.2 The state transition of the energy level at transmitter *i* under the constant EH rate model.

block can be determined. To simplify our analysis, the transmission block is treated as one time slot with length L for the purpose of counting battery state transitions. In addition, we assume that the battery works in half-duplex mode, i.e. it cannot be charged when the transmitter transmits data.

For transmitter i with EH rate E^i, $1 \leq i \leq I$, the set of its energy states is given by $B_t^i \in \boldsymbol{\Delta}_i = \left\{0, E^i l, 2E^i l, \dots, \left\lfloor \frac{B_{\max}\delta}{E^i l} \right\rfloor E^i l, B_{\max}\delta \right\}$, where $t \geq 1$ is the slot index. The state transition is depicted in Figure 5.2. In addition, we denote the distribution of the energy level for transmitter i at time t as $\Pi_t^i = [\pi_{t,0}^i \cdots \pi_{t,B_{\max}}^i]$.

Next, we consider the state transition probability. Supposing that transmitter i is at energy level $u_i \in \boldsymbol{\Delta}_i$, there are three events that may happen at time slot t:

(i) It occupies the channel and transmits. According to Section 5.1.1, transmitter i consumes all the energy for the transmission and transfers to the energy level 0 after the transmission. Thus, the transition probability is given by

$$p_{u_i,0}^i = Q_i p_{\text{tr}}^i(u_i), \tag{5.31}$$

where Q_i is the probability that the ith transmitter occupies the channel and $p_{\text{tr}}^i(u_i)$ is the probability that it successfully transmits with the energy level u_i. Furthermore, according to (5.24), $p_{\text{tr}}^i(u_i)$ can be computed as

$$\begin{aligned} p_{\text{tr}}^i(u_i) &= \mathbb{P}\{R^i(V^*) \geq \lambda^*\} \\ &= \mathbb{P}\left\{ \log\left(1 + |h^i|^2 \frac{u_i + V^* l E^i}{(L/l - V^*) l \sigma_i^2}\right) \geq \frac{\lambda^*}{1 - \frac{V^* l}{L}} \right\}, \end{aligned} \tag{5.32}$$

where V^* is defined by (5.21) in Proposition 5.4. Note that in (5.32), $|h^i|^2$ is the only random variable and its distribution is known.

(ii) Other transmitters occupy the channel and transmit. If anyone among the other $I-1$ transmitters sends data, transmitter i will harvest $E^i L$ units of energy during this period and then attain level $v_i = \min\{u + E^i L, B_{\max}\delta\}$. Suppose that the jth transmitter transmits. Similar to the first case, the probability of transmission performed by the jth transmitter is given by $Q_j \sum_{b=0}^{B_{\max}} \pi_{t,b}^j p_{\text{tr}}^j(bE^j l)$, where $bE^j l \in \boldsymbol{\Delta}_j$ and thus $b \in \left\{0, 1, 2, \dots, \left\lfloor \frac{B_{\max}\delta}{E^j l} \right\rfloor, B_{\max} \right\}$. Since there are in total $I-1$ transmitters, the transition probability for the transmitter i from level u_i to v_i is given by

$$p_{u_i,v_i}^i = \sum_{j \neq i} Q_j \sum_{b=0}^{B_{\max}} \pi_{t,b}^j p_{\text{tr}}^j(bE^j l). \tag{5.33}$$

(iii) No transmission happens. In this case, transmitter i just harvests $E^i l$ units of the energy and goes into state $w_i = \min\{u_i + E^i l, B_{\max}\delta\}$. The probability of this case happening can be directly obtained as

$$p^i_{u_i,w_i} = 1 - p^i_{u_i,0} - p^i_{u_i,v_i}. \tag{5.34}$$

Note that when $\tilde{u}_i = v_i = w_i$, the transition probability is just given by

$$\begin{aligned} p^i_{u_i,\tilde{u}_i} &= p^i_{u_i,v_i} + p^i_{u_i,w_i} = p^i_{u,v_i} + 1 - p^i_{u_i,0} - p^i_{u_i,v_i} \\ &= 1 - p^i_{u_i,0}. \end{aligned} \tag{5.35}$$

In this way, we can compute all $\{p^i_{u_i,\tilde{u}_i}\}$ for $1 \le i \le I$, where $u_i \in \mathbf{\Delta}_i$ and $\tilde{u}_i \in \{0, v_i, w_i, B_{\max}\delta\}$. The transition probability matrix is nothing but $\mathbf{P}^i_t = \{p^i_{u_i,\tilde{u}_i}\}$ with dimension $\left(\left\lceil \frac{B_{\max}\delta}{E^i l} \right\rceil + 1\right) \times \left(\left\lceil \frac{B_{\max}\delta}{E^i l} \right\rceil + 1\right)$. Obviously, $\mathbf{P}^i_t$ is a stochastic matrix, i.e. a square matrix in which all elements are nonnegative and the row sum is 1. However, $\mathbf{P}^i_t$ depends on t since $p^i_{u_i,v_i}$ depends on the state distribution Π^j_t for all $j \neq i$. Therefore, $\{B^i_t\}_{t\ge 0}$ is a nonhomogeneous Markov chain, whose state evolution is given by

$$\Pi^i_{t+1} = \Pi^i_t \mathbf{P}^i_t, \quad t \ge 0. \tag{5.36}$$

We propose Algorithm 5.1, to compute the steady-state distribution for all transmitters. Here, the infinity norm is applied, which is defined as $\|\mathbf{a}\|_\infty = \max_{1\le i\le n} |a_i|$ for $\mathbf{a} = [a_1 \cdots a_n]$.

Algorithm 5.1: Compute the steady-state distribution for all transmitters.

Input: $t = 0, \varepsilon$
Output: $\mathbf{P}^i_t, \Pi^i_{t+1}$

Initialize Π^i_0 for $1 \le i \le I, \varepsilon$
Compute $p^i_{u_i,0}$ by (5.31) for all $u_i \in \mathbf{\Delta}_i$ and $1 \le i \le I$
Compute $\mathbf{P}^i_0$ by (5.33)–(5.35) for all $1 \le i \le I$
Compute Π^i_1 by (5.36) for all $1 \le i \le I$
if $\max_{1\le i\le I} \| \Pi^i_{t+1} - \Pi^i_t \|_\infty > \varepsilon$ **then**
  Let $t = t + 1$
  Update $\mathbf{P}^i_t$ by (5.33)–(5.35) for all $1 \le i \le I$
  Compute Π^i_{t+1} by (5.36) for all $1 \le i \le I$
end if

Proposition 5.10 *For any given initial state distribution* Π^i_0,

$$\Pi^i_t = [\pi^i_{t,0}, \dots, \pi^i_{t,B_{\max}}] \tag{5.37}$$

that is generated by Algorithm 5.1 converges to a unique steady-state distribution Π^i *for all* $1 \le i \le I$.

Proof: Please see [13]. □

Remark 5.5 The steady-state distribution for all transmitters can be obtained by the iterative computation $\mathbf{\Pi}_{t+1} = \mathbf{\Pi}_t\mathbf{P}$ over the "super" Markov system as well, which is constructed in [13]. However, this is not as efficient as Algorithm 5.1. From the computational complexity point of view, suppose that each transmitter has m energy levels and there are n transmitters in total. The number of the states in the "super" Markov chain is m^n. If there is only one processer, the floating-point calculation for one iteration of the state distribution for the "super" Markov chain is approximately on the order of $O(2m^{2n})$. On the contrary, by using Algorithm 5.1, (5.33) requires n^2m^2 calculations, and updating $\{\mathbf{P}_t^i\}$ requires about nm calculations according to (5.34). In addition, $\{\Pi_t^i\mathbf{P}_t^i\}$ requires $2nm^2$ calculations. Overall, one iteration for all transmitters is approximately on the order of $O(n^2m^2)$, which is more efficient than the case for the "super" Markov chain especially when m and n are large. Moreover, our algorithm can also be operated in a parallel way, i.e. computing $\Pi_{t+1}^i = \Pi_t^i\mathbf{P}_t^i$ for $1 \le i \le n$ at the same time over different cores.

5.1.3.2 Battery with i.i.d. EH Model

The argument that the battery state evolves as a Markov process for the random case is analogous to that of the constant case in the previous subsection. The main difference is that the probability $p_{\text{tr}}^i(u_i)$ defined by (5.32) is changed, which needs to be further developed under the i.i.d. EH rate model.

We now consider the calculation of $p_{\text{tr}}^i(u_i)$. When transmitter i grabs the channel with energy level u_i, according to the stopping rule M^* (5.28) and N^* (5.29), the transmitter checks the condition $\max\{(R(0)-\lambda)L, -\lambda l + \mathbb{E}[U_1(\mathcal{F}_1) \mid \mathcal{F}_0]\} \ge 0$. If it is true, the transmitter starts EP until the M^*th slot and transmits when $(R(M^*)-\lambda^*)L \ge -\lambda^* M^* l$ according to (5.29). Specifically, given $U_0(u_i, |h^i|^2) \ge 0$, the transmitter continues EP at slot k for $0 \le k \le M^*-1$, which is equivalent to $\max\{(R(k)-\lambda^*)L, -\lambda^* kl\} < \mathbb{E}[U_{k+1}(\mathcal{F}_{k+1}) \mid \mathcal{F}_k]$, where $\mathcal{F}_k = \{u_i + l\sum_{j=0}^k E_j^i, |h^i|^2\}$. Then, at slot $M^* = m \le L/l$, the transmitter stops EP and transmits when $(R(m)-\lambda^*)L \ge \max\{-\lambda^* ml, \mathbb{E}[U_{m+1}(\mathcal{F}_{m+1}) \mid \mathcal{F}_m]\}$. Thus, we obtain

$$\begin{aligned} p_{\text{tr}}^i(u_i) = \int_0^\infty & \mathbb{P}\{\text{Transmits at } M^* \mid U_0(u_i, d|h^i|^2) \ge 0\}\cdot \\ & \mathbb{P}\{U_0(u_i, d|h^i|^2) \ge 0\} f(|h^i|^2) d|h^i|^2, \end{aligned} \tag{5.38}$$

where $f(|h^i|^2)$ is the PDF of the channel power gain. The probability

$$\mathbb{P}\{U_0(u_i, d|h^i|^2) \ge 0\} \tag{5.39}$$

can be computed based on Proposition 5.7. For notation simplicity, we omit the condition $U_0(u_i, d|h^i|^2) \ge 0$, and the first term in the integral of (5.38) can be expanded as

$$\mathbb{P}\{\text{Transmits at } M^*\} = \sum_{m=0}^{L/l}\left(\prod_{k=0}^{m-1}\mathbb{P}\{\alpha_k < 0\}\right)\mathbb{P}\{\beta_m \le 0\} \tag{5.40}$$

where $\alpha_k = \max\{(R(k)-\lambda^*)L, -\lambda kl\} - \mathbb{E}[U_{k+1}(\mathcal{F}_{k+1}) \mid \mathcal{F}_k]$ and $\beta_m = \max\{-\lambda ml, \mathbb{E}[U_{m+1}(\mathcal{F}_{m+1}) \mid \mathcal{F}_m]\} - (R(m)-\lambda^*)L$. Note that in $\mathbb{P}\{\alpha_k < 0\}$, $R(k)$ and $\mathbb{E}[U_{k+1}(\mathcal{F}_{k+1}) \mid \mathcal{F}_k]$ are random since they are the functions of $\sum_{j=0}^k E_j^i$, where $\{E_j^i\}_{1\le j\le k}$ are i.i.d. with a known distribution and $E_0^i = 0$. Thus, $\mathbb{P}\{\alpha_k < 0\}$ can be computed. Using the similar argument, it is easy to see that $\mathbb{P}\{\beta_m \le 0\}$ can be computed as well. Therefore, the probability given in (5.40) is computable. Overall, we could obtain $p^i_{\text{tr}}(u_i)$ after plugging (5.40) into (5.38).

After obtaining $p^i_{\text{tr}}(u_i)$, the transition probability $\{p^i_{u_i,\tilde{u}_i}\}$, where $u_i \in \boldsymbol{\Delta}$ and $\tilde{u}_i \in \{0, u_i, u_i + \delta, \ldots, B_{\max}\delta\}$, can be calculated similarly as the case of constant EH rate. In addition, Algorithm 5.1 and Proposition 5.10 could be modified, such that they could suit the i.i.d. EH model, which is omitted in this chapter.

5.1.4 Computation of the Optimal Throughput

The optimal throughput λ^* hinges upon the optimal stopping rules in (5.24) and (5.29). Thus, to fully obtain the optimal scheduling policy of the proposed DOS, we next turn our attention to computing the value of λ^*.

By Propositions 5.5 and 5.8, λ^* can be obtained by solving (5.25) or (5.30) under the constant or i.i.d. EH model, respectively. Next, we briefly introduce the idea why there exists λ^* such that Eq. (5.25) or (5.30) holds and how to search λ^*. For brevity, we focus the constant EH rate case.

Note that $R(V^*)$ is a function of random variables h^i and B^i_0; we could calculate the expectation on the left-hand side of (5.25) for each given $\lambda \geq 0$. Such expectation requires the distribution of B^i_0, i.e. the steady-state distribution Π^i, which could be approximately computed as shown in Section 5.1.3. In addition, for a given λ, an upper bound of this expectation can be obtained by fixing $\Pi^i = [0, \ldots, 0, 1]$. As λ increases from zero to infinity, this upper bound decreases to zero at some $\tilde{\lambda} < \infty$. Since the right-hand side of (5.25) is strictly increasing over λ within the range $[0, +\infty)$, there at least exists one λ^* satisfying (5.25). Therefore, an exhaustive one-dimensional search can be applied to obtain the optimal throughput over the range $[0, \tilde{\lambda}]$. Note that during each iteration of the exhaustive search, Algorithm 5.1 (given in Section 5.1.3) is used to obtain the steady-state distribution for a given $\lambda \in [0, \tilde{\lambda}]$, and then we check if Eq. (5.25) or (5.30) holds. Finally, λ^* should be the largest one in $[0, \tilde{\lambda}]$ that makes Eq. (5.25) or (5.30) hold.

In summary, the above search can characterize the optimal stopping rules given in Propositions 5.5 and 5.8, which completes the proposed DOS framework.

5.1.5 Numerical Results

In this section, we first validate Propositions 5.5 and 5.8 to show that the optimal throughput λ^* exists and can be found via one-dimensional search. Second, we investigate the throughput gain of our proposed DOS with two-level probing over the best -effort delivery method, where the data is transmitted whenever the channel contention is successful. Note that such a method can be realized in the proposed DOS framework by fixing $M = 0$ and setting $N = 1$ in (5.24) and (5.29). Let λ_0 denote the throughput obtained by the best-effort scheme, which can be calculated as

$$\lambda_0 = \frac{\sum_{i=1}^{I} \frac{Q_i}{Q} \mathbb{E}\left[L \log\left(1 + |h^i_n|^2 \frac{B^i_{n,0}}{L\sigma^2}\right)\right]}{\frac{l}{Q} + L}. \tag{5.41}$$

In general, a typical button cell battery has the capacity of 150 mAh with the end-point voltage of 0.9 V, which is equal to 150 mAh × 3600 s h^{-1} × 0.9 V = 486 J. A thin-film rechargeable battery can offer 50 μAh with 3.3 V, which is equal to 0.594 J. Since a typical transmission time interval is on the time scale of milliseconds, we let the energy unit be $\delta = 10^{-3}$ J in the simulation. Accordingly, we set the capacity of the battery $B_{\max}\delta = 10^5\delta$, which falls between the capacity volume of a thin-film battery and that of a button cell battery. Also, the current commercial solar panel can provide power from 1 to about 400 W, which is equivalent to $1\delta\cdot\text{ms}^{-1} \sim 400\delta\cdot\text{ms}^{-1}$. According to this fact, in our simulation, we let the EH rate vary within the range $[0, 40\delta]$. In addition, the channel gains are i.i.d for different links, and the channel power gains follow an exponential distribution with mean 5. The variance of the noise is set to be 10 mW. The length of one time slot is unified as $l = 1$ ms, and the length of a transmission block is $L = 100l$.

Validation of Propositions 5.5 and 5.8: In Figure 5.3, we illustrate the variation of the average throughput as the "threshold" λ changes. Without loss of generality, we first consider a homogeneous network with 10 user pairs, i.e. all pairs are identical. For the constant EH model, the EH rate is set to be $E = 10\delta$ for all transmitters. For the i.i.d. EH case, we choose the Bernoulli model [14, 15]: the EH rate is either zero or of a finite value with probability 0.5. In our simulation, we consider three cases for the mean values in i.i.d. EH model: 7.5δ, 10δ, and 20δ.

First, we observe in Figure 5.3 that as λ increases from zero, the average throughput is increasing and then decreasing. Then, the optimal point is achieved at λ^*, where the average throughput is at its apex that is also approximately of the same value as λ^*. Taking the case of i.i.d. EH model with mean 20δ as an example in Figure 5.3, the value of the optimal throughput $\lambda = \lambda^*$ is approximately 4.5, and the actual optimal average throughput is about 4.5 as well. Therefore, this observation validates our Propositions 5.5 and 5.8 and discussions in Section 5.1.4. Second, we observe that the average throughput is almost the same when the mean of the EH rate in the i.i.d. EH model is equal to the EH rate in the constant EH model. Thus, the type of EH rate models does not directly determine the average throughput performance.

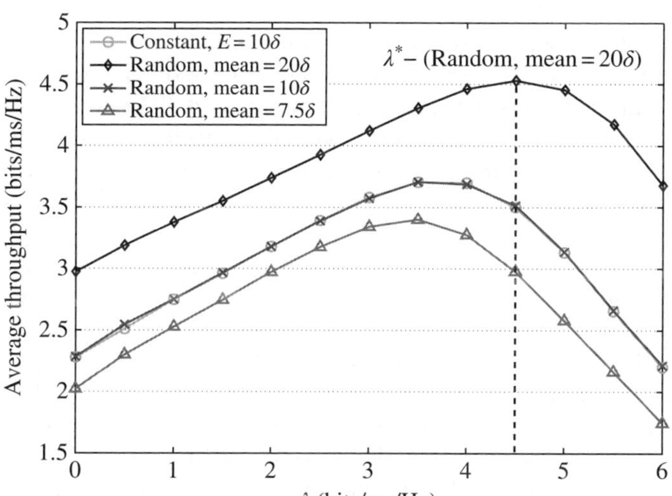

Figure 5.3 λ versus the average throughput.

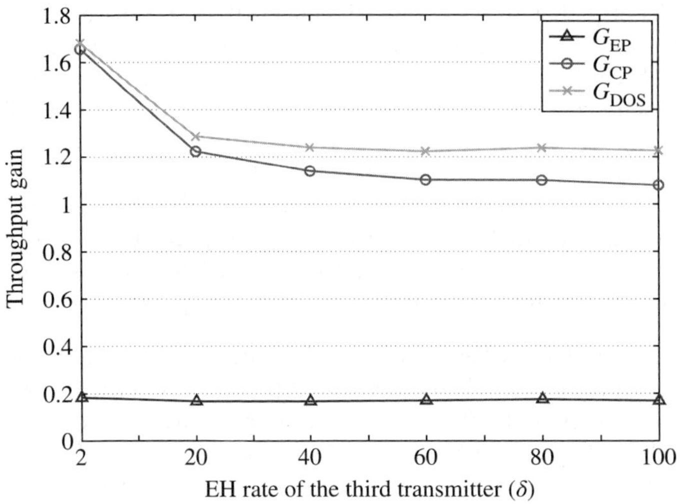

Figure 5.4 The throughput gain versus EH rate of the third transmitter.

Throughput Gain: We use λ_{EP} to denote the throughput where only EP is adopted, i.e. setting $N = 1$ and $M = M^*$, and λ_{CP} to denote the throughput where only CP is adopted, i.e. setting $N = N^*$ and $M = 0$. Thus, the throughput gains are defined as

$$\begin{cases} G_{EP} = \dfrac{\lambda_{EP} - \lambda_0}{\lambda_0} & \text{gain from EP} \\ G_{CP} = \dfrac{\lambda_{CP} - \lambda_0}{\lambda_0} & \text{gain from CP} \\ G_{DOS} = \dfrac{\lambda^* - \lambda_0}{\lambda_0} & \text{gain from CP + EP.} \end{cases} \tag{5.42}$$

In Figure 5.4, we evaluate the above throughput gains for the network with $I = 3$ user pairs. Although the number of pairs is small, such a scenario is still able to illustrate the variation of the throughput gains. Since the constant and i.i.d. EH rate models could attain the same throughput performance over λ, we only consider the constant EH model in this case. Particularly, we study a heterogeneous case where the first two transmitters have the same EH rates 2δ, while the EH rate of the third transmitter varies from 2δ to 100δ.

We observe in Figure 5.4 that as the EH rate of the third transmitter increases, G_{EP} almost keeps constant and can achieve a gain about 19%. It implies that after the channel contention, the successful transmitter with any EH rate could do EP to enhance its average transmission rate over the transmission block. Thus, the ESI of the successful transmitter does not have obvious impact on the throughput. However, we notice that G_{CP} achieves its maximum when all transmitters are identical (with the same EH rate 2δ) and then decreases slowly as the EH rate of the third transmitter increases. The intuition is that when the difference among EH rates becomes larger, the stopping rule of CP will more likely let the transmitter with relatively low energy level to give up the channel, which results in a longer time on CP, and then the throughput gain is lower

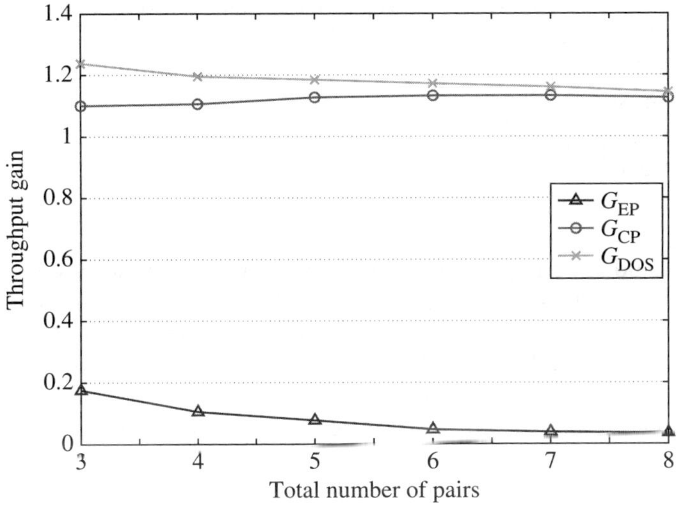

Figure 5.5 The throughput gain versus the size of the network.

than the case when all transmitters are identical. Regarding G_{DOS}, our proposed DOS with two-stage probing can achieve the highest throughput gain among three schemes. It is worth noticing that as the EH rate of the third transmitter increases, the efficiency of DOS becomes more apparent, although slowly, than the scheme with pure CP, which implies that the second stage probing brings more benefits. Our intuition is that a larger difference among the EH rates leads to a bigger difference of energy levels. Since EP allows the successful transmitter with relatively lower energy level to possibly harvest more energy after CP, EP will play a more important role as the difference among the EH rates increases.

In Figure 5.5, we illustrate how the size of the network influences the throughput gains. In this scenario, we start from a three-pair network with EH rates 2δ, 2δ, and 80δ, respectively. Then, we keep adding pairs with EH rate 2δ at the transmitter side. We observe that the throughput gain G_{CP} is increasing a little as the size of the network is increasing. It is reasonable since CP could utilize the multiuser diversity of both channel gains and energy levels. We see that G_{CP} increases slowly, since we only add a low-EH-rate transmitter at each time. We also observe that G_{EP} is decreasing. The reason is that the more transmitters in the network, the less probability to transmit for each transmitter, and then more transmitters would maintain a high energy level. Thus, EP is rarely triggered after a channel contention. For the same reason, G_{DOS} would approach G_{CP} as the size of the network increases.

5.2 Multiuser Gain Analysis

5.2.1 System Model

We assume that the system is slotted over time and the length of the time slot is unified for convenience. Regarding the variation time scale of energy sources, it is assumed that the EH rate (i.e. the harvested energy per time slot) is constant over T consecutive time

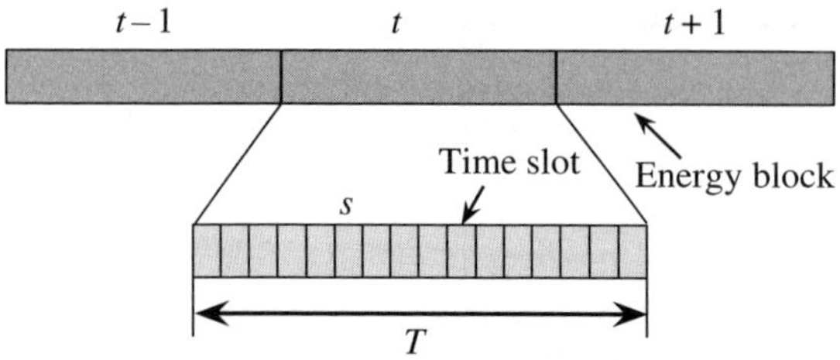

Figure 5.6 The time-slotted system with a two-time-scale structure.

slots. We call one T-slot period an "energy block." As shown in Figure 5.6, we use t as the index for the energy block and use s for the time slot within one energy block to differentiate these two time scales. Then, an arbitrary time slot can be indexed by a tuple (t, s), with $1 \leq s \leq T$. In our considered system, if two or more transmitters transmit at the same time slot, collisions occur, and no data get through.[2]

Supposing that at time slot (t, s), only the nth transmitter transmits, the received signal $y_{t,s}^{(n)}$ is given by[3]

$$y_{t,s}^{(n)} = \sqrt{P_{t,s}^{(n)}} x_{t,s}^{(n)} + z_{t,s}^{(n)}, \tag{5.43}$$

where $P_{t,s}^{(n)}$ is the transmission power, $x_{t,s}^{(n)}$ is the transmit signal of unit power, and $z_{t,s}^{(n)}$ is the CSCG noise with zero mean and unit variance. Then, the instantaneous transmission rate over one time slot could be expressed as $\log(1 + P_{t,s}^{(n)})$ under the assumption of unit noise power [17].

Let $E_t^{(n)}$ denote the EH rate (harvested energy per time slot) of the nth transmitter, which remains constant in energy block t (which contains T consecutive time slots). Furthermore, $\{E_t^{(n)}\}$ are assumed to be Bernoulli random variables[4] with probability p across different energy blocks and i.i.d. among different transmitters. Let $B_{t,s}^{(n)}$ denote the energy level of the nth user at the beginning of time slot (t, s). For notation simplicity, we let $B_t^{(n)} = B_{t,1}^{(n)}$. We also call $E_t^{(n)}$ and $B_{t,s}^{(n)}$ ESI. Here, we adopt the following setups for the energy model:

- Each transmitter has a battery with infinite battery capacity. Given that the slot-based operations are usually on the level of microseconds, an infinite battery capacity is a reasonable assumption when the battery capacity is large enough compared with the magnitude of the EH rate, e.g. an AA-sized NiMH battery with a capacity of 7.7 kJ usually requires couple of hours to be fully charged by some commercial solar panels [18].
- The transmitter is able to work in an energy-full-duplex fashion [19], i.e. it can withdraw the energy from and harvest energy to the same battery at the same time.

5.2.2 Centralized Access

In this section, we investigate the multiuser gain from the EH effect under two different centralized access schemes: fixed time division multiple access (TDMA) and energy

2 This is a typical channel model for studying random medium access protocols [16].
3 Again, to better illustrate the EH effect, we only consider Gaussian channel.
4 The analysis in this chapter can be extended to the cases of different distributions by following our analytical technique, and the scaling law has the similar form. We choose Bernoulli distribution due to its generality and notation simplicity.

greedy. Note that in this chapter, we are not focusing on optimizing the network throughput. Instead, we mainly aim to analyze the throughput scaling under specific feasible access schemes, which could serve as guidelines for future system design.

5.2.2.1 Fixed TDMA

The first centralized scheme considered is the fixed TDMA, where users transmit in turn periodically. This scheme does not require ESI from transmitters, and simply assign each transmitter the same opportunity for transmissions. The transmission of each user (within each operation period) would last for T time slots.[5] Moreover, each transmitter uses a constant transmission power and consumes all the available energy over the T time slots to fully utilize the transmission opportunity. It may be intuitively expected that this scheme may not explore any diversity with respect to energy since no ESI is utilized. In fact, as we will show next, this scheme can still achieve energy diversity gain.

It is easy to prove that under the fixed TDMA, the energy level at each transmitter is stable, i.e. $\lim_{t\to\infty}\mathbb{P}\{B_{t,s}^{(n)}=\infty\}=0$, for $1\le n\le N$ and $1\le s\le T$. Thus, when $t\to\infty$, $B_t^{(n)}$ has a stationary distribution that is the same as some random variable $B^{(n)}$. Then, the expected throughput per time slot can be written and upper-bounded by Jensen's inequality as

$$\mathbb{E}[R^{\mathrm{FT}}]=\mathbb{E}\left[\frac{1}{T}\sum_{s=1}^{T}\log\left(1+\frac{B^{(n)}}{T}\right)\right] \tag{5.44}$$

$$\le \log\left(1+\mathbb{E}\left[\frac{B^{(n)}}{T}\right]\right)$$

$$=\log(1+N\mathbb{E}[E^{(n)}]). \tag{5.45}$$

It can be observed that the upper-bound (5.45) increases as N gets large. Next, we show that the gap between $\mathbb{E}[R^{\mathrm{FT}}]$ and the upper-bound (5.45) gets close as $N\to\infty$, which implies that when N is sufficiently large, $\mathbb{E}[R^{\mathrm{FT}}]$ and the upper-bound (5.45) scale on the same order.

Proposition 5.11 *The expected throughput $\mathbb{E}[R^{\mathrm{FT}}]$ approaches its upper bound given in (5.45). That is, as $N\to\infty$, there is*

$$\mathbb{E}[R^{\mathrm{FT}}]-\log(1+N\mathbb{E}[E^{(n)}])\to 0. \tag{5.46}$$

Proof: First, it is worth pointing out that for $k\le N_{\mathrm{p}}$, the probability

$$\mathbb{P}\{B^{(n)}/T\le k\}\le \exp\left(-2\frac{(N_{\mathrm{p}}-k)^2}{N}\right) \tag{5.47}$$

gets smaller and smaller as N increases. Thus, given some $k>0$, it follows that for any $\epsilon>0$, there exists a large number a such that when $N\ge a$, $\mathbb{P}\{B^{(n)}\le k\}\le\epsilon$.

---

5 In general, the transmission could take multiple energy blocks. Here, it is without loss of generality that we restrict the transmission to last a duration of one block.

Therefore, we obtain

$$\mathbb{E}[R^{\rm FT}] = \sum_{i=1}^{k} \mathbb{P}\{B^{(n)} = i\} \log\left(1+\frac{i}{T}\right) \tag{5.48}$$

$$+ \sum_{i=k+1}^{N} \mathbb{P}\{B^{(n)} = i\} \log\left(1+\frac{i}{T}\right)$$

$$\leq \mathbb{P}\{B^{(n)} \leq k\} \log\left(1+\frac{k}{T}\right)$$

$$+ \sum_{i=k+1}^{N} \mathbb{P}\{B^{(n)} = i\} \log\left(1+\frac{i}{T}\right)$$

$$\leq \epsilon_1 + \sum_{i=k+1}^{N} \mathbb{P}\{B^{(n)} = i\} \log\left(1+g_s^{(n)}\frac{i}{T}\right), \tag{5.49}$$

where $\epsilon_1 = \epsilon \log\left(1+\frac{k}{T}\right)$.

Moreover, the concavity of the function $f(x) = \log(1+x)$ is weak on the region where x is large. In other words, for any $\epsilon_2 > 0$, there exist large numbers b and c such that $f''(x) \leq \epsilon_2$ for $x \in [b, c]$. Then, when $N \geq b$, we obtain

$$(5.49) \approx \epsilon_1 + \log\left(1 + \frac{\sum_{i=k+1}^{N} \mathbb{P}\{B^{(n)} = i\} i}{T}\right)$$

$$\leq \epsilon_1 + \log(1 + N\mathbb{E}[E^{(n)}]). \tag{5.50}$$

Therefore, when $N > \max\{a, b\}$, we have

$$\mathbb{E}[R^{\rm FT}] - \log(1 + N\mathbb{E}[E^{(n)}]) \lessapprox \epsilon_1, \tag{5.51}$$

which proves the proposition. □

We obtain from the above proposition that the upper-bound (5.45) is a reasonable estimation of $\mathbb{E}[R^{\rm FT}]$ for large N. From (5.45), we observe that the average throughput grows on a scale of $\log(pN)$. Compared with the single-user case, where the transmitter adopts a greedy power utilization strategy with transmission power E_t, we see that such scaling law mainly results from the increase of total harvested energy, which is in the nature of multiuser effect.

To further discover how energy diversity plays, we unify the EH rate in the single-user and multiuser scenarios. Then, the expected rate $\mathbb{E}[R^{\rm FT}]$ in (5.44) becomes

$$\mathbb{E}[R^{\rm FT}] = \mathbb{E}\left[\frac{1}{T}\sum_{s=1}^{T} \log\left(1 + \sum_{i=1}^{N} \frac{E_i^{(n)}}{N}\right)\right] \tag{5.52}$$

$$\leq \log(1 + \mathbb{E}[E^{(n)}]). \tag{5.53}$$

We observe that the upper-bound (5.53) is no longer scaling over N. By using the similar argument in Proposition 5.11, it can be shown that the power-normalized $\mathbb{E}[R^{\rm FT}]$ increases and approaches this fixed upper-bound (5.53) as N increases. Thus, without the power gain and without knowing ESI, the fixed TDMA can still achieve certain throughput gain as N increases, although may be limited.

In fact, the energy diversity is explored over time naturally by the fixed TDMA since users may harvest different amounts of energy over different energy blocks. For the conventional (non-EH-based) system, on the other hand, no energy diversity can be exploited due to the constant power supply. Thus, the effective transmission power is improved to $\frac{B^{(n)}}{NT} = \frac{\sum_{i=1}^{N} E_i}{N}$ according to (5.44) and (5.52).

5.2.2.2 Energy-Greedy Access

In this scheme, the central controller chooses the transmitter with the maximum available energy to transmit, and thus the ESI is explored. In addition, the transmitter uses up all the stored energy in the next T slots, and then the transmission power is $\max_{1\leq n\leq N}\left\{\frac{B_t^{(n)}}{T}\right\}$ for $t \geq 1$, where the $B_t^{(n)}$ is the energy level at the beginning of energy block t. If this transmission power is stable, the expected throughput can be written as

$$\mathbb{E}[R^{\mathrm{EG}}] = \mathbb{E}\left[\frac{1}{T}\sum_{s=1}^{T}\log\left(1+\frac{B^{(n^*)}}{T}\right)\right] \tag{5.54}$$

$$\leq \frac{1}{T}\sum_{s=1}^{T}\log\left(1+\frac{1}{T}\mathbb{E}[B^{(n^*)}]\right), \tag{5.55}$$

where

$$n^* = \arg\max_{1\leq n\leq N}\{B^{(n)}\}. \tag{5.56}$$

By applying similar argument in Proposition 5.11, it can be shown that the upper-bound (5.54) is tight for $\mathbb{E}[R^{\mathrm{EG}}]$ when N is large.

Thus, in the following, we will first investigate how $\max_{1\leq n\leq N}\{B_t^{(n)}\}$ behaves when $t \to \infty$ and then study how the upper-bound (5.54) scales over N. We use $\mu = p$ to denote the mean of EH rate and $\sigma^2 = p(1-p)$ to denote the variance. Note that our analysis in this subsection is not limited to the Bernoulli energy arrival model; it also works for any arrival model with finite mean and variance.

First, the following lemma provides a clue to find out the distribution of the battery levels.

Lemma 5.2 *Under the energy-greedy scheduling policy, the energy levels of all the transmitters are stable, i.e.* $\lim_{t\to\infty}\mathbb{P}\{B_{t,s}^{(n)} = \infty\} = 0$ *for* $1 \leq n \leq N$.

Proof: We prove this lemma by contradiction. Suppose that transmitter 1 violates the condition, i.e. $\lim_{t\to\infty}\mathbb{P}\{B_{t,s}^{(1)} = \infty\} > 0$. Note that this event would happen only if transmitter 1 keeps saving for infinite time slots from, say, the (t_1, s_1)th time slot, given that the EH rate has a finite nonnegative mean μ and variance σ^2. That is, as $t \to \infty$, we have

$$\{B_{t,s}^{(1)} = \infty\}$$

$$\Leftrightarrow \left\{(T-s_1)E_{t_1}^{(1)} + T\sum_{i=t_1+1}^{t-1}E_i^{(1)} + (s-1)E_t^{(1)} = \infty\right\}$$

$$\Leftrightarrow \left\{T\sum_{i=t_1+1}^{t-1}E_i^{(1)} = \infty\right\}.$$

Moreover, if the event $\left\{T\sum_{i=t_1+1}^{t-1} E_i^{(1)} = \infty\right\}$ happens as $t \to \infty$, according to the energy-greedy scheme, it is equivalent to the event that the energy level of transmitter 1 is never the highest among those of all transmitters after slot (t_1, s_1), i.e.

$$\{B_{t,s}^{(1)} = \infty\} \Leftrightarrow \{B_{t,s}^{(1)} < \max_{1\leq n\leq N}\{B_{t,s}^{(n)}\} = \infty\} \tag{5.57}$$

as $t \to \infty$. Hence, there must exist at least one transmitter, say, the second transmitter, such that the event $\{B_{t,s}^{(2)} > B_{t,s}^{(1)}\}$ happens. Transmitter 2 needs to start saving from some time slot (t_2, s_2) until forever in order to make the event $\{B_{t,s}^{(2)} = \infty\}$ happen with a nonzero probability as $t \to \infty$.

Similar to the case of transmitter 1, if the second transmitter also saves for infinitely many time slots, there must be

$$\left\{T\sum_{i=t_2+1}^{t-1} E_i^{(2)} \leq \max_{n\neq 1,2}\{B_{t,s}^{(n)}\} = \infty\right\} \quad \text{as} \quad t \to \infty.$$

Analogously, it directly implies that all N transmitters must keep saving energy for infinite numbers of times slots. However, this cannot happen since with the energy-greedy access, some transmitter must be chosen to fulfill a transmission for each time. Therefore, all N transmitters cannot keep saving energy forever, which contradicts the assumption at the beginning, and thus the lemma is proved. □

Note that this lemma also holds for the case when EH rates are only i.i.d. across time, but not i.i.d. across different transmitters.

Following Lemma 5.2, we have the next proposition.

Proposition 5.12 *Suppose that $\{E_t^{(n)}\}$ are i.i.d. across different transmitters and over time. Then, under the energy-greedy scheduling policy, each transmitter is equally likely to be chosen for channel access, i.e. there is*

$$\lim_{t\to\infty} \mathbb{P}\{B_t^{(i)} = \max_{1\leq n\leq N}\{B_t^{(n)}\}\} = \frac{1}{N}, \tag{5.58}$$

for any $i \in \{1, 2, \dots, N\}$.

Proof: Since the energy levels of all transmitters are stable as $t \to \infty$ by Lemma 5.2, it follows that each transmitter could be chosen to transmit with a nonzero probability. Also, given that EH rates are respectively i.i.d. across different transmitters and over time, we obtain by symmetry that

$$\lim_{t\to\infty} \mathbb{P}\{B_t^{(i)} = \max_{1\leq n\leq N}\{B_t^{(n)}\}\} = \frac{1}{N}, \tag{5.59}$$

for any $1 \leq i \leq N$. □

Remark 5.6 This proposition also implies that the stationary probability that a transmitter achieves the highest energy level among all transmitters is $1/N$. Then, the waiting time for a transmitter to fulfill a transmission satisfies a geometric distribution with parameter $1/N$.

In the following, we only keep the transmitter index n when it is necessary for the presentation; otherwise we remove it since all transmitters are identical to our interests.

Based on Proposition 5.12 and Remark 5.6, we obtain the distribution of energy levels at an arbitrary transmitter, which is given as

$$B_t/T \to^d B/T = \begin{cases} E_1 & \frac{1}{N} \\ E_1 + E_2 & \frac{1}{N}\frac{N-1}{N} \\ E_1 + E_2 + E_3 & \frac{1}{N}\left(\frac{N-1}{N}\right)^2 \\ \cdots & \cdots, \end{cases} \tag{5.60}$$

as $t \to \infty$, where the notation $\xrightarrow{d}$ denotes the convergence in distribution. In other words, we have

$$B/T = \sum_{i=1}^{S} E_i, \tag{5.61}$$

where $S \sim \text{Geo}(\frac{1}{N})$. Then, we obtain

$$\max_{1 \le n \le N} \{B_t^{(n)}\} \xrightarrow{d} \max_{1 \le n \le N} \{B^{(n)}\} \quad \text{as} \quad t \to \infty, \tag{5.62}$$

where $B^{(n)}$ is given in (5.61) with transmitter index n.

Scaling of the transmission power: First, in the next lemma, we present the strong law of large numbers (SLLN) for the random sum B.

Lemma 5.3 *Given $\mu = \mathbb{E}[E] < \infty$, the stationary energy level B satisfies*

$$\frac{B/T - \mu S}{S} \to 0, \quad a.s., \tag{5.63}$$

as $N \to \infty$, where $S \sim Geo(\frac{1}{N})$.

Proof: It is efficient to show that for $\forall \epsilon > 0$,

$$\mathbb{P}\left\{ \lim_{N\to\infty} \left| \frac{B/T - \mu S}{S} \right| > \epsilon \right\} = 0. \tag{5.64}$$

Let $X_i = E_i - \mu$. Note that SLLN holds for $X_1, X_2, \ldots, X_k$, i.e. $\sum_{i=1}^{k} X_i/k \to 0$ as $k \to \infty$ with probability 1, which implies

$$\sum_{k=1}^{\infty} \mathbb{P}\left\{ \left| \sum_{i=1}^{k} X_i \right| > k\epsilon \right\} < \infty. \tag{5.65}$$

Define

$$A_k = \left\{ \left| \sum_{i=1}^{S} X_i \right| > S\epsilon, S = k \right\}, \quad F_N = \bigcup_{k \ge N} A_k.$$

Then, we have

$$\mathbb{P}\left\{ \lim_{N\to\infty} \left| \frac{B/T - \mu S}{S} \right| > \epsilon \right\} = \mathbb{P}\left\{ \bigcap_{N=1}^{\infty} F_N \right\} = \mathbb{P}\{A_k \text{ i.o.}\},$$

where i.o. stands for "infinitely often." Next, we need to show $\mathbb{P}\{A_k \text{ i.o.}\} = 0$:

$$\sum_{k=1}^{\infty} \mathbb{P}\{A_k\} = \sum_{k=1}^{\infty} \mathbb{P}\left\{\left|\sum_{i=1}^{k} X_i\right| > k\epsilon \mid S = k\right\} \mathbb{P}\{S = k\}$$
$$\leq \sum_{k=1}^{\infty} \mathbb{P}\left\{\left|\sum_{i=1}^{k} X_i\right| > k\epsilon\right\} < \infty.$$

Therefore, $\mathbb{P}\{A_k \text{ i.o.}\} = 0$, which implies that the convergence in (5.63) holds by the Borel–Cantelli lemma [20]. □

Lemma 5.3 also implies $\mathbb{E}[B] = \mu TN$. Next, we present the central limit theorem for the random sum B.

Proposition 5.13 *Given $\{E_t\}$ are i.i.d. and $\mu = \mathbb{E}[E_t] < \infty$, B satisfies*

$$\frac{B/T - \mu S}{\sigma\sqrt{S}} \xrightarrow{d} X, \tag{5.66}$$

as $N \to \infty$, where $X \sim \mathcal{N}(0, 1)$.

Proof: Let

$$\frac{B/T - \mu S}{\sigma\sqrt{S}} = \sum_{i=1}^{S} \frac{E_i - \mu}{\sigma\sqrt{S}} = \sum_{i=1}^{S} \frac{Y_i}{\sqrt{S}}.$$

Then, we calculate its characteristic function as

$$\mathbb{E}\left[\exp\left(t\sum_{i=1}^{S-1} \frac{Y_i}{\sqrt{S}}\right)\right] = \mathbb{E}\left[\prod_{i=1}^{S} \exp\left(\frac{Y_i}{\sqrt{S}}\right)\right]$$
$$= \sum_{s=1}^{\infty} \mathbb{E}\left[\prod_{i=1}^{s} \exp\left(\frac{Y_i}{\sqrt{s}}\right)\middle| S = s\right] \mathbb{P}\{S = s\}$$
$$= \sum_{s=1}^{\infty} \left(\mathbb{E}\left[\exp\left(\frac{Y_i}{\sqrt{s}}\right)\right]\right)^s \mathbb{P}\{S = s\}$$
$$= \sum_{s=1}^{\infty} \left(1 - \frac{t^2}{2s} + o\left(\frac{t^2}{s}\right)\right)^s \mathbb{P}\{S = s\}. \tag{5.67}$$

Note that for a large s, we have the approximation $\left(1 - \frac{t^2}{2s} + o\left(\frac{t^2}{s}\right)\right)^s \approx e^{-\frac{t^2}{2}}$ when $s \geq K$. Thus, we obtain

$$(5.67) = \sum_{s=1}^{K-1} \left(1 - \frac{t^2}{2s} + o\left(\frac{t^2}{s}\right)\right)^s \mathbb{P}\{S = s\} + \sum_{s \geq K} e^{-\frac{t^2}{2}} \mathbb{P}\{S = s\}. \tag{5.68}$$

Further, by letting $N \to \infty$, we have

$$\lim_{N\to\infty} (5.68) = \lim_{N\to\infty} \sum_{s=1}^{K-1} \left(1 - \frac{t^2}{2s} + o\left(\frac{t^2}{s}\right)\right)^s \mathbb{P}\{S = s\}$$
$$+ e^{-\frac{t^2}{2}} \lim_{N\to\infty} \sum_{s\geq K} \mathbb{P}\{S = s\}$$
$$= e^{-\frac{t^2}{2}} \lim_{N\to\infty} \mathbb{P}\{S \geq K\} = e^{-\frac{t^2}{2}}.$$

Thus, we obtain that the characteristic function of $\frac{B/T-\mu S}{\sigma\sqrt{S}}$ converges to $e^{-\frac{t^2}{2}}$ as $N \to \infty$. Finally, by the Lévy's continuity theorem (Chapter 18 in [20]), we can obtain the conclusion. □

Based on Proposition 5.13, we obtain that the transmission power under the energy-greedy policy satisfies

$$\max_{1\leq n\leq N} \left\{ \frac{B^{(n)}/T - \mu S}{\sigma\sqrt{S}} \right\} \xrightarrow{d} \max_{1\leq n\leq N} \{X_n\}, \quad \text{as} \quad N \to \infty,$$

where $X_n \sim \mathcal{N}(0, 1)$. Moreover, we can further approximate the distribution of $\max_{1\leq n\leq N}\{X_n\}$ according to the next lemma [21, 22].

Lemma 5.4 *If* $X_n \sim \mathcal{N}(0, 1)$ *for* $1 \leq n \leq N$, *the distribution of* $Z_N = \max\{X_1, X_2, \ldots, X_N\}$ *satisfies*

$$\mathbb{P}\{a_N(Z_N - b_N) < x\} \to \exp(-e^{-x}) \tag{5.69}$$

as $N \to \infty$, *where* a_N *and* b_N *are normalizing factors given as*

$$a_N = \sqrt{2\ln N},$$
$$b_N = \sqrt{2\ln N} - \frac{\ln\ln N + \ln 4\pi}{2\sqrt{2\ln N}}.$$

Based on Lemma 5.4, we obtain the following proposition.

Proposition 5.14 *Under the energy-greedy access policy, the transmission power satisfies*

$$a_N\left(\max_{1\leq n\leq N} \left\{ \frac{B^{(n)}/T - \mu S}{\sigma\sqrt{S}} \right\} - b_N\right) \xrightarrow{d} Y, \quad \text{as} \quad N \to \infty,$$

where a_N *and* b_N *are given in Lemma 5.4 and the CDF of* Y *is* $\exp(-e^{-x})$ *for* $x \in (-\infty, +\infty)$.

Proof: It can be directly proved by using Proposition 5.13 and Lemma 5.4. □

Scaling of the expected throughput: We are now ready to investigate how the upper-bound (5.54) of $\mathbb{E}[R^{\mathrm{EG}}]$ behaves. The next lemma [23] is used to bound the mean of the transmission power $\max_{1\leq n\leq N}\{B^{(n)}\}$.

Lemma 5.5 *If $X_n \sim \mathcal{N}(0,1)$ for $1 \le n \le N$, the mean of $Z_N = \max\{X_1, X_2, \dots, X_N\}$ satisfies*

$$\mathbb{E}[Z_N] \le \sqrt{2\ln N} + o(1)$$

for large N, where $o(1)$ denotes a function such that $\lim_{N\to\infty} o(1) < \epsilon$ for any $\epsilon > 0$.

By Lemma 5.5, when N is large, we have

$$\mathbb{E}\left[\max_{1\le n\le N}\left\{\frac{B^{(n)}/T - \mu S}{\sigma\sqrt{S}}\right\}\right] = \mathbb{E}\left[\max_{1\le n\le N}\{X_n\}\right] \le \sqrt{2\ln N} + 1.$$

Therefore, we obtain an approximated upper bound, i.e. for large N,

$$\mathbb{E}\left[\max_{1\le n\le N}\{B^{(n)}\}\right] \lessapprox \mu NT + \sigma T\mathbb{E}\left[\sqrt{S}\right]\left(\sqrt{2\ln N} + 1\right) = \mu NT + o(TN). \tag{5.70}$$

Note that this approximation is more accurate if the variable S is deterministic such that $S = N$. Furthermore, we could bound $\mathbb{E}[\max_{1\le n\le N}\{B^{(n)}\}]$ from below by

$$\mathbb{E}\left[\max_{1\le n\le N}\{B^{(n)}\}\right] \ge \mathbb{E}[B] = \mu TN.$$

Finally, it follows that

$$\mathbb{E}[R^{\mathrm{EG}}] = O\left(\log(\mu N)\right), \tag{5.71}$$

where $O(\log(\mu N))$ denotes a function such that $\lim_{N\to\infty} \frac{O(\log(\mu N))}{\log(\mu N)} < \infty$.

Comparing with the fixed TDMA, we observe that the expected throughput achieved by the energy-greedy scheduling grows on the same order as the fixed TDMA. It tells us that the utilization of ESI does not contribute too much on the throughput performance. Besides the power gain, the energy-greedy scheme explores the energy diversity over both time and space, and thus the effective transmission power is improved from E_t to $\frac{1}{TN}\max_{1\le n\le N}\{B^{(n)}\}$.

5.2.3 Distributed Access

Suppose that the nth user contends for the channel use with probability q_n at the very beginning of each time slot; then the successful contention probability of the nth user is

$$Q_n = q_n \prod_{j\ne n}(1 - q_j). \tag{5.72}$$

Here, we assume that channel contention consumes negligible energy as we focus on investigating the order-wise throughput performance. If the nth transmitter successfully occupies the channel, it transmits for T time slots and uses up all its available energy. Note that similar to TDMA, this access and power control scheme does not use any ESI.[6]

6 Exploring ESI can improve the throughput performance, e.g. like the distributed scheme in [13]. Again, in this chapter, we only focus on the scaling law of throughput instead of optimizing it. Thus, we only consider some basic access schemes.

It is easy to see that the energy level at each transmitter is stable under this contention-based access scheme. Then, the average throughput across the whole system is given by

$$\begin{aligned}\mathbb{E}[R] &= \sum_{n=1}^{N} q_n \prod_{j\neq n}(1-q_j)\mathbb{E}\left[\frac{1}{T}\sum_{s=1}^{T}\log\left(1+\frac{B^{(n)}}{T}\right)\right] \\ &\leq \sum_{n=1}^{N} Q_n \log\left(1+\frac{1}{T}\mathbb{E}[B^{(n)}]\right). \end{aligned} \tag{5.73}$$

Again, it is worth noticing that when $\mathbb{E}[B^{(n)}]$ is large, the upper bound is tight.

As finding the distribution of the energy level is challenging, we turn to estimate its mean value. Note that for the nth user, the successful transmission probability is given by Q_n, and it follows that the waiting time is a geometric random variable with parameter Q_n. If we let $W^{(n)}$ denote the number of waiting time slots, it can be written as

$$W^{(n)} = aT + b, \tag{5.74}$$

for some $a, b \geq 0$, and thus the energy level is given by

$$B^{(n)} = \sum_{t=1}^{a} TE_t^{(n)} + bE_{a+1}^{(n)}. \tag{5.75}$$

When user n does not transmit, there are two possible situations for the system: (i) none of the users transmit, or (ii) one user other than user n transmits. Mathematically, there is a probability $\frac{1-\sum_{i=1}^{N} Q_i}{1-Q_n}$ that no user transmits and a probability $\frac{\sum_{i\neq n}^{N} Q_i}{1-Q_n}$ that another user transmits. Therefore, the average waiting time is

$$\mathbb{E}[W^{(n)}] = \frac{1}{Q_n}\left(\frac{1-\sum_{i=1}^{N} Q_i}{1-Q_n} + T\frac{\sum_{i\neq n}^{N} Q_i}{1-Q_n}\right),$$

and the expected energy level is

$$\begin{aligned}\mathbb{E}[B^{(n)}] &= \mathbb{E}[W^{(n)}]\mathbb{E}[E^{(n)}] \\ &= \frac{p}{Q_n}\left(\frac{1-\sum_{i=1}^{N} Q_i}{1-Q_n} + T\frac{\sum_{i\neq n}^{N} Q_i}{1-Q_n}\right).\end{aligned}$$

For simplification, suppose that all transmitters are identical. Then, the expected energy level is given by

$$\begin{aligned}\mathbb{E}[B] &= \frac{p}{Q}\left(\frac{1-NQ}{1-Q} + T\frac{(N-1)Q}{1-Q}\right) \\ &= p\frac{T-1}{1-Q}N + p\frac{1-TQ}{Q(1-Q)}.\end{aligned}$$

It can be observed that the energy level satisfies

$$\mathbb{E}[B] = O\left(p\frac{T-1}{1-Q}N\right).$$

Next, we consider the expected throughput. It can be obtained from (5.73) that

$$\mathbb{E}[R] \leq NQ\log\left(1+\frac{1}{T}\mathbb{E}[B]\right).$$

When $Q = O(N^{-\alpha})$, where $\alpha > 0$, there is $\mathbb{E}[B] = O(pN^{\alpha} + p(T-1)N)$. Thus, we obtain

$$\mathbb{E}[R] \approx O\left(\frac{\log\left(1+\frac{p}{T}N^{\alpha}+pN\right)}{N^{\alpha-1}}\right). \tag{5.76}$$

If α is too large, the expected rate will surprisingly decrease. To see this point, we consider the limit of (5.76). When $\alpha > 1$, we have

$$\begin{aligned}
&\lim_{N\to\infty}\frac{\log\left(1+\frac{p}{T}N^{\alpha}+pN\right)}{N^{\alpha-1}}\\
&=\lim_{N\to\infty}\frac{\frac{p\alpha}{T}N^{\alpha-1}+p}{(\alpha-1)N^{\alpha-2}\left(1+\frac{p}{T}N^{\alpha}+pN\right)}\\
&=\lim_{N\to\infty}\frac{\frac{p\alpha}{T}N^{\alpha-1}+p}{pN^{\alpha-1}+N^{\alpha-2}+\frac{p}{T}N^{2\alpha-2}}\cdot\frac{1}{\alpha-1}.
\end{aligned} \tag{5.77}$$

It can be found that (5.77) $= 0$ when $\alpha > 1$. Thus, the parameter α should be chosen from $(0, 1]$. At last, we consider a specific random access strategy and study how the throughput scales.

Example 5.1 *(ALOHA [uniform contention]) When transmitters contend with probability $q_n = 1/N^{\alpha}$, for $0 < \alpha \leq 1$, we obtain $Q_n = \frac{1}{N^{\alpha}}\left(1-\frac{1}{N^{\alpha}}\right)^{N-1}$. Note that there is*

$$\begin{aligned}
\lim_{N\to\infty}\left(1-\frac{1}{N^{\alpha}}\right)^{N-1} &= \lim_{N\to\infty} e^{(N-1)\log\left(1-\frac{1}{N^{\alpha}}\right)}\\
&\approx \lim_{N\to\infty} e^{-\frac{N-1}{N^{\alpha}}}\\
&= \lim_{N\to\infty} e^{-N^{1-\alpha}}\\
&= \begin{cases} 0 & 0<\alpha<1\\ e^{-1} & \alpha=1. \end{cases}
\end{aligned}$$

Thus, the only choice is $\alpha = 1$, and we arrive at

$$\begin{aligned}
\mathbb{E}[B] &= \frac{pN^{\alpha}}{\left(1-\frac{1}{N^{\alpha}}\right)^{N-1}}\left(\frac{1-\frac{1}{N^{\alpha-1}}\left(1-\frac{1}{N^{\alpha}}\right)^{N-1}}{1-\frac{1}{N^{\alpha}}\left(1-\frac{1}{N^{\alpha}}\right)^{N-1}}+T\frac{\frac{N-1}{N^{\alpha}}\left(1-\frac{1}{N^{\alpha}}\right)^{N-1}}{1-\frac{1}{N^{\alpha}}\left(1-\frac{1}{N^{\alpha}}\right)^{N-1}}\right)\\
&= O\left(peN\left(1+\frac{T-1}{e}\right)\right),
\end{aligned}$$

which implies that the average throughput is

$$\mathbb{E}[R] \approx O\left(e^{-1} \log\left(\frac{pe}{T}\left(1 + \frac{T-1}{e}\right)N\right)\right). \tag{5.78}$$

In this case, the average throughput scales on the same order as that in the fixed TDMA scheme, while there is a loss factor e^{-1} *due to the nature of random access.*

5.2.4 Numerical Analysis and Discussions

In this section, we provide more insights on the average throughput based on some numerical analysis and discuss where the multiuser gain in EH-based communication system comes from. Here, the benchmark is the throughput achieved by a point-to-point EH communication system over a Gaussian channel, where the transmitter adopts a greedy power utilization strategy.[7] The multiuser effect is expected to provide certain throughput gain against this benchmark.

The average throughputs under different schemes are numerically compared in Figure 5.7. We observe that all the access schemes discussed in the previous two sections can achieve certain throughput gain. The energy-greedy one slightly outperforms TDMA while their performance gets close when N is large, which agrees with our discussion in Section 5.2.2. For ALOHA, it has a throughput loss against centralized schemes, which results from the time cost on channel contentions. This observation is similar to the case with conventional multiuser diversity in fading channels, where the ALOHA has a throughput loss $\frac{1}{e}$ compared with the centralized protocol [16].

To highlight the effect of energy diversity in the total multiuser gain, we also show the throughput performance in Figure 5.8, where the total energy arrival rate is unified

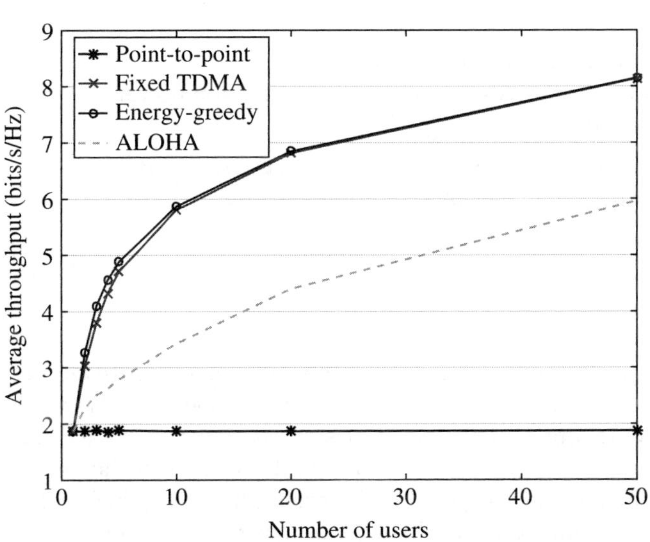

Figure 5.7 The average throughput with different access schemes.

7 Note that the greedy power utilization strategy is not a capacity-achieving strategy in general; we use it here to unify the power allocation for single- and multiple-user scenarios. The capacity-achieving power allocation strategy is discussed in [19, 24].

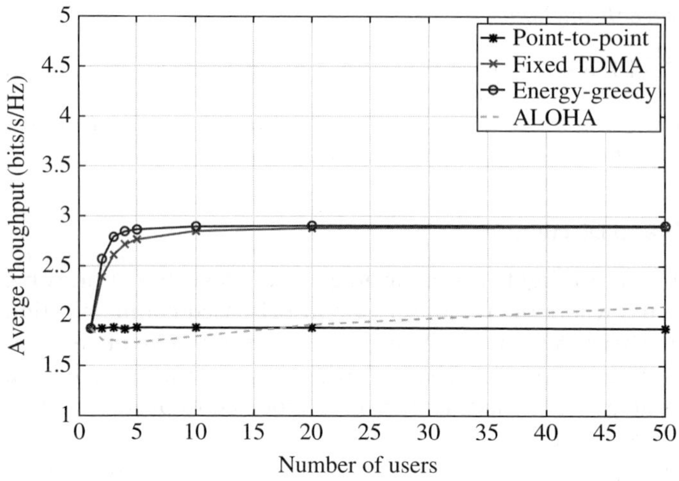

Figure 5.8 The average throughput with different access schemes at unified total EH rate.

such that the power gain over space is eliminated. We observe that compared with the single-user benchmark, the fixed TDMA, the energy-greedy, and the ALOHA schemes are able to achieve certain multiuser gain. However, such diversity gain does not always scale with the number of users when N gets large, which agrees with our discussions in Section 5.2.2.

In short, our theoretical and numerical results imply that compared with the point-to-point EH communication system, the multiuser gain due to EH effects mainly comes from two aspects:

(1) The power gain from the increase of total energy arrivals.
(2) The energy diversity gain that improves the effective transmission power:
   - For fixed TDMA, improved from $E_t^{(1)}$ to $\frac{1}{NT}B_t^{(1)}$.
   - For energy-greedy scheduling, improved from $E_t^{(1)}$ to $\max_{1\le n\le N}\frac{1}{NT}B_t^{(n)}$.
   - For ALOHA, improved from $E_t^{(1)}$ to $\frac{1}{W}B_t^{(n)}$.

5.3 Summary

In this chapter, we discuss a DOS framework for a heterogeneous single-hop *ad hoc* network, in which each transmitter is powered by a renewable energy source and accesses the channel randomly. Our DOS framework includes two successive processes: all transmitters first probe the channel via random access, and then the successful transmitter decides whether to give up the channel or to optimally probe the energy before data transmission. The optimal scheduling policy of the DOS framework is obtained as follows: First, assuming that the battery state is stationary at each transmitter, the expected throughput maximization problem is formulated as a rate-of-return optimal stopping problem, which is solved for both the constant and i.i.d. EH rate models. Second, by fixing the stopping rule, the stored energy level at each transmitter is shown to own a steady-state distribution as time goes to infinity, where we also propose an efficient iterative algorithm for its computation. Finally, the optimal throughput and the scheduling policy are obtained via one-dimensional search with the above two steps (i.e. finding the form of the optimal stopping rule and calculating the steady-state distribution) repeated in each iteration. Numerical results are also provided to validate our analysis; the proposed DOS with two-level probing is shown to outperform the best-effort delivery method.

Next, the multiuser gain is investigated for EH-based communication systems. The centralized access schemes are shown to have a scaling law of $\log(\mu N)$ for the average throughput, and the distributed one is shown to have the same scaling law with a constant loss factor e^{-1}. Besides the increase of total energy arrivals, the multiuser gain also benefits from the energy diversity over time and space, which improves the effective transmission power. The access schemes discussed in this chapter capture the "energy-accumulating" feature in EH communications, and the corresponding throughput scaling sets the guidelines for the future transmission design of EH wireless systems.

References

1 Ho, C.K. and Zhang, R. (2012). Optimal energy allocation for wireless communications with energy harvesting constraints. *IEEE Transactions on Signal Processing* 60 (9): 4808–4818. https://doi.org/10.1109/TSP.2012.2199984.

2 Huang, C., Zhang, R., and Cui, S. (2013). Throughput maximization for the Gaussian relay channel with energy harvesting constraints. *IEEE Journal on Selected Areas in communications* 31 (8): 1469–1479. https://doi.org/10.1109/JSAC.2013.130811.

3 Ozel, O., Tutuncuoglu, K., Yang, J. et al. (2011). Transmission with energy harvesting nodes in fading wireless channels: optimal policies. *IEEE Journal on Selected Areas in communications* 29 (8): 1732–1743. https://doi.org/10.1109/JSAC.2011.110921.

4 Iannello, F., Simeone, O., and Spagnolini, U. (2012). Medium access control protocols for wireless sensor networks with energy harvesting. *IEEE Transactions on Communications* 60 (5): 1381–1389. https://doi.org/10.1109/TCOMM.2012.030712.110089.

5 Abramson, N. (1977). The throughput of packet broadcasting channels. *IEEE Transactions on Communications* 25 (1): 117–128. https://doi.org/10.1109/TCOM.1977.1093713.

6 Zheng, D., Ge, W., and Zhang, J. (2009). Distributed opportunistic scheduling for Ad Hoc networks with random access: an optimal stopping approach. *IEEE Transactions on Information Theory* 55 (1): 205–222. https://doi.org/10.1109/TIT.2008.2008137.

7 Zheng, D. (2007). Physical-layer aware control and optimization in stochastic wireless networks. Ph.D. thesis. Arizona State University.

8 Thejaswi, P.S., Zhang, J., Pun, M. et al. (2010). Distributed opportunistic scheduling with two-level probing. *IEEE/ACM Transactions on Networking* 18 (5): 1464–1477. https://doi.org/10.1109/TNET.2010.2042610.

9 Ferguson, T.S. (2006). Optimal stopping and applications. www.math.ucla.edu/tom/Stopping/Contents.html (accessed 02 August 2018).

10 Ferguson, T.S. and MacQueen, J.B. (1992). Some time-invariant stopping rule problems. *Optimization* 23 (2): 155–169. https://doi.org/10.1080/02331939208843754.

11 Peskir, G. and Shiryaev, A. (2006). *Optimal Stopping and Free-Boundary Problems.* Boston, MA: Springer.

12 Boyd, S. and Vandenberghe, L. (2004). *Convex Optimization.* Cambridge: Cambridge University Press.

13 Li, H., Huang, C., Zhang, P. et al. (2016). Distributed opportunistic scheduling for energy harvesting based wireless networks: a two-stage probing approach. *IEEE/ACM Transactions on Networking* 24 (3): 1618–1631. https://doi.org/10.1109/TNET.2015.2421320.

14 Seyedi, A. and Sikdar, B. (2010). Energy efficient transmission strategies for body sensor networks with energy harvesting. *IEEE Transactions on Communications* 58 (7): 2116–2126. https://doi.org/10.1109/TCOMM.2010.07.080544.

15 Kashef, M. and Ephremides, A. (2012). Optimal packet scheduling for energy harvesting sources on time varying wireless channels. *Journal of Communications and Networks* 14 (2): 121–129. https://doi.org/10.1109/JCN.2012.6253060.

16 Qin, X. and Berry, R.A. (2006). Distributed approaches for exploiting multiuser diversity in wireless networks. *IEEE Transactions on Information Theory* 52 (2): 392–413. https://doi.org/10.1109/TIT.2005.862103.

17 Tse, D. and Viswanath, P. (2005). *Fundamentals of Wireless Communication.* Cambridge: Cambridge University Press.

18 Sharma, V., Mukherji, U., Joseph, V., and Gupta, S. (2010). Optimal energy management policies for energy harvesting sensor nodes. *IEEE Transactions on Wireless Communications* 9 (4): 1326–1336. https://doi.org/10.1109/TWC.2010.04.080749.

19 Ulukus, S., Yener, A., Erkip, E. et al. (2015). Energy harvesting wireless communications: a review of recent advances. *IEEE Journal on Selected Areas in communications* 33 (3): 360–381. https://doi.org/10.1109/JSAC.2015.2391531.

20 Williams, D. (1991). *Probability with Martingales.* Cambridge: Cambridge University Press.

21 Bovier, A. (2005). *Extreme Values of Random Processes.* Bonn: Lecture Notes Technische Universität Berlin.

22 Embrechts, P., Klüppelberg, C., and Mikosch, T. (2013). *Modelling Extremal Events: For Insurance and Finance.* Boston, MA: Springer Science & Business Media.

23 Romberg, J. (2012). Maximum of a sequence of Gaussian random variables. OpenStax CNX, February 27. http://cnx.org/contents/8bd316d8-6442-4f5a-a597-aef1d6202f87@1 (accessed 02 August 2018).

24 Rajesh, R., Sharma, V., and Viswanath, P. (2014). Capacity of Gaussian channels with energy harvesting and processing cost. *IEEE Transactions on Information Theory* 60 (5): 2563–2575. https://doi.org/10.1109/TIT.2014.2311822.

6

Cost-Aware Design for Energy Harvesting Powered Cellular Networks

6.1 Introduction

To meet the dramatic wireless data traffic growth driven by the popularity of new mobile devices and mobile applications, the fifth-generation (5G) cellular technology has recently attracted a lot of research interests from both academia and industry. As compared with its fourth-generation (4G) counterpart, 5G is expected to achieve roughly 1000 times data rate increases via dense base station (BS) deployments and advanced physical layer communication techniques [1]. However, the large number of BSs will lead to large energy consumption and high electricity bills for cellular operators, which amount to a large portion of their operational expenditure (OPEX). For example, China Mobile owns around 920 000 BSs in 2011, and the total energy cost per year is almost US$ 3 billion, given that the annual cost for each BS is about US$ 3000 [2]. Therefore, in the 5G era, it is emerging for these cellular operators to reduce their energy costs by employing new cost-saving solutions in the design of cellular BSs.

In general, the cost-saving solutions can be categorized into two classes that manage the energy supply and the communication demand of cellular BSs, respectively [2–5]. On the supply side, one commonly adopted solution is to use energy harvesting devices (e.g. solar panels and wind turbines) at cellular BSs, which can harvest cheap and clean renewable energy to reduce or even substitute the energy purchased from the grid [5]. However, since the renewable energy is often randomly distributed in both time and space and cellular BSs are very energy hungry, it is very difficult to solely use different BSs' individually harvested energy to power their operations. As a result, the power grid is still needed to provide reliable energy to BSs. Besides serving as a reliable energy supply, power grid also provides new opportunities for the BSs' cost saving with its ongoing paradigm shift from traditional grid to smart grid. Unlike traditional grid, which uses one-way energy flow to deliver power from central generators to electricity users, smart grid deploys smart meters at end users to enable both two-way information and energy flows between the grid and end users [6, 7]. The two-way energy flow in smart grid motivates another form of *energy cooperation* in cellular networks, which is different from that in Chapter 3, as we will elaborate later in this chapter, and allows the BSs to trade and share their unevenly harvested renewable energy through the smart grid to support the nonuniform wireless traffic in a cost-effective way.

On the demand side, various techniques have been proposed in cellular networks across different layers of communication protocols for reducing the energy

Energy Harvesting Wireless Communications, First Edition.
Chuan Huang, Sheng Zhou, Jie Xu, Zhisheng Niu, Rui Zhang and Shuguang Cui.

consumption [2]. Among them, communication cooperation (e.g. traffic offloading [8], spectrum sharing [9], and coordinated multipoint (CoMP) [10]) is particularly appealing, which allows the BSs to share the wireless resources and shift the traffic loads with each other for energy saving. However, the introduction of renewable energy at BSs imposes new challenges on the existing communication cooperation design: the conventional energy-saving design may not be cost effective any longer. This is due to the fact that the renewable energy (though unreliable in supplying) is in general much cheaper than the energy purchased from the grid, and therefore BSs should maximally use it to save cost, whereas under the energy-saving design the harvested renewable energy at BSs may not be efficiently utilized when serving the time- and space-varying wireless traffic. To overcome this problem, it is desirable to design new *cost-aware communication cooperation* approaches by taking into account the cost differences between renewable and conventional energy.

In this chapter, we first overview the recent advances in energy cooperation and cost-aware communication cooperation. Figure 6.1 illustrates the general energy and communication cooperation concept for cellular networks at both the energy supply layer and the communication demand layer, respectively. Then we present the new

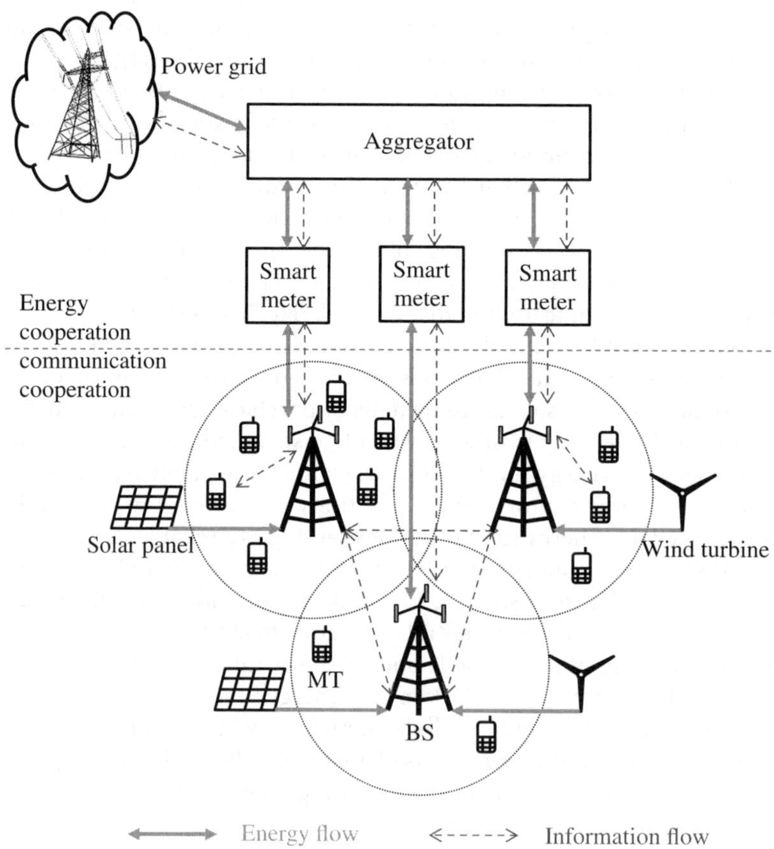

Figure 6.1 A general model of cellular networks with energy and communication cooperation among BSs. Source: Xu et al. 2015 [11]. Reproduced with permission of IEEE.

joint energy and communication cooperation to exploit both benefits. Specifically, we present the following three approaches [11]:

- *Approach I – Energy cooperation on the supply side*: Cellular systems or BSs use the two-way energy flow in smart grid to trade or share renewable energy by taking the energy demands for communications as given.
- *Approach II – Communication cooperation on the demand side*: Cellular systems or BSs perform cost-aware communication cooperation to share wireless resources and reshape wireless load over space and time by taking the energy supply (renewable and/or conventional) as given.
- *Approach III – Joint energy and communication cooperation on both sides*: Cellular systems or BSs jointly cooperate on both the supply and demand sides to maximally reduce their total energy cost.

In the rest of this chapter, we first introduce the energy supply and demand models of cellular systems. Then we present the essential ideas of energy, communication, and joint cooperation approaches, respectively. Next, we introduce two joint cooperation schemes with CoMP in more detail. Finally, we point out several future research directions and conclude this chapter.

6.2 Energy Supply and Demand of Cellular Systems

In this section, we introduce the energy supply and demand models for cellular systems. The models will be used to motivate various cooperation schemes introduced later. For notational convenience, we focus on one particular time slot and thus skip the time index of variables (such as the energy demand and supply at BSs as well as the energy prices), as will be shown later. Note that these variables can vary over time in practice. In addition, we normalize the length of each time slot into unity and thus use the terms energy and power interchangeably throughout this chapter.

We consider a single cellular system with $N > 1$ BSs, in which each BS i is connected to the smart grid and also deployed with a renewable energy harvesting device that has a harvesting rate of $E_i \geq 0, i = 1, \dots, N$. The value of E_i at a given time instance depends on the type of the renewable energy source (e.g. solar or wind), the harvesting capacity of the device (e.g. size of the solar panel), and the weather condition at that location. As shown in the upper subfigure of Figure 6.2, the E_i's are generally different among BSs at different locations.

On the demand side, the power consumption of each cellular BS i, denoted by $Q_i \geq 0$, is composed of two main parts: the dynamic power consumption related to the transmission and reception of wireless signals for serving the mobile terminals (MTs) and the constant power consumption (e.g. at the circuits and air conditioners) for maintaining other operations. In reality, the value of Q_i varies according to the traffic load over the service coverage area of BS i. Due to MTs' mobility across cells and their time-varying service requests, the traffic loads (and thus the Q_i's) are different among BSs and change over time, as shown in the middle subfigure of Figure 6.2.

By combining the supply and demand sides, we denote the *net load* at BS i as $\delta_i = Q_i - E_i$, where $\delta_i > 0$ shows the deficit status of renewable energy and $\delta_i < 0$ indicates the energy surplus status. Since Q_i's and E_i's are usually independent (see Figure 6.2),

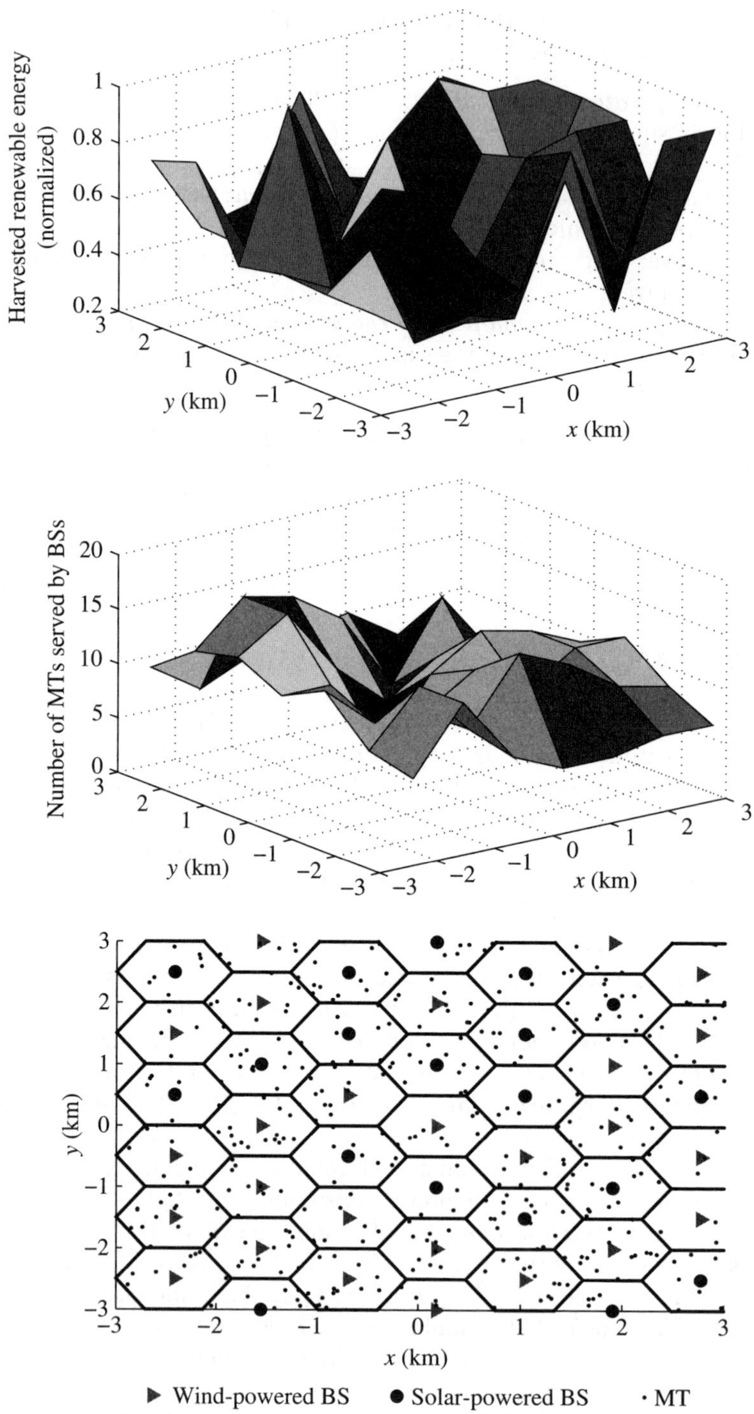

Figure 6.2 An example of a cellular system with BSs having spatially distributed traffic load and harvested energy at a given time instance. It is assumed that the (normalized) energy harvesting capacity of all solar-powered BSs is 1 and that of all wind-powered BSs is 0.5. Source: Xu et al. 2015 [11]. Reproduced with permission of IEEE.

it is likely that some BSs are short of renewable energy to match demand (i.e. $\delta_i > 0$), while the other BSs are adequate in renewable energy (i.e. $\delta_i < 0$). Such a geographical diversity in net load requires some BSs to purchase energy from the grid (e.g. a δ_i amount of energy purchase for BS i with $\delta_i > 0$)[1] while the other BSs to waste the extra renewable energy (i.e. a $|\delta_j|$ amount of energy waste for BS j with $\delta_j < 0$).[2] Overall, the total purchased energy amount from the grid by all the N BSs is the total renewable energy deficit, denoted by $\Delta_+ := \sum_{i=1}^{N} [\delta_i]^+ \geq 0$ with $[x]^+ := \max(x, 0)$, while the total wasted renewable energy amount by them is the total renewable energy surplus, given by $\Delta_- := -\sum_{j=1}^{N} [\delta_j]^- \geq 0$ with $[x]^- := \min(x, 0)$. By denoting the price for BSs to purchase energy from the grid as $\pi > 0$, then the total energy cost of the cellular system is

$$C_1 = \pi\Delta_+, \tag{6.1}$$

which is independent of Δ_-. This fact motivates us to use the wasted renewable energy surplus (Δ_-) to compensate the deficit (Δ_+) for cost saving. To this end, we will implement the energy and communication cooperation on the supply and demand sides, respectively, to reschedule and balance E_i's and Q_i's.

6.3 Energy Cooperation

Energy cooperation is a cost-saving approach on the supply side, in which the cellular BSs are allowed to employ a two-way energy trading or sharing for better utilization of their otherwise wasted renewable energy surplus (Δ_-). Although the idea of energy cooperation has been mentioned in smart grid for microgrids' energy trading [7], it is new to cellular networks. Particularly, since it is too complex for the grid to directly control a large number of BSs, the energy trading and sharing in cellular networks should be enabled by using aggregators [12] (see the upper energy cooperation layer in Figure 6.1). With aggregators, we can cluster BSs into a finite number of groups, and an aggregator can serve as an intermediary party to control each group of BSs for the grid, thus helping realize the two-way energy flow between the grid and BS groups. The implementation of energy cooperation is not difficult in smart grid: it only requires the two-way energy flow and aggregators and does not change the existing infrastructure of cellular networks.

6.3.1 Aggregator-Assisted Energy Trading

Aggregator-assisted energy trading is an energy cooperation scheme in which the aggregator performs a two-way energy trading with the BSs by deciding buying and selling prices. In this scheme, the BSs adequate in renewable energy can sell their extra energy to the aggregator, from which the selling revenue can be gained to compensate the total energy cost; at the same time, the other BSs short of renewable energy can obtain such cheap energy from the aggregator at a lower price than the regular price π to purchase from the grid directly. As the coordinator in this trading market, the aggregator can also

1 By purchasing energy from the grid, all the BSs can maintain their routine operation all the time.

2 Another possible solution to reduce such renewable energy waste is to equip storage devices at BSs to store extra energy for future use. The energy storage and energy cooperation techniques can compensate with each other to further improve the energy efficiency at cellular BSs.

obtain some revenue by properly deciding the energy selling and buying prices. Here, the energy selling and buying at each BS is managed with the help of the smart meter in real time, which can decide the sold/purchased energy amount at any time slot *a priori* based on the energy harvesting rates, the power demand, and the energy prices.

Let $\pi_{\text{buy}} > 0$ and $\pi_{\text{sell}} > 0$ denote the unit prices for each BS to buy and sell energy from and to the aggregator, respectively.[3] Here, $\pi_{\text{sell}} < \pi_{\text{buy}}$ holds to avoid the trivial case where a BS can benefit by reselling its bought energy from the aggregator, and $\pi_{\text{buy}} < \pi$ is also true, since otherwise all BSs short of energy will buy cheaper energy from the grid directly. With the two-way energy trading, the BSs adequate in renewable energy will sell their total Δ_- amount of energy surplus to the aggregator at the price π_{sell}, and accordingly an energy quota is set by the aggregator as Δ_-. The BSs short of renewable energy will first purchase a $\min(\Delta_+, \Delta_-)$ amount of cheap energy from the aggregator at the price π_{buy} (with the quota limitation of Δ_-) to maximally use this resource and (if not enough) will buy a $\Delta_+ - \min(\Delta_+, \Delta_-)$ amount from the grid at the price π. Depending on the relationship between Δ_+ and Δ_-, the total cost of all the N BSs is

$$C_2 = \begin{cases} \pi_{\text{buy}}\Delta_+ - \pi_{\text{sell}}\Delta_- & \text{if} \quad \Delta_+ \leq \Delta_- \\ \pi_{\text{buy}}\Delta_- + \pi(\Delta_+ - \Delta_-) - \pi_{\text{sell}}\Delta_- & \text{if} \quad \Delta_+ > \Delta_-. \end{cases} \tag{6.2}$$

Note that C_2 can be even negative, which is the case when Δ_- is sufficiently larger than Δ_+ such that $\pi_{\text{buy}}\Delta_+ < \pi_{\text{sell}}\Delta_-$. By comparing (6.1) and (6.2), it follows that $C_2 \leq C_1$.

6.3.2 Aggregator-Assisted Energy Sharing

Aggregator-assisted energy sharing is another energy cooperation scheme that allows BSs in a BS group to mutually negotiate and share renewable energy by simultaneously injecting and drawing energy to and from the aggregator, respectively. By matching the local renewable energy deficit (positive δ_i's) and surplus (negative δ_i's) between any two BSs, this scheme helps the group of BSs reduce their aggregate renewable energy deficit. The practical implementation of the energy sharing requires this group of BSs to sign a *contract* with the aggregator at a contract fee that motivates the aggregator to support the energy sharing. Note that the contract still requires BSs to commit to not interfere in the common operation of the aggregator by equating their total injected energy (into the aggregator) to their total drawn energy (from the aggregator) at any given time instance.[4] Unlike in the aggregator-assisted energy trading scheme, the aggregator does not need to be actively involved in this energy sharing scheme and endures limited coordination complexity.

Specifically, suppose that BS i wants to transfer an $e_{ij} \geq 0$ amount of energy to BS j, $i \neq j$. This is accomplished at an appointed time by BS i injecting an e_{ij} amount of energy into the aggregator and at the same time BS j drawing the same e_{ij} amount from the

3 The energy prices may vary over time based on the time-varying relationship between aggregate energy demand and supply at the aggregator. Depending on information and energy exchange frequencies, the aggregator can decide such prices either day ahead or in real time.

4 The aggregator can alternatively provide a long-term contract to the BSs, such that the BSs only need to ensure the energy sharing balance over a longer time period (e.g. one day or even a month), thus offering more flexibility for the BSs' energy sharing. In this case, however, a higher contract fee may be required by the aggregator due to the short-term perturbation to the aggregator.

aggregator. Note that the energy transfer between two BSs through the aggregator may lead to a certain amount of energy loss (e.g. less than 5% of the transferred energy) [13], which will be explicitly considered in Section 6.7. For the purpose of exposition, here we ignore such an effect unless otherwise mentioned. Thanks to the mutual sharing of e_{ij}'s among the N BSs, the total energy deficit Δ_+ and surplus Δ_- can be effectively matched. When $\Delta_+ \leq \Delta_-$, the N BSs can maintain their operation without purchasing any energy from the grid; otherwise, a total $\Delta_+ - \Delta_-$ amount of energy should be purchased from the grid at the price π. By denoting the contract fee to the aggregator as $\bar{C}$, the total cost of all the N BSs is given by

$$C_3 = \begin{cases} \bar{C} & \text{if} \quad \Delta_+ \leq \Delta_- \\ \pi(\Delta_+ - \Delta_-) + \bar{C} & \text{if} \quad \Delta_+ > \Delta_-. \end{cases} \tag{6.3}$$

By comparing (6.3) and (6.1), it generally follows that $C_3 \leq C_1$, i.e. the total energy cost is reduced, as long as $\bar{C}$ is sufficiently small.

The aggregator-assisted energy trading and sharing schemes may have different cost-saving performances depending on the energy buying and selling prices in the former scheme, and the contract fee in the latter one, both of which incentivize the aggregator to help.

6.4 Communication Cooperation

Communication cooperation refers to a cost-saving approach on the demand side that exploits the broadcast nature of wireless channels and uses wireless resource sharing to reshape BSs' wireless load and energy consumption. Different from conventional communication cooperation (e.g. [8–10]) aiming to maximize data throughput or minimize energy consumption, the communication cooperation of our interest here seeks to minimize the total energy cost by optimally utilizing both the cheap renewable energy and reliable on-grid energy. In the so-called cost-aware communication cooperation, the rescheduling of BSs' traffic load and energy consumption should follow their given renewable energy supply, such that the renewable energy can be maximally used to support the quality of service (QoS) requirements of the MTs, and the on-grid energy purchase is thus minimized. To implement this approach, BSs should share with each other the communication information (e.g. channel state information [CSI] and QoS requirements of MTs) and the energy information (e.g. the energy harvesting rates) through the backhaul links connecting them, as shown in Figure 6.1. This may require the cellular operator to install new infrastructures (e.g. high-capacity and low-latency backhaul links) and/or coordinate and standardize the communication protocols. It may incur more implementation complexity than the energy cooperation in Section 6.3.

In this section, we discuss three different cost-aware communication cooperation schemes, namely, traffic offloading [8], spectrum sharing [9], and CoMP [10], which are implemented in different time scales. For the purpose of illustration, we consider a simple cellular system setup with two BSs as shown in Figure 6.3a, in which BS 1 has sufficient harvested renewable energy but light traffic load (serving two MTs), thus having the net load $\delta_1 < 0$, while BS 2 has insufficient renewable energy but heavy traffic load (serving four MTs), leading to the net load $\delta_2 > 0$. The corresponding

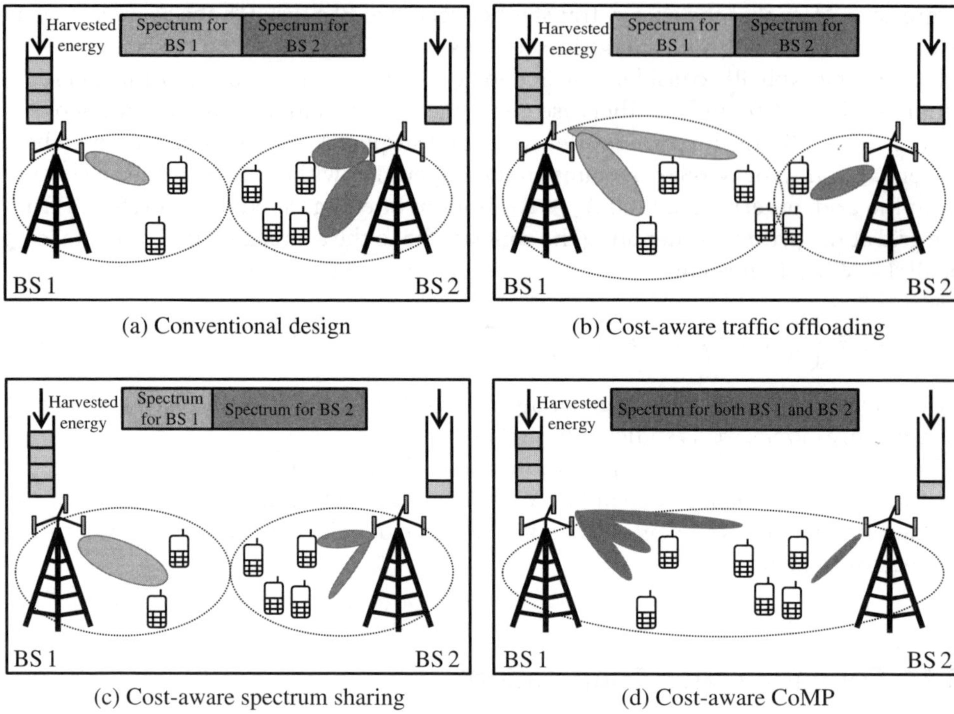

Figure 6.3 An example of different communication cooperation designs in a simple cellular network with two BSs. (a) Conventional design, (b) cost-aware traffic offloading, (c) cost-aware spectrum sharing, and (d) cost-aware CoMP. Source: Xu et al. 2015 [11]. Reproduced with permission of IEEE.

spectrum and power allocation and user–BS association for the two BSs are shown in Figure 6.3b–d for different cost-aware communication cooperation schemes.

6.4.1 Cost-Aware Traffic Offloading

Traffic offloading is traditionally designed to shift the traffic load (or served MTs) of heavily loaded BSs to the lightly loaded ones, for the purpose of avoiding the traffic congestion and improving the QoS of the MTs. Differently, the cost-aware traffic offloading here focuses on the new issue of energy cost reduction, i.e. BSs short of renewable energy can offload their MTs to neighboring BSs with abundant renewable energy (even if they have more or similar traffic loads), thus reducing the whole energy drawn from the grid to save cost. As shown in the example of Figure 6.3b, it is cost effective for BS 2 to offload two MTs (at its cell edge) to BS 1, such that the renewable energy at BS 1 is better utilized. Traffic offloading is often employed in a time scale of several seconds.

6.4.2 Cost-Aware Spectrum Sharing

Besides energy, spectrum is another scarce resource in cellular networks, and spectrum sharing has been considered as a solution to improve the spectrum utilization efficiency [9]. Different from the conventional spectrum sharing, the cost-aware spectrum sharing is based on the fact that the energy and spectrum resources can partially substitute each other to support the wireless transmission, and sharing spectrum to a BS short of energy can better save the energy cost of that BS.[5] As shown in the example of Figure 6.3c, BS 1 shares part of its available spectrum to BS 2. Under the same QoS requirements of MTs, BS 2 can decrease its transmission power purchased from the grid, while BS 1 uses more renewable energy for its transmission. Hence, the total cost is reduced. The implementation of spectrum sharing requires the BSs (and perhaps even the MTs) to have the capability of aggregating different frequency bands in transmission and reception, e.g. with the advanced carrier aggregation technique. Spectrum sharing can be realized in a time scale of minutes.

6.4.3 Cost-Aware Coordinated Multipoint (CoMP)

Traditionally, CoMP is considered as a technique to improve the spectral efficiency in cellular networks, by which BSs can implement coordinated baseband signal processing to cooperatively serve multiple MTs over the same time-frequency resources, transforming the harmful intercell interference (ICI) into useful information signals [10]. Differently, the cost-aware CoMP is motivated by the following observation: since different BSs can cooperatively send information signals to the MTs (in the downlink), their transmission power can be compensated by each other for satisfying the QoS requirements at MTs. Therefore, by adaptively adjusting the BSs' transmit signals, the cost-aware CoMP helps match the BSs' transmission power with their harvested renewable energy, thus minimizing the total energy drawn from the grid to save cost. For example, in Figure 6.3d, BS 1 adequate in renewable energy should use a high transmission power for providing strong wireless signals to the MTs, while BS 2 short of renewable energy should transmit at a low power level in their CoMP transmission.

CoMP should be performed at a symbol or frame level in the time scale of microseconds/milliseconds, which is more complex than the aforementioned traffic offloading and spectrum sharing schemes but can generally achieve higher cost saving. In practical cost-aware cellular networks, the three communication cooperation schemes can be employed depending on the tradeoff between cost saving and implementation complexity.

6.5 Joint Energy and Communication Cooperation

Joint energy and communication cooperation can maximally save cost by applying both the energy cooperation on the supply side and communication cooperation on the demand side. To realize the joint operation, the BSs should share the energy information by using the two-way information flow supported by the smart grid (through the smart meters) and also exchange the communication information through their backhaul connections (see Figure 6.1). Here, the exact required information sharing among BSs depends on the specific energy and communication cooperation schemes employed.

---

5 Considering a point-to-point additive white Gaussian noise (AWGN) channel, the relationship between the transmit power $P \geq 0$ and the bandwidth $W \geq 0$ is given by $P = N_0 W(2^{r/W} - 1)$, where r and N_0 denote the transmission rate and the noise power spectral density at the receiver, respectively. In this case, assuming fixed r, as the system bandwidth W increases, the transmit power P will decrease, and vice versa.

The joint energy and communication cooperation is more complex than energy or communication cooperation only, due to the implementation complexity for solving the cost minimization problem by optimizing both the supply (e.g. energy trading/sharing among BSs) and demand (e.g. spectrum and power allocations at BSs) sides, as well as the signaling overhead for sharing both the energy and communication information among BSs. The complexity increases significantly as the network size or the number of BSs becomes large. One potential solution to resolve this problem is to dynamically group the huge number of BSs into different BS clusters, where BSs within each cluster can implement the joint cooperation in a centralized manner, and different clusters can perform limited coordination in a decentralized way. In this chapter, we focus our study on the joint cooperation among a limited number of BSs in one single cluster.

As there are two energy cooperation schemes (aggregator-assisted energy trading and energy sharing) in Section 6.3 and three communication cooperation schemes (traffic offloading, spectrum sharing, and CoMP) in Section 6.4, there are totally six combinations of joint cooperation designs. In this section, we introduce a joint energy and spectrum sharing scheme and use a case study to illustrate the benefit of joint energy and communication cooperation. In the next two sections, we present two joint energy cooperation and CoMP designs in more detail.

The joint energy and spectrum sharing [14] is a scheme that allows neighboring BSs to share energy and spectrum with each other through the aggregator-assisted energy sharing in Section 6.3.2 and the spectrum sharing in Section 6.4.2, respectively. In this scheme, the BSs share their energy harvesting rates, energy prices, available bandwidth, and CSI (e.g. channel gains), as well as the QoS requirements of MTs among each other. Accordingly, the BSs exchange energy and spectrum to take advantage of resource complementarity.

Building upon the spectrum sharing in Figure 6.3c, [14] considered joint energy and spectrum sharing between two BSs to minimize their total energy cost while ensuring the QoS requirements for all the MTs. It is shown that at the optimality, it is possible that one BS adequate in both energy and spectrum shares these two resources to the other (in unidirectional cooperation) or one BS exchanges its energy for spectrum with the other (in bidirectional cooperation).

6.5.1 A Case Study

Now we present a case study to compare the energy cooperation in Section 6.3, the cost-aware spectrum sharing in Section 6.4, and the three joint energy and communication cooperation schemes proposed in this section. Also, we consider the conventional design without energy or communication cooperation as the performance benchmark, where each BS first individually minimizes its energy consumption on the demand side while ensuring the QoS requirements at MTs and then (if the energy demand exceeds the renewable energy supply) purchases the additional energy from the grid. For the purpose of illustration, as shown in Figure 6.4, we consider the downlink of a cellular system with two single-antenna BSs (i.e. BS 1 and BS 2), each applying orthogonal frequency-division multiple access (OFDMA) to serve $K_1 = 5$ and $K_2 = 15$ single-antenna MTs (denoted by the MT sets $\mathcal{K}_1$ and $\mathcal{K}_2$), respectively. Each BS uses an orthogonal frequency band with the same bandwidth ($W_1 = W_2 = 10$) initially. For simplicity, we randomly generate the channels based on the independent and

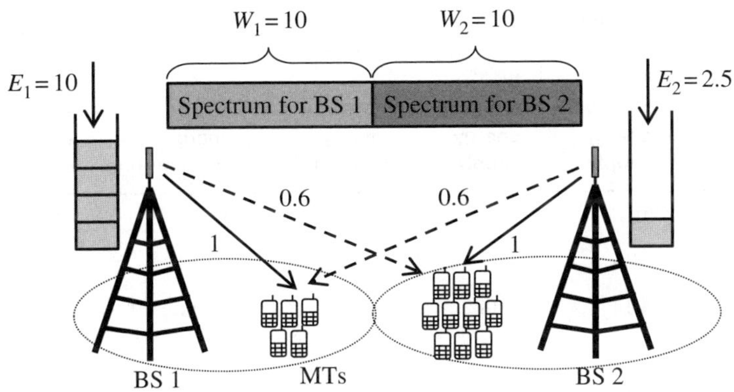

Figure 6.4 The case study model for comparing energy cooperation, communication cooperation, and joint cooperation.

identically distributed (i.i.d.) Rayleigh fading with the average channel powers from each BS to its own associated MTs (i.e. from BS 1 to any MT in $\mathcal{K}_1$ and from BS 2 to any MT in $\mathcal{K}_2$) being 1 and that from each BS to the other BS's associated MTs (i.e. from BS 1 to any MT in $\mathcal{K}_2$ and from BS 2 to any MT in $\mathcal{K}_1$) being 0.6. We set the noise power spectral density at each MT to be 1 and the QoS requirement of each MT to be a minimum data rate 1. On the demand side, we set the power consumptions Q_1 and Q_2 at the two BSs as their transmission power only; on the supply side, we set their harvested renewable energy as $E_1 = 10$ and $E_2 = 2.5$, respectively, and their energy buying price from the grid as $\pi = 1$. Additionally, for the aggregator-assisted energy trading, the BSs' energy buying and selling prices from and to the aggregator are $\pi_{\text{buy}} = 0.5$ and $\pi_{\text{sell}} = 0.4$, respectively, and for the aggregator-assisted energy sharing, the contract fee paid to the aggregator is $\bar{C} = 0.1$. Furthermore, in each scheme, the BSs employ the equal bandwidth allocation among MTs, and there is only one MT served in each sub-band. Note that all units are normalized for simplicity here.

Based on the above setting, we summarize the results in Table 6.1, from which we have the following observations:

- For the conventional design, the two BSs' energy demands for communications are computed to be $Q_1 = 4.14$ and $Q_2 = 18.28$. Their total energy cost is 15.78.
- For both energy cooperation approaches, it is observed that the renewable energy supplies at BS 1 and BS 2 are changed to 4.14 and 8.36, respectively, by BS 1 trading or sharing its excessive renewable energy to BS 2 through the aggregator. Since the new renewable energy supplies better match with the given energy consumptions at the two BSs, their total energy cost reduces to 10.51 and 10.03, respectively, where the different cost reductions are related to the different service fees charged by the aggregator.
- Regarding communication cooperation, it is observed that the energy consumptions at the two BSs are respectively changed to $Q_1 = 10.00$ and $Q_2 = 14.04$ for the spectrum sharing scheme. Compared with the conventional design, communication cooperation increases the transmission power at BS 1 to partially substitute that at BS 2 (together with certain wireless resource sharing) while satisfying the MTs' QoS

Table 6.1 Energy cost performance comparison.

	BS 1's renewable energy supply	BS 2's renewable energy supply	BS 1's energy consumption	BS 2's energy consumption	Total energy cost
Conventional design without energy or communication cooperation	10	2.5	4.14	18.28	15.78
Energy cooperation via aggregator-assisted energy trading	4.14	8.36	4.14	18.28	10.51
Energy cooperation via aggregator-assisted energy sharing	4.14	8.36	4.14	18.28	10.03
Communication cooperation via spectrum sharing	10	2.5	10.00	14.04	11.54
Joint energy and spectrum sharing	5.00	7.50	5.00	15.00	7.60

requirements. The transmission power adaptation matches and better uses the given cheap renewable energy supplies at the two BSs. Consequently, the resulting total energy costs reduce to 11.54.

- For joint energy and communication cooperation, it is observed that by exploiting both supply- and demand-side management, the joint scheme outperforms the schemes with only energy or communication cooperation. In particular, the total energy cost of joint energy and spectrum sharing (7.6) is less than that of aggregator-assisted energy sharing only (10.03) and spectrum sharing only (11.54).

6.6 Joint Aggregator-Assisted Energy Trading and CoMP

In the next two sections, we present two joint energy cooperation and CoMP designs. In such designs, different BSs implement CoMP-based transmission/reception in Section 6.4.3 to serve one or more MTs over the same time-frequency resources and at the same time perform aggregator-assisted energy trading in Section 6.3.1 or aggregator-assisted energy sharing in Section 6.3.2. To implement this, the BSs need to share their energy harvesting rates, energy prices, and instantaneous CSI (both channel gains and phases), as well as the QoS requirements of MTs among each other. In particular, this section focuses on the joint aggregator-assisted energy trading and CoMP, and the next section focuses on the joint aggregator-assisted energy sharing and CoMP.

6.6.1 System Model

We consider one BS cluster, in which $N > 1$ BSs each equipped with $M \geq 1$ antennas cooperatively send independent messages to K single-antenna MTs. For convenience, we denote the set of MTs and that of BSs as $\mathcal{K} = \{1, \ldots, K\}$ and $\mathcal{N} = \{1, \ldots, N\}$, respectively. We assume that each BS is locally equipped with one or more energy harvesting devices (wind turbines and/or solar panels) and is also connected to the aggregator for implementing the two-way energy trading. We also assume that there is a central unit deployed for each CoMP cluster that coordinates the cooperative energy trading as well as the cooperative communication within the cluster. To this end, the central unit needs to collect both the communication data (i.e. the transmit messages and CSI) from each of the BSs through the cellular backhaul links and the energy information (i.e. the energy harvesting rates and energy buying/selling prices) via the smart meters installed at BSs and the grid-deployed communication/control links connecting them. The central unit can be one of the N BSs that serves as the cluster head or a dedicatedly deployed entity in the network.

As assumed above, each BS is equipped with energy harvesting devices and is also connected to the grid for two-way energy trading. We denote the harvested energy at each BS $i \in \mathcal{N}$ as $E_i \geq 0$, which is a given constant for one block of our interest. We also denote the energy purchased (sold) from (to) the grid at BS i as $G_{b,i} \geq 0$ ($G_{s,i} \geq 0$). When each BS i buys (sells) one unit energy from (to) the grid, we denote the price that it needs to pay to (or will be paid by) the grid as $\alpha_{b,i} > 0$ ($\alpha_{s,i} > 0$). Then we have the net energy cost at BS i as

$$C_i = \alpha_{b,i} G_{b,i} - \alpha_{s,i} G_{s,i}, \quad i \in \mathcal{N}. \tag{6.4}$$

Note that C_i can be positive (e.g. $G_{b,i} > 0, G_{s,i} = 0$), negative (e.g. $G_{s,i} > 0, G_{b,i} = 0$), or zero. In practice, to prevent any BS from buying the energy from the grid and then selling back to it to make unjustifiable profit that leads to energy inefficiency, the grid operator should set $\alpha_{s,i} \leq \alpha_{b,i}, \forall i \in \mathcal{N}$; as a result, we can induce that at most one of $G_{b,i}$ and $G_{s,i}$ can be strictly positive, i.e. $G_{b,i} \cdot G_{s,i} = 0$ (otherwise, the cost in (6.4) can be further reduced by setting $G_{b,i} \leftarrow G_{b,i} - G_{s,i}$ and $G_{s,i} \leftarrow 0$ if $G_{b,i} \geq G_{s,i} > 0$ or $G_{s,i} \leftarrow G_{s,i} - G_{b,i}$ and $G_{b,i} \leftarrow 0$ if $G_{s,i} \geq G_{b,i} > 0$). Moreover, the energy selling price is usually subject to a minimum value, given by $\alpha_{\min} > 0$, to encourage the renewable generation investment at the BSs, while the energy buying price cannot exceed the maximum electricity price in the grid, given by $\alpha_{\max} > 0$. Thus, we have

$$0 < \alpha_{\min} \leq \alpha_{s,i} \leq \alpha_{b,i} \leq \alpha_{\max}, \quad \forall i \in \mathcal{N}. \tag{6.5}$$

In cellular systems, the power consumption at each BS typically includes both the transmission power due to radio frequency (RF) power amplifiers (PAs) and the non-transmission power due to other components such as cooling systems, baseband units (BBUs) for data processing, and circuits of RF chains. We denote the radiated transmit power of each BS i by $P_{t,i} \geq 0$ and generally model the non-transmission power as a constant denoted by $P_{c,i} > 0$. By combining them, we obtain the total power consumption at BS i, denoted by P_i, which should be no larger than the total energy available at BS i, i.e.

$$P_i = P_{t,i}/\eta + P_{c,i} \leq E_i + G_{b,i} - G_{s,i}, \quad i \in \mathcal{N}, \tag{6.6}$$

where $0 < \eta \leq 1$ denotes the PA efficiency. Since η is a constant, we normalize it as $\eta = 1$ unless stated otherwise.

Next, we present the downlink CoMP transmission among the N BSs in one cluster. We denote the channel vector from BS i to MT k as $\boldsymbol{h}_{ik} \in \mathbb{C}^{M\times 1}, i \in \mathcal{N}, k \in \mathcal{K}$, and the channel vector from all N BSs in the cluster to MT k as $\boldsymbol{h}_k = [\boldsymbol{h}_{1k}^T \quad \cdots \quad \boldsymbol{h}_{Nk}^T]^T \in$

$\mathbb{C}^{MN\times 1}, k \in \mathcal{K}$. We consider linear transmit beamforming applied at the BSs. Let the information signal for MT $k \in \mathcal{K}$ be denoted by s_k and its associated beamforming vector across the N BSs by $\boldsymbol{w}_k \in \mathbb{C}^{MN\times 1}$. Then the transmitted signal for MT k can be expressed as

$$\boldsymbol{x}_k = \boldsymbol{w}_k s_k,$$

where s_k is assumed to be a complex random variable with zero mean and unit variance. Thus, the received signal at MT k is given by

$$y_k = \boldsymbol{h}_k^H \boldsymbol{x}_k + \sum_{l\in\mathcal{K},l\neq k} \boldsymbol{h}_k^H \boldsymbol{x}_l + v_k, \quad k \in \mathcal{K},$$

where $\boldsymbol{h}_k^H \boldsymbol{x}_k$ is the desired signal for MT k, $\sum_{l\in\mathcal{K},l\neq k} \boldsymbol{h}_k^H \boldsymbol{x}_l$ is the inter-user interference within the same cluster, and v_k denotes the background AWGN at MT k, which may also include the downlink interference from other BSs outside this cluster. We assume that v_k's are independent circularly symmetric complex Gaussian (CSCG) random variables, each with zero mean and variance σ_k^2, i.e. $v_k \sim \mathcal{CN}(0, \sigma_k^2)$. Thus, the signal-to-interference-plus-noise ratio (SINR) at MT k can be expressed as

$$\texttt{SINR}_k(\{\boldsymbol{w}_k\}) = \frac{|\boldsymbol{h}_k^H \boldsymbol{w}_k|^2}{\sum_{l\in\mathcal{K},l\neq k} |\boldsymbol{h}_k^H \boldsymbol{w}_l|^2 + \sigma_k^2}, \quad k \in \mathcal{K}.$$

The transmit power at each BS i, i.e. $P_{t,i}$ in (6.6), can be expressed as

$$P_{t,i} = \sum_{k\in\mathcal{K}} \boldsymbol{w}_k^H \boldsymbol{B}_i \boldsymbol{w}_k, \quad \forall i \in \mathcal{N}, \tag{6.7}$$

where $\boldsymbol{B}_i \triangleq \texttt{Diag}(\underbrace{0,\ldots,0}_{(i-1)M},\underbrace{1,\ldots,1}_{M},\underbrace{0,\ldots,0}_{(N-i)M})$, with $\texttt{Diag}(\boldsymbol{a})$ denoting a diagonal matrix with the diagonal elements given in the vector $\boldsymbol{a}$. In addition, we assume that the maximum transmit power at each BS i is denoted by $P_{\max,i} > 0$, and thus we have $P_{t,i} \leq P_{\max,i}, \forall i \in \mathcal{N}$.

We aim to jointly optimize the N BSs' purchased/sold energy from/to the grid, $\{G_{b,i}\}$ and $\{G_{s,i}\}$, and their cooperative transmit beamforming vectors, $\{\boldsymbol{w}_k\}$, so as to minimize the total energy cost of all N BSs, i.e. $\sum_{i\in\mathcal{N}} C_i$ with C_i given in (6.4), subject to each MT's QoS constraint that is specified by a minimum SINR requirement γ_k for MT $k \in \mathcal{K}$. Here, the value of γ_k should be set based on the service type (e.g. video call or online game) requested by each MT k. Mathematically, we formulate the joint energy trading and beamforming optimization problem as

$$\min_{\{\boldsymbol{w}_k\},\{G_{b,i}\},\{G_{s,i}\}} \sum_{i\in\mathcal{N}} (\alpha_{b,i} G_{b,i} - \alpha_{s,i} G_{s,i}) \tag{6.8}$$

$$\text{s.t.} \quad \texttt{SINR}_k(\{\boldsymbol{w}_k\}) \geq \gamma_k, \quad \forall k \in \mathcal{K} \tag{6.9}$$

$$\sum_{k\in\mathcal{K}} \boldsymbol{w}_k^H \boldsymbol{B}_i \boldsymbol{w}_k + P_{c,i} \leq E_i + G_{b,i} - G_{s,i}, \quad \forall i \in \mathcal{N} \tag{6.10}$$

$$\sum_{k\in\mathcal{K}} \boldsymbol{w}_k^H \boldsymbol{B}_i \boldsymbol{w}_k \leq P_{\max,i}, \quad \forall i \in \mathcal{N} \tag{6.11}$$

$$G_{b,i} \geq 0, G_{s,i} \geq 0, \quad \forall i \in \mathcal{N}, \tag{6.12}$$

where (6.9) denotes the set of QoS constraints for the K MTs, (6.10) specifies the power constraints at the N BSs by combining (6.6) and (6.7), and (6.11) is for the individual maximum transmit power constraint at each of the N BSs. In problem (6.8), we do not explicitly add the constraint that at most one of $G_{b,i}$ and $G_{s,i}$ can be strictly positive for BS i, i.e. $G_{b,i} \cdot G_{s,i} = 0, \forall i \in \mathcal{N}$; however, it will be shown that the optimal solution to problem (6.8) always satisfies such constraints, and thus there is no loss of optimality by removing these constraints. Notice that problem (6.8) is in general non-convex due to the non-convex QoS constraints in (6.9).

Before solving problem (6.8), we first check its feasibility as follows. Note that given any transmit beamforming vectors $\{\boldsymbol{w}_k\}$ satisfying (6.11), each BS can always purchase sufficiently large amount of energy from the grid to satisfy the constraints in (6.10) and (6.12), e.g. by setting $G_{b,i} = (P_{\max,i} + P_{c,i} - E_i)^+$ and $G_{s,i} = 0, \forall i \in \mathcal{N}$. As a result, to check the feasibility of (6.8), we only need to check whether the constraints in (6.9) and (6.11) can be ensured at the same time. This is equivalent to solving the following feasibility problem to determine whether the N BSs can use their individual power to meet the QoS constraints for all the K MTs:

$$\begin{aligned} &\text{find}\{\boldsymbol{w}_k\} \\ &\text{s.t. } \mathtt{SINR}_k(\{\boldsymbol{w}_k\}) \geq \gamma_k, \quad \forall k \in \mathcal{K} \\ &\quad \sum_{k\in\mathcal{K}} \boldsymbol{w}_k^H \boldsymbol{B}_i \boldsymbol{w}_k \leq P_{\max,i}, \quad \forall i \in \mathcal{N}. \end{aligned} \tag{6.13}$$

Problem (6.13) has been solved by the standard convex optimization techniques via reformulating it as a second-order cone program (SOCP) [15] or by the uplink–downlink duality-based fixed point iteration algorithm [16, 17]. Later in this section, we focus on the case that problem (6.8) (or equivalently problem (6.13)) is feasible unless otherwise stated.

6.6.2 Optimal Solution

In this section, we present the optimal solution to problem (6.8) by proposing an algorithm based on the Lagrange duality method [18] and the uplink–downlink duality technique [17].

Let the dual variable associated with the ith power constraint in (6.10) be denoted by $\mu_i \geq 0$ and that corresponding to the ith individual maximum transmit power constraint in (6.11) by $\nu_i \geq 0, i \in \mathcal{N}$. Then we can express the partial Lagrangian of problem (6.8) as

$$\begin{aligned} &\mathcal{L}(\{\boldsymbol{w}_k\}, \{G_{b,i}\}, \{G_{s,i}\}, \{\mu_i\}, \{\nu_i\}) \\ &= \sum_{k\in\mathcal{K}} \boldsymbol{w}_k^H \boldsymbol{B}_{\mu,\nu} \boldsymbol{w}_k + \sum_{i\in\mathcal{N}} (\alpha_{b,i} - \mu_i) G_{b,i} + \sum_{i\in\mathcal{N}} (\mu_i - \alpha_{s,i}) G_{s,i} \\ &\quad + \sum_{i\in\mathcal{N}} (P_{c,i} - E_i)\mu_i - \sum_{i\in\mathcal{N}} P_{\max,i}\nu_i, \end{aligned} \tag{6.14}$$

where $\boldsymbol{B}_{\mu,\nu} \triangleq \sum_{i\in\mathcal{N}} (\mu_i + \nu_i)\boldsymbol{B}_i$. Accordingly, the dual function is given by

$$\begin{aligned} &g(\{\mu_i\}, \{\nu_i\}) = \\ &\min_{\{\boldsymbol{w}_k\}, \{G_{b,i}\geq 0\}, \{G_{s,i}\geq 0\}} \quad \mathcal{L}(\{\boldsymbol{w}_k\}, \{G_{b,i}\}, \{G_{s,i}\}, \{\mu_i\}, \{\nu_i\}) \\ &\quad \text{s.t.} \quad \mathtt{SINR}_k(\{\boldsymbol{w}_k\}) \geq \gamma_k, \quad \forall k \in \mathcal{K}, \end{aligned} \tag{6.15}$$

and thus the dual problem is expressed as

$$\max_{\{\mu_i \geq 0, \nu_i \geq 0\}} g(\{\mu_i\}, \{\nu_i\}). \tag{6.16}$$

Since problem (6.8) itself is non-convex, in general only weak duality holds between problem (6.8) and its dual problem (6.16), that is, the optimal value achieved by problem (6.16) is generally a lower bound on that of problem (6.8) [18]. However, due to the specific structure of problem (6.8), we can show that strong duality indeed holds between problems (6.8) and (6.16), as stated in the following proposition.

Proposition 6.1 *The optimal value achieved by problem (6.16) is equal to that by problem (6.8).*

Proof: This proposition relies on the fact that problem (6.8) can be recast as a convex optimization problem. Please refer to [19] for details. □

According to Proposition 6.1, we can solve problem (6.8) by equivalently solving problem (6.16). In the following, we first solve the problem in (6.15) for obtaining $g(\{\mu_i\}, \{\nu_i\})$ under any given $\{\mu_i\}$ and $\{\nu_i\}$, satisfying $\mu_i \geq 0$ and $\nu_i \geq 0, \forall i \in \mathcal{N}$, and then minimize $g(\{\mu_i\}, \{\nu_i\})$ over $\{\mu_i\}$ and $\{\nu_i\}$.

6.6.2.1 Solve Problem (6.15) for Obtaining $g(\{\mu_i\}, \{\nu_i\})$

First, we have the following lemma.

Lemma 6.1 *In order for $g(\{\mu_i\}, \{\nu_i\})$ to be bounded from below, i.e. $g(\{\mu_i\}, \{\nu_i\}) > -\infty$, it must hold that*

$$\alpha_{s,i} \leq \mu_i \leq \alpha_{b,i}, \quad \forall i \in \mathcal{N}. \tag{6.17}$$

Proof: Please refer to [19]. □

According to Lemma 6.1, we only need to derive $g(\{\mu_i\}, \{\nu_i\})$ for the given $\{\mu_i\}$ and $\{\nu_i\}$ satisfying $0 < \alpha_{s,i} \leq \mu_i \leq \alpha_{b,i}$ and $\nu_i \geq 0, \forall i \in \mathcal{N}$. In this case, we have $\boldsymbol{B}_{\mu,\nu} \succ \boldsymbol{0}$. Furthermore, we observe that the problem in (6.15) can be decomposed into the following $2N + 1$ subproblems by dropping the irrelative term $\sum_{i \in \mathcal{N}} (P_{c,i} - E_i)\mu_i - \sum_{i \in \mathcal{N}} P_{\max,i}\nu_i$:

$$\min_{\{\boldsymbol{w}_k\}} \sum_{k \in \mathcal{K}} \boldsymbol{w}_k^H \boldsymbol{B}_{\mu,\nu} \boldsymbol{w}_k$$
$$\text{s.t. } \mathtt{SINR}_k(\{\boldsymbol{w}_k\}) \geq \gamma_k, \quad \forall k \in \mathcal{K}, \tag{6.18}$$

$$\min_{G_{b,i} \geq 0} (\alpha_{b,i} - \mu_i) G_{b,i}, \quad \forall i \in \mathcal{N}, \tag{6.19}$$

$$\min_{G_{s,i} \geq 0} (\mu_i - \alpha_{s,i}) G_{s,i}, \quad \forall i \in \mathcal{N}, \tag{6.20}$$

where (6.19) and (6.20) each correspond to N subproblems (one for each BS i). For the subproblems in (6.19) and (6.20), it is easy to show that the optimal solutions are given by

$$G_{b,i}^{\star} = G_{s,i}^{\star} = 0, \quad \forall i \in \mathcal{N}. \tag{6.21}$$

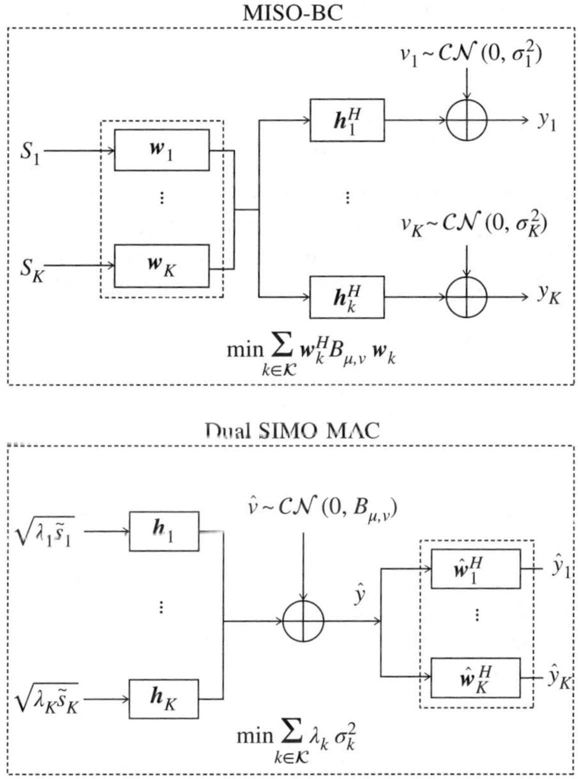

Figure 6.5 Uplink–downlink duality for MISO-BC and SIMO-MAC.

Note that if $\mu_i = \alpha_{b,i}$ or $\mu_i = \alpha_{s,i}$ for any $i \in \mathcal{N}$, then the corresponding optimal solution of $G^{\star}_{b,i}$ or $G^{\star}_{s,i}$ in (6.21) is generally not unique and can take any nonnegative value. In this case, $G^{\star}_{b,i} = 0$ or $G^{\star}_{s,i} = 0$ is employed here for the purpose of solving problem (6.19) or (6.20) to obtain the dual function only, while they may not be the optimal solution to the original problem (6.8), as will be discussed later in Section 6.6.2.2.

Now, it only remains to solve problem (6.18) with $\boldsymbol{B}_{\mu,\nu} \succ \boldsymbol{0}$ for obtaining $g(\{\mu_i\}, \{\nu_i\})$. To this end, we exploit the uplink–downlink duality as follows.

Problem (6.18) can be viewed as a transmit beamforming problem for a multiple-input single-output broadcast channel (MISO-BC), as shown in the left subfigure of Figure 6.5, with the goal of minimizing the weighted sum power $\sum_{k\in\mathcal{K}} \boldsymbol{w}_k^H \boldsymbol{B}_{\mu,\nu} \boldsymbol{w}_k$ at the transmitter subject to a set of SINR constraints $\{\gamma_k\}$. For the MISO-BC, its dual single-input multiple-output multiple-access channel (SIMO-MAC) is shown in the right subfigure of Figure 6.5 by conjugating and transposing the channel vectors, where K single-antenna transmitters send independent information to one common receiver with MN antennas. For transmitter $k \in \mathcal{K}$, let λ_k be its transmit power, $\tilde{s}_k$ denote its transmitted information signal with zero mean and unit variance, and $\boldsymbol{h}_k$ be its channel vector to the receiver in the dual SIMO-MAC. Then the received signal is expressed as $\hat{\boldsymbol{y}} = \sum_{k\in\mathcal{K}} \boldsymbol{h}_k \sqrt{\lambda_k}\tilde{s}_k + \hat{\boldsymbol{v}}$, where $\hat{\boldsymbol{v}}$ is a CSCG random vector with zero mean and covariance matrix $\boldsymbol{B}_{\mu,\nu}$ denoting the equivalent noise vector at the receiver, i.e. $\hat{\boldsymbol{v}} \sim \mathcal{CN}(\boldsymbol{0}, \boldsymbol{B}_{\mu,\nu})$. By applying receive beamforming vector $\hat{\boldsymbol{w}}_k$'s, the SINRs of different users in the dual SIMO-MAC are then given by

$$\texttt{SINR}_k^{\mathrm{MAC}}(\{\hat{\boldsymbol{w}}_k, \lambda_k\}) = \frac{\lambda_k |\boldsymbol{h}_k^H \hat{\boldsymbol{w}}_k|^2}{\sum_{l\in\mathcal{K}, l\neq k} \lambda_l |\boldsymbol{h}_l^H \hat{\boldsymbol{w}}_k|^2 + \hat{\boldsymbol{w}}_k^H \boldsymbol{B}_{\mu,\nu} \hat{\boldsymbol{w}}_k}, \quad \forall k \in \mathcal{K}. \tag{6.22}$$

The design objective for the dual SIMO-MAC is to minimize the weighted sum transmit power $\sum_{k\in\mathcal{K}} \lambda_k \sigma_k^2$ by jointly optimizing the power allocation $\{\lambda_k\}$ and receive beamforming vectors $\{\hat{\boldsymbol{w}}_k\}$ subject to the same set of SINR constraints $\{\gamma_k\}$ as in the original MISO-BC given by (6.18). We thus formulate the dual uplink problem as

$$\min_{\{\hat{\boldsymbol{w}}_k\},\{\lambda_k\geq 0\}} \sum_{k\in\mathcal{K}} \lambda_k \sigma_k^2$$
$$\text{s.t.} \quad \texttt{SINR}_k^{\mathrm{MAC}}(\{\hat{\boldsymbol{w}}_k, \lambda_k\}) \geq \gamma_k, \quad \forall k \in \mathcal{K}. \tag{6.23}$$

With $\boldsymbol{B}_{\mu,\nu} \succ \boldsymbol{0}$, it has been shown in [17] that problems (6.18) and (6.23) are equivalent. Thus, we can solve the downlink problem (6.18) by first solving the uplink problem (6.23) and then mapping its solution to that of problem (6.18), shown as follows.

First, consider the uplink problem (6.23). Since it can be shown that the optimal solution of (6.23) is always achieved when all the SINR constraints are met with equality [17], it follows that the optimal uplink transmit power $\{\lambda_k^\star\}$ must be a fixed point solution of the following equations and thus can be found via an iterative function evaluation procedure [15].

$$\lambda_k^\star = \frac{1}{\left(1+\frac{1}{\gamma_k}\right) \boldsymbol{h}_k^H \left(\sum_{l\in\mathcal{K}} \lambda_l^\star \boldsymbol{h}_l \boldsymbol{h}_l^H + \boldsymbol{B}_{\mu,\nu}\right)^{-1} \boldsymbol{h}_k}, \quad \forall k \in \mathcal{K}. \tag{6.24}$$

With $\{\lambda_k^\star\}$ at hand, the optimal receive beamforming vector $\{\hat{\boldsymbol{w}}_k^\star\}$ can then be obtained based on the minimum mean squared error (MMSE) principle as

$$\hat{\boldsymbol{w}}_k^\star = \frac{\left(\sum_{l\in\mathcal{K}} \lambda_l^\star \boldsymbol{h}_l \boldsymbol{h}_l^H + \boldsymbol{B}_{\mu,\nu}\right)^{-1} \boldsymbol{h}_k}{\left\|\left(\sum_{l\in\mathcal{K}} \lambda_l^\star \boldsymbol{h}_l \boldsymbol{h}_l^H + \boldsymbol{B}_{\mu,\nu}\right)^{-1} \boldsymbol{h}_k\right\|}, \quad \forall k \in \mathcal{K}. \tag{6.25}$$

After obtaining the optimal solution of $\{\hat{\boldsymbol{w}}_k^\star\}$ and $\{\lambda_k^\star\}$ for the uplink problem (6.23), we then map the solution to $\{\boldsymbol{w}_k^\star\}$ for the downlink problem (6.18). As shown in [17], $\{\boldsymbol{w}_k^\star\}$ and $\{\hat{\boldsymbol{w}}_k^\star\}$ are identical up to a certain scaling factor. Using this argument together with the fact that the optimal solution of (6.18) is also attained with all the SINR constraints being tight similarly to that in problem (6.23), it follows that $\{\boldsymbol{w}_k^\star\}$ can be obtained as $\boldsymbol{w}_k^\star = \sqrt{p_k^\star}\hat{\boldsymbol{w}}_k^\star, \forall k \in \mathcal{K}$, with $\boldsymbol{p}^\star = [p_1^\star, \ldots, p_K^\star]^T$ given by

$$\boldsymbol{p}^\star = \left(\boldsymbol{I} - \boldsymbol{D}(\{\hat{\boldsymbol{w}}_k^\star, \gamma_k\})\right)^{-1} \boldsymbol{u}(\{\hat{\boldsymbol{w}}_k^\star, \gamma_k\}), \tag{6.26}$$

where $[\boldsymbol{D}]_{kl}(\{\hat{\boldsymbol{w}}_k, \gamma_k\}) = \begin{cases} 0, & k = l \\ \dfrac{\gamma_k |\boldsymbol{h}_k^H \hat{\boldsymbol{w}}_l|^2}{|\boldsymbol{h}_k^H \hat{\boldsymbol{w}}_k|^2}, & k \neq l \end{cases}$ and $\boldsymbol{u}(\{\hat{\boldsymbol{w}}_k, \gamma_k\}) = \left[\dfrac{\gamma_1 \sigma_1^2}{|\boldsymbol{h}_1^H \hat{\boldsymbol{w}}_1|^2}, \ldots, \dfrac{\gamma_K \sigma_K^2}{|\boldsymbol{h}_K^H \hat{\boldsymbol{w}}_K|^2}\right]^T$.

6.6.2.2 Minimize $g(\{\mu_i\},\{\nu_i\})$ over $\{\mu_i\}$ and $\{\nu_i\}$

Up to now, we have obtained the optimal solution of $\{\boldsymbol{w}_i^\star\}$, $\{G_{\mathrm{b},i}^\star\}$, and $\{G_{\mathrm{s},i}^\star\}$ to the problem in (6.15) with given $\{\mu_i\}$ and $\{\nu_i\}$. Accordingly, the dual function $g(\{\mu_i\},\{\nu_i\})$ has been obtained. Next, we solve problem (6.16) by minimizing $g(\{\mu_i\},\{\nu_i\})$ over $\{\mu_i\}$ and $\{\nu_i\}$. Since $g(\{\mu_i\},\{\nu_i\})$ is convex but not necessarily differentiable, we can employ the ellipsoid method [20] to obtain the optimal $\{\mu_i^*\}$ and $\{\nu_i^*\}$ for problem (6.16) by using the fact that the subgradients of $g(\{\mu_i\},\{\nu_i\})$ at the given μ_i and ν_i can be shown to be $\sum_{k\in\mathcal{K}}\boldsymbol{w}_k^{\star H}\boldsymbol{B}_i\boldsymbol{w}_k^\star + P_{\mathrm{c},i} - E_i - G_{\mathrm{b},i}^\star + G_{\mathrm{s},i}^\star = \sum_{k\in\mathcal{K}}\boldsymbol{w}_k^{\star H}\boldsymbol{B}_i\boldsymbol{w}_k^\star + P_{\mathrm{c},i} - E_i$ and $\sum_{k\in\mathcal{K}}\boldsymbol{w}_k^{\star H}\boldsymbol{B}_i\boldsymbol{w}_k^\star - P_{\max,i}$, $i\in\mathcal{N}$, respectively.

With the obtained $\{\mu_i^*\}$ and $\{\nu_i^*\}$, the corresponding $\{\boldsymbol{w}_i^\star\}$ becomes the optimal transmit beamforming vectors for problem (6.8), denoted by $\{\boldsymbol{w}_k^*\}$. However, the solutions of $\{G_{\mathrm{b},i}^\star\}$ and $\{G_{\mathrm{s},i}^\star\}$ given by (6.21) in general may not be the optimal solution to problem (6.8), since they are not unique if $\alpha_{\mathrm{b},i} - \mu_i^* = 0$ or $\mu_i^* - \alpha_{\mathrm{s},i} = 0$, for any $i\in\mathcal{N}$. Nevertheless, it can be easily checked that the optimal solution to problem (6.8) is achieved when the constraints in (6.10) are all met with equality. As a result, the optimal solution of $\{G_{\mathrm{b},i}^*\}$ and $\{G_{\mathrm{s},i}^*\}$ for problem (6.8) can be obtained as

$$G_{\mathrm{b},i}^* = \left(\sum_{k\in\mathcal{K}}\boldsymbol{w}_k^{*H}\boldsymbol{B}_i\boldsymbol{w}_k^* + P_{\mathrm{c},i} - E_i\right)^+, \quad \forall i\in\mathcal{N}, \tag{6.27}$$

$$G_{\mathrm{s},i}^* = \left(E_i - \sum_{k\in\mathcal{K}}\boldsymbol{w}_k^{*H}\boldsymbol{B}_i\boldsymbol{w}_k^* - P_{\mathrm{c},i}\right)^+, \quad \forall i\in\mathcal{N}. \tag{6.28}$$

This result is intuitive, since if the amount of harvested energy by BS i, E_i, is smaller (or larger) than that of its consumed energy, $\sum_{k\in\mathcal{K}}\boldsymbol{w}_k^{*H}\boldsymbol{B}_i\boldsymbol{w}_k^* + P_{\mathrm{c},i}$, then BS i should purchase the insufficient energy (or sell the excess energy) from (or to) the grid.

To summarize, our proposed algorithm to solve problem (6.8) is given in Algorithm 6.2.

Algorithm 6.1: Algorithm for solving problem (6.8).

Input: $\{\mu_i\}$ and $\{\nu_i\}$ with $\alpha_{\mathrm{s},i} \le \mu_i \le \alpha_{\mathrm{b},i}$ and $\nu_i \ge 0$, $\forall i\in\mathcal{N}$
Output: $\{G_{\mathrm{b},i}^*\}$ and $\{G_{\mathrm{s},i}^*\}$

1: **repeat**
2: Compute $\{\lambda_k^\star\}$ as a fixed point solution of (6.24) by iterative function evaluation [15], and compute the uplink receive beamforming vectors $\{\hat{\boldsymbol{w}}_k^\star\}$ by (6.25)
3: Compute the downlink beamforming vectors as $\boldsymbol{w}_k^\star = \sqrt{p_k^\star}\hat{\boldsymbol{w}}_k^\star$, $\forall k\in\mathcal{K}$, with $\{p_k^\star\}$ given by (6.26)
4: Compute the subgradients of $g(\{\mu_i\},\{\nu_i\})$ associated with μ_i and ν_i as $\sum_{k\in\mathcal{K}}\boldsymbol{w}_k^{\star H}\boldsymbol{B}_i\boldsymbol{w}_k^\star - E_i + P_{\mathrm{c},i}$ and $\sum_{k\in\mathcal{K}}\boldsymbol{w}_k^{\star H}\boldsymbol{B}_i\boldsymbol{w}_k^\star - P_{\max,i}$, respectively, $i\in\mathcal{N}$, and then update $\{\mu_i\}$ and $\{\nu_i\}$ accordingly based on the ellipsoid method [20], subject to $\alpha_{\mathrm{s},i} \le \mu_i \le \alpha_{\mathrm{b},i}$ and $\nu_i \ge 0$, $\forall i\in\mathcal{N}$
5: **until** $\{\mu_i\}$ and $\{\nu_i\}$ all converge within a prescribed accuracy
6: $\boldsymbol{w}_k^* \leftarrow \boldsymbol{w}_k^\star$, $\forall k\in\mathcal{K}$
7: Compute $\{G_{\mathrm{b},i}^*\}$ and $\{G_{\mathrm{s},i}^*\}$ given by (6.27) and (6.28)

Remark 6.1 For problem (6.8), it follows that the optimal dual solution $\{\mu_i^*\}$ must satisfy that

$$\mu_i^* \begin{cases} = \alpha_{s,i} & \text{if} \quad G_{b,i}^* = 0, G_{s,i}^* > 0 \\ \in [\alpha_{s,i}, \alpha_{b,i}] & \text{if} \quad G_{b,i}^* = G_{s,i}^* = 0 \\ = \alpha_{b,i} & \text{if} \quad G_{b,i}^* > 0, G_{s,i}^* = 0, \end{cases} \quad \forall i \in \mathcal{N}, \tag{6.29}$$

which can be explained intuitively by interpreting μ_i^* as the marginal energy cost for BS $i \in \mathcal{N}$. In other words, the marginal energy cost for BS i (i.e. μ_i^*) is equal to the unit energy selling/buying price $\alpha_{s,i}/\alpha_{b,i}$, if the BS sells/purchases energy to/from the grid (i.e. $G_{s,i}^* > 0$ or $G_{b,i}^* > 0$). Note that the set of marginal energy costs $\{\mu_i^*\}$ plays an important role in adjusting the cooperative transmit beamforming vectors at the N BSs (cf. (6.25)). They allow the cooperative BSs to reallocate their power consumption pattern to follow the corresponding renewable generation profile, such that the BSs can better utilize the cheap renewable energy and therefore reduce the total energy cost, as will be validated in our simulation results.

6.6.3 Numerical Results

In this subsection, we provide numerical results to evaluate the performance of the proposed design as compared with the following three alternative schemes, which are briefly introduced as follows and more details can be found in [13]:

- *Suboptimal solution to problem* (6.8) *with ZF beamforming*: This scheme corresponds to a lower-complexity suboptimal solution to problem (6.8) based on cooperative ZF precoding/beamforming at the BSs [21, 22]. In particular, the transmit beamforming vectors should be designed to precancel any inter-user interference among different MTs, i.e. $\boldsymbol{h}_k^H \boldsymbol{w}_l = 0, \forall l, k \in \mathcal{K}, l \neq k$. Note that in order for the ZF beamforming to be feasible, it is assumed that all the channel vector $\boldsymbol{h}_k$'s are linearly independent in this scheme.
- *Conventional design with optimal beamforming*: First, the N BSs design the cooperative transmit beamforming so as to minimize their transmit sum power subject to the QoS constraints given in (6.9). Next, each of the N BSs sets the energy trading solution with the grid independently.
- *Conventional design with ZF beamforming*: This scheme performs similarly as for the conventional design with optimal beamforming; expect that ZF beamforming is employed at the N BSs.

In the simulation, we consider a practical three-BS cluster (with $N = 3$), where the cells are hexagonal with the inter-BS distance of 1 km, the number of transmit antennas at each BS is $M = 4$, and the total number of MTs is $K = 8$. We assume that BS 1 and BS 2 are deployed with solar and wind generators, respectively, while BS 3 has both of them deployed. Then based on a real-world solar and wind energy production data,[6] we model the energy harvesting rates at the three BSs as shown in Figures 6.6–6.8, where the harvested energy at each BS has been averaged over 15 minutes, and thus there are 384 energy harvesting rate samples for each BS over 96 hours (i.e. four days). For

6 See http://www.elia.be/en/grid-data/power-generation/.

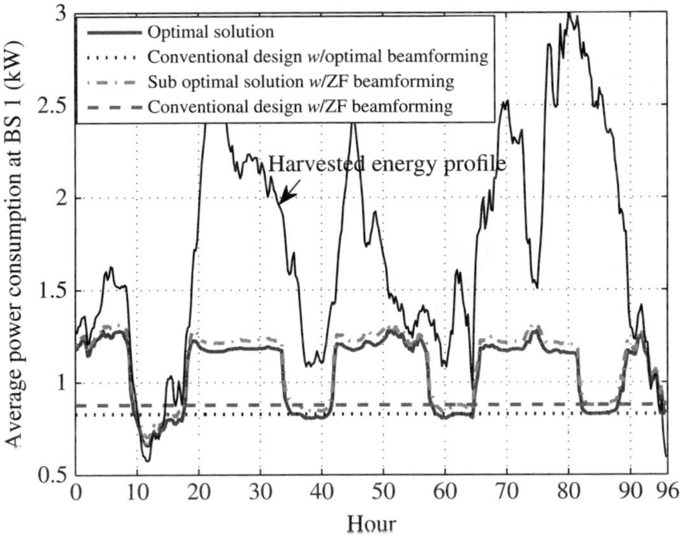

Figure 6.6 The average power consumption at BS 1 over time.

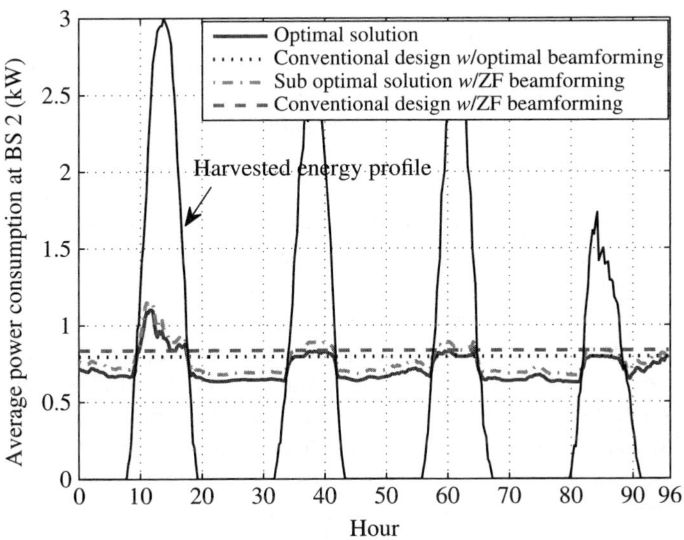

Figure 6.7 The average power consumption at BS 2 over time.

each renewable energy sample, we apply the same set of 100 randomly generated user channels in order to focus our study on the impact of renewable generation variation. For each channel realization, we randomly generate the eight MTs in the three cells following a uniform distribution and model each channel by a superposition of path loss and short-term Rayleigh fading. We assume that the background noise and the QoS requirement at each MT receiver are $\sigma_k^2 = -85$ dBm and $\gamma_k = 10$ dB, $\forall k \in \mathcal{K}$, respectively, while the PA efficiency, the maximum transmit power, and the non-transmission constant power at each BS are $\eta = 0.1$ and $P_{\max,i} = 100$ W, and $P_{c,i} = 500$ W, $\forall i \in \mathcal{N}$,

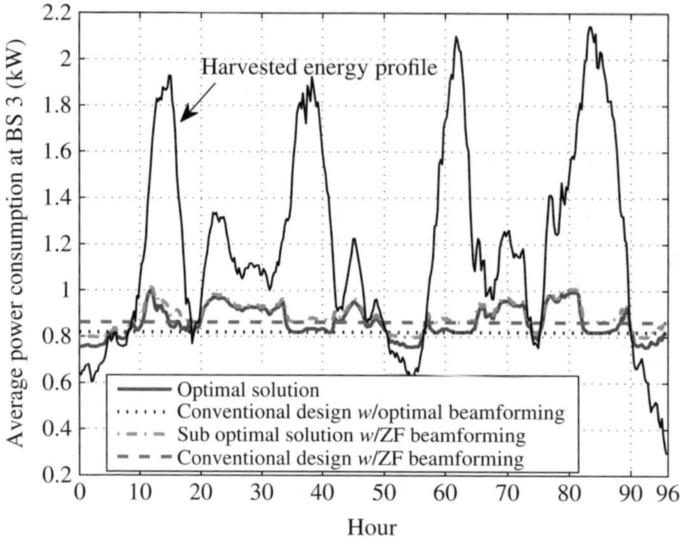

Figure 6.8 The average power consumption at BS 3 over time.

respectively. We consider the prices for buying (selling) energy from (to) the grid as $\alpha_{b,i} = 1\ \text{kW}^{-1}$ ($\alpha_{s,i} = 0.1\ \text{kW}^{-1}$), $\forall i \in \mathcal{N}$, where the price unit is normalized without loss of generality. Note that among the randomly generated channels, we have only chosen the realizations such that all the four schemes are both feasible for the ease of performance comparison.

Figures 6.6–6.8 show the average power consumption at the three BSs over time, together with their harvested energy profiles. For the two conventional designs with optimal and ZF beamforming, it is observed that the average power consumption at each BS remains constant over time, regardless of the fluctuations in energy harvesting rates at each BS. This is because the two conventional designs obtain the respective transmit beamforming solutions to minimize the sum power by ignoring the two-way energy trading prices and the energy harvesting rates at all the BSs; as a result, the average power consumption at each BS is constant for this purposefully designed setup with fixed number of users and the same set of wireless channels over the time. In contrast, for the proposed optimal and suboptimal solutions, the resulting average power consumption at each BS is observed to vary following a similar pattern as the corresponding energy harvesting rates. For example, during hours 0–9, BS 1 with large locally generated wind energy increases its transmission power and accordingly decreases the excess energy sold to the grid, while BS 2 and BS 3 with zero/smaller locally generated solar energy reduce their transmission power that need to be purchased from the grid, in order to minimize the total energy cost of the three BSs, given that $\alpha_{s,i} < \alpha_{b,i}, \forall i \in \mathcal{N}$. Finally, it is observed that the optimal solution leads to small power consumption than the suboptimal solution with ZF beamforming for each of the three BSs over the 96 hours. The energy (thus cost) reduction, however, is achieved at a cost of higher algorithm implementation complexity.

Figures 6.9 and 6.10 show the total purchased energy from the grid (i.e. $\sum_{i\in\mathcal{N}} G_{b,i}$) and the total energy cost of the three BSs over time, respectively, based on the results in Figures 6.6–6.8. It is observed that the proposed optimal and suboptimal solutions

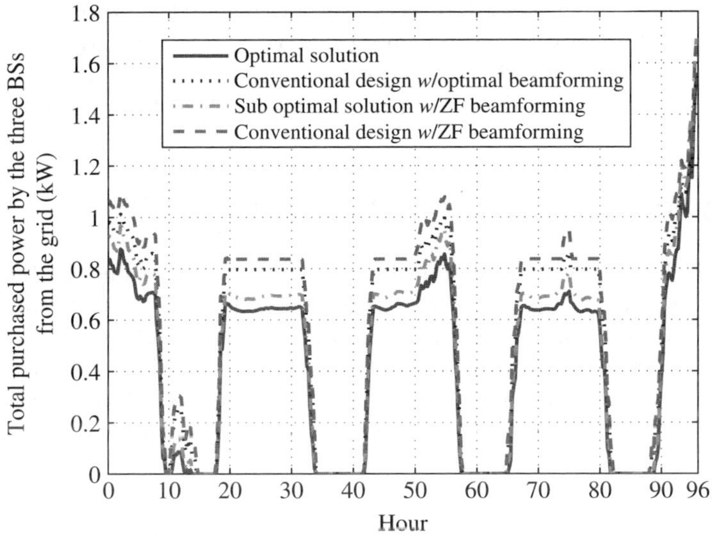

Figure 6.9 The total purchased power by the three BSs from the grid over time.

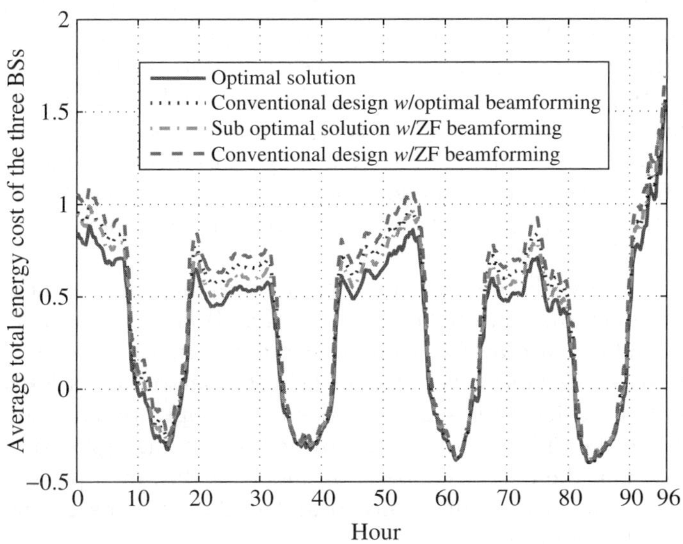

Figure 6.10 The total energy cost of the three BSs over time.

reduce the expensive grid energy purchase (see Figure 6.9) and accordingly the total energy cost (see Figure 6.10) at all time, as compared with the conventional designs with optimal and ZF beamforming, respectively. The grid energy purchase and the energy cost reductions are also observed to be more substantial when the energy harvesting rates at different BSs are more unevenly distributed, e.g. during hours 0–9 and 20–30. The average total energy costs over the entire period of 96 hours achieved by the four schemes of optimal solution, conventional design with optimal beamforming, suboptimal solution with ZF beamforming, and conventional design with ZF beamforming are obtained as 0.3110, 0.3993, 0.3729, and 0.4601, respectively. As a result, the optimal solution and the suboptimal solution with ZF beamforming achieve 22.12% and 6.61% total energy cost reductions over the conventional design with optimal beamforming and 32.41% and 18.96% cost reductions over the conventional design with ZF beamforming, respectively.

6.7 Joint Aggregator-Assisted Energy Sharing and CoMP

In this section, we study a purely renewable powered cellular system enabled by the aggregator-assisted energy sharing, in which the BSs do not purchase any energy from the grid to minimize the cost, but use the harvested renewable energy together with the energy sharing to maintain their operations. Under this system, we consider the joint aggregator-assisted energy sharing and CoMP design by taking into account the energy loss during the aggregator-assisted energy sharing and ignoring the contract fee.

6.7.1 System Model

We consider a CoMP cluster, in which N BSs each equipped with M antennas cooperatively send independent messages to K single-antenna MTs. The BSs are assumed to be locally deployed with solar panels and/or wind turbines for energy harvesting from the environment and also equipped with smart meters to enable their energy cooperation through the aggregator in smart grid. We consider a narrow-band system with ZF-based precoding, which requires that the number of active MTs in each cluster is no larger than the total number of transmitting antennas at all N BSs, i.e. $K \leq MN$, while the results of this paper can be readily generalized to more practical setups with broadband transmission and arbitrary number of MTs by applying time-frequency user scheduling and/or other precoding schemes. For convenience, we denote the set of MTs and that of BSs as $\mathcal{K} = \{1, \dots, K\}$ and $\mathcal{N} = \{1, \dots, N\}$, respectively, with k, l indicating MT index and i, j for BS index.

We assume that all BSs within each cluster can perfectly share their communication information (including both the transmit messages and CSI) via high-capacity low-latency backhaul links and can also perfectly measure and exchange their energy information (i.e. the energy harvesting rates over time) by using the smart meters. Such information sharing helps BSs in each cluster to implement energy cooperation to share energy with each other by locally injecting/drawing power to/from the aggregator. In the following, we first explain the energy cooperation model at BSs, then present the downlink CoMP transmission model, and finally formulate the joint energy and communication cooperation design problem.

We consider that the BS does not purchase any expensive on-grid energy to minimize the cost, but instead only uses its locally harvested energy and the transferred energy from other BSs (if not zero) in the same cluster. The energy harvesting, consumption, and sharing at each BS is coordinated by a smart meter. For energy harvesting, let the harvested renewable energy at BS $i \in \mathcal{N}$ be denoted as $E_i \geq 0$. Regarding energy consumption, we only consider the transmit power consumption at each BS by ignoring its non-transmission power due to air conditioner, data processor, etc. for simplicity,[7] and we denote the transmit power of BS $i \in \mathcal{N}$ as $P_i \geq 0$.

7 Since the non-transmission power is generally modeled as a constant term, our results are readily extendible to the case with non-transmission power included by simply modifying the harvested energy as that offset by the non-transmission related energy consumption at each BS.

Next, we introduce the energy cooperation among BSs. Let the transferred energy from BS i to BS j be denoted as $e_{ij} \geq 0$, $i, j \in \mathcal{N}, i \neq j$. Practically, this can be implemented by BS i injecting power e_{ij} to the blackaggregator and at the same time BS j drawing power $e_{ij} - q_{ij}(e_{ij})$ from the blackaggregator, where $q_{ij}(e_{ij})$ denotes the resulting power loss in the power network with $0 < q_{ij}(e_{ij}) < e_{ij}$. In practical power systems, the power transfer loss $q_{ij}(e_{ij})$ is normally characterized by a linear model, i.e. $q_{ij}(e_{ij}) = \zeta_{ij} e_{ij}$, where $0 < \zeta_{ij} < 1$ is termed the "incremental loss" that represents the incremental power loss in the power network caused by blackboth the power injected by BS i and power drawn by BS j. For notational convenience, we define $\beta_{ij} \triangleq 1 - \zeta_{ij}$ as the energy transfer efficiency from BS i to BS j, where $0 < \beta_{ij} < 1, \forall i, j \in \mathcal{N}, i \neq j$. Thus, when BS i transfers power e_{ij} to BS j, the effective energy drawn at BS j needs to be $\beta_{ij} e_{ij}$, in order to have a zero net energy exchanged with the aggregator to maintain its stability. For the ease of analysis, we also consider the special case of $\beta_{ij} = 0, \forall i, j \in \mathcal{N}, i \neq j$, which occurs when no energy transfer among BSs is implemented, as well as another special case of $\beta_{ij} = 1, \forall i, j \in \mathcal{N}, i \neq j$ for the ideal scenario of no energy transfer loss in the network.

With the aforementioned energy cooperation model, the available transmit power of BS i, P_i, should satisfy the following constraint:

$$0 \leq P_i \leq E_i + \sum_{j \in \mathcal{N}, j \neq i} \beta_{ji} e_{ji} - \sum_{j \in \mathcal{N}, j \neq i} e_{ij}, \quad i \in \mathcal{N}. \tag{6.30}$$

It is worth noting from (6.30) that to implement the energy exchange with the other $N - 1$ BSs, each BS $i \in \mathcal{N}$ only needs to either draw the total amount of energy, $\sum_{j \in \mathcal{N}} \beta_{ji} e_{ji} - \sum_{j \in \mathcal{N}} e_{ij}$, from the aggregator if this term is positive or otherwise inject the opposite amount (i.e. $-\sum_{j \in \mathcal{N}} \beta_{ji} e_{ji} + \sum_{j \in \mathcal{N}} e_{ij}$) of energy into the aggregator. Furthermore, since the total power injected to the grid by all BSs, i.e. $\sum_{i \in \mathcal{N}} \sum_{j \in \mathcal{N}, j \neq i} e_{ij}$, is equal to that drawn from the grid by all BSs, i.e. $\sum_{i \in \mathcal{N}} \sum_{j \in \mathcal{N}, j \neq i} \beta_{ij} e_{ij}$, plus the total power loss, i.e. $\sum_{i \in \mathcal{N}} \sum_{j \in \mathcal{N}, j \neq i} \zeta_{ij} e_{ij}$, the net energy exchanged with the aggregator is zero.

We denote the channel vector from BS i to MT k as $\boldsymbol{h}_{ik} \in \mathbb{C}^{1 \times M}, i \in \mathcal{N}, k \in \mathcal{K}$ and the channel vector from all N BSs in one particular cluster to MT k as $\boldsymbol{h}_k = [\boldsymbol{h}_{1k} \quad \cdots \quad \boldsymbol{h}_{Nk}] \in \mathbb{C}^{1 \times MN}, k \in \mathcal{K}$. It is assumed that the channel vector $\boldsymbol{h}_k$'s are drawn from a certain set of independent continuous distributions. Since we consider cooperative downlink transmission by N BSs, the downlink channel in each cluster can be modeled as a K-user MISO-BC with a total number of MN transmitting antennas from all N BSs.

We consider cooperative ZF precoding at BSs [22], with $K \leq MN$, although the cases of other precoding schemes can also be studied similarly. Let the information signal for MT $k \in \mathcal{K}$ be denoted by s_k and its associated precoding vector across N BSs denoted by $\boldsymbol{t}_k \in \mathbb{C}^{MN \times 1}$. Accordingly, the transmitted signal for MT k can be expressed as $\boldsymbol{x}_k = \boldsymbol{t}_k s_k$. Thus, the received signal at MT k is given by

$$y_k = \boldsymbol{h}_k \boldsymbol{x}_k + \sum_{l \in \mathcal{K}, l \neq k} \boldsymbol{h}_k \boldsymbol{x}_l + v_k, \quad k \in \mathcal{K}, \tag{6.31}$$

where $\boldsymbol{h}_k \boldsymbol{x}_k$ is the desired signal for MT k, $\sum_{l \in \mathcal{K}, l \neq k} \boldsymbol{h}_k \boldsymbol{x}_l$ is the inter-user interference within the cluster, and v_k denotes the background AWGN at MT k, which is assumed to be of zero mean and variance σ_k^2. Note that the background noise v_k at each receiver k may include the downlink interference from other BSs outside the cluster, although this effect can be neglected if proper frequency assignments over different clusters have been designed to minimize any intercluster interference. It is also assumed that Gaussian signaling is employed at BSs, i.e. s_k's are CSCG random variables with zero mean and unit variance. Thus, the covariance matrix of the transmitted signal for MT k can be expressed as $\boldsymbol{S}_k = \mathbb{E}(\boldsymbol{x}_k\boldsymbol{x}_k^H) = \boldsymbol{t}_k\boldsymbol{t}_k^H \succeq \boldsymbol{0}$. Accordingly, the transmit power at BS i can be expressed as [22]

$$P_i = \sum_{k\in\mathcal{K}} \mathtt{tr}(\boldsymbol{B}_i\boldsymbol{S}_k), \quad i \in \mathcal{N}, \tag{6.32}$$

where

$$\boldsymbol{B}_i \triangleq \mathtt{Diag}(\underbrace{0,\dots,0}_{(i-1)M},\underbrace{1,\dots,1}_{M},\underbrace{0,\dots,0}_{(N-i)M}).$$

By combining (6.32) and (6.30), we obtain the transmit power constraints under BSs' energy cooperation, given by

$$\sum_{k\in\mathcal{K}} \mathtt{tr}(\boldsymbol{B}_i\boldsymbol{S}_k) \le E_i + \sum_{j\in\mathcal{N},j\neq i} \beta_{ji}e_{ji} - \sum_{j\in\mathcal{N},j\neq i} e_{ij}, \quad i \in \mathcal{N}. \tag{6.33}$$

Specifically, the cooperative ZF precoding is described as follows. Define $\boldsymbol{H}_{-k} = [\boldsymbol{h}_1^T,\dots,\boldsymbol{h}_{k-1}^T,\boldsymbol{h}_{k+1}^T,\dots,\boldsymbol{h}_K^T]^T$, $k \in \mathcal{K}$, where $\boldsymbol{H}_{-k} \in \mathbb{C}^{(K-1)\times MN}$. Let the (reduced) singular value decomposition (SVD) of $\boldsymbol{H}_{-k}$ be denoted as $\boldsymbol{H}_{-k} = \boldsymbol{U}_k\boldsymbol{\Sigma}_k\boldsymbol{V}_k^H$, where $\boldsymbol{U}_k \in \mathbb{C}^{(K-1)\times(K-1)}$ with $\boldsymbol{U}_k\boldsymbol{U}_k^H = \boldsymbol{U}_k^H\boldsymbol{U}_k = \boldsymbol{I}$, $\boldsymbol{V}_k \in \mathbb{C}^{MN\times(K-1)}$ with $\boldsymbol{V}_k^H\boldsymbol{V}_k = \boldsymbol{I}$, and $\boldsymbol{\Sigma}_k$ is a $(K-1)\times(K-1)$ diagonal matrix. Define the projection matrix $\boldsymbol{P}_k = \boldsymbol{I} - \boldsymbol{V}_k\boldsymbol{V}_k^H$. Without loss of generality, we can express $\boldsymbol{P}_k = \tilde{\boldsymbol{V}}_k\tilde{\boldsymbol{V}}_k^H$, where $\tilde{\boldsymbol{V}}_k \in \mathbb{C}^{MN\times(MN-K+1)}$ satisfies $\tilde{\boldsymbol{V}}_k^H\tilde{\boldsymbol{V}}_k = \boldsymbol{I}$ and $\boldsymbol{V}_k^H\tilde{\boldsymbol{V}}_k = \boldsymbol{0}$. Note that $[\boldsymbol{V}_k, \tilde{\boldsymbol{V}}_k]$ forms an $MN \times MN$ unitary matrix. We then consider the cooperative ZF precoding vector given by

$$\boldsymbol{t}_k = \sqrt{p_k}\tilde{\boldsymbol{V}}_k\frac{(\boldsymbol{h}_k\tilde{\boldsymbol{V}}_k)^H}{\|\boldsymbol{h}_k\tilde{\boldsymbol{V}}_k\|}, \tag{6.34}$$

and accordingly the transmit covariance matrices can be expressed as

$$\boldsymbol{S}_k = p_k\frac{\tilde{\boldsymbol{V}}_k\tilde{\boldsymbol{V}}_k^H\boldsymbol{h}_k^H\boldsymbol{h}_k\tilde{\boldsymbol{V}}_k\tilde{\boldsymbol{V}}_k^H}{\|\boldsymbol{h}_k\tilde{\boldsymbol{V}}_k\|^2}, \tag{6.35}$$

where $p_k \ge 0, \forall k \in \mathcal{K}$. Note that for simplicity, in this paper we consider separately designed ZF precoding and power allocation as in (6.34) and (6.35), while our results can be extended to the cases with optimal joint ZF precoding and power allocation (see [22]) as well as other precoder designs. Under the above ZF precoding design, the inter-user interference within each cluster can be completely eliminated, i.e. we have $\boldsymbol{h}_k\boldsymbol{t}_l = 0$ or equivalently $\boldsymbol{h}_k\boldsymbol{S}_l\boldsymbol{h}_k^H = 0, \forall k, l \in \mathcal{K}, k \neq l$. As a result, the achievable data rate by the kth MT can be expressed as

$$r_k = \log_2\left(1 + \frac{\boldsymbol{h}_k\boldsymbol{S}_k\boldsymbol{h}_k^H}{\sigma_k^2}\right) = \log_2(1 + a_kp_k), \tag{6.36}$$

where $a_k = \frac{\|\boldsymbol{h}_k\tilde{\boldsymbol{V}}_k\|^2}{\sigma_k^2}, \forall k \in \mathcal{K}$. Meanwhile, given $\{\boldsymbol{S}_k\}$ in (6.35), the power constraints in (6.33) can be rewritten as

$$\sum_{k\in\mathcal{K}} b_{ik}p_k \le E_i + \sum_{j\in\mathcal{N},j\neq i} \beta_{ji}e_{ji} - \sum_{j\in\mathcal{N},j\neq i} e_{ij}, \quad i \in \mathcal{N}, \tag{6.37}$$

where $b_{ik} = \frac{\mathrm{tr}(\boldsymbol{B}_i \tilde{\boldsymbol{V}}_k \tilde{\boldsymbol{V}}_k^H \boldsymbol{h}_k^H \boldsymbol{h}_k \tilde{\boldsymbol{V}}_k \tilde{\boldsymbol{V}}_k^H)}{\|\boldsymbol{h}_k \tilde{\boldsymbol{V}}_k\|^2}, \forall i \in \mathcal{N}, k \in \mathcal{K}$.[8]

We aim to jointly optimize the transmit power allocations $\{p_k\}$ at all N BSs, as well as their transferred energy $\{e_{ij}\}$, so as to maximize the weighted sum-rate throughput (in bps Hz^{-1}) for all K MTs given by $\sum_{k\in\mathcal{K}} \omega_k r_k$, where r_k denotes the achievable rate by the kth MT given in (6.36) and $\omega_k > 0$ denotes the given weight for MT $k, k \in \mathcal{K}$. Here, larger weight value of ω_k indicates higher priority of transmitting information to MT k as compared with other MTs; hence, by properly designing the weight ω_k's, rate fairness among different MTs can be ensured. Next, we formulate the joint communication and energy cooperation problem as

$$\max_{\{p_k\},\{e_{ij}\}} \sum_{k\in\mathcal{K}} \omega_k \log_2(1 + a_k p_k) \tag{6.38}$$

$$\text{s.t.} \quad \sum_{k\in\mathcal{K}} b_{ik} p_k \le E_i + \sum_{j\in\mathcal{N}, j\neq i} \beta_{ji} e_{ji} - \sum_{j\in\mathcal{N}, j\neq i} e_{ij}, \quad \forall i \in \mathcal{N} \tag{6.39}$$

$$p_k \ge 0, \quad e_{ij} \ge 0, \qquad \forall k \in \mathcal{K}, \quad i, j \in \mathcal{N}, i \neq j. \tag{6.40}$$

Before we proceed to solving problem (6.38), we first consider the following two special cases for problem (6.38) with $\beta_{ij} = 0, \forall i, j \in \mathcal{N}, i \neq j$ and $\beta_{ij} = 1, \forall i, j \in \mathcal{N}, i \neq j$ to draw some insights. The former case of $\beta_{ij} = 0, \forall i, j \in \mathcal{N}, i \neq j$ corresponds to the scenario of no energy transfer between any two BSs in the cluster, for which it is equivalent to setting $e_{ij} = 0, \forall i, j \in \mathcal{N}, i \neq j$. In this case, the power constraints in (6.39) reduce to

$$\sum_{k\in\mathcal{K}} b_{ik} p_k \le E_i, \quad \forall i \in \mathcal{N}. \tag{6.41}$$

Accordingly, problem (6.38) reduces to the conventional MISO-BC weighted sum-rate maximization problem with N per BS power constraints, given by $E_i, \forall i \in \mathcal{N}$ [22]. On the other hand, if $\beta_{ij} = 1, \forall i, j \in \mathcal{N}, i \neq j$, the energy of any two BSs can be shared ideally without any loss. It can then be easily verified that in this case BSs' individual power constraints in (6.39) can be combined into one single sum power constraint as

$$\sum_{k\in\mathcal{K}} p_k \le \sum_{i\in\mathcal{N}} E_i. \tag{6.42}$$

Thus, problem (6.38) reduces to the conventional MISO-BC weighted sum-rate maximization problem under one single sum power constraint [22].

As an example, we consider a two-cell system with $N = 2$, where $\beta_{ij} \triangleq \beta, \forall i, j \in \{1, 2\}, i \neq j$. We depict in Figure 6.11 the feasible power region (shown as the shaded area) consisting of the all available transmit power pairs (P_1, P_2) at BS 1 and BS 2 under energy cooperation in the case of $0 < \beta < 1$, as compared with the other two extreme cases of $\beta = 0$ and $\beta = 1$. It is observed that the boundary of the power region when $\beta > 0$ is always attained either by $e_{12} \ge 0$, $e_{21} = 0$ or $e_{12} = 0$, $e_{21} \ge 0$. Similarly, it can be shown that for the general multicell case with $N > 2$, the boundary of the N-dimension power region should also be attained either by $e_{ij} \ge 0$, $e_{ji} = 0$ or $e_{ij} = 0$, $e_{ji} \ge 0, \forall i, j \in \mathcal{N}, i \neq j$ (for any pair of two BSs i and j). This result together with the fact that the weighted sum rate in (6.38) always increases with the transmit power

8 Since it is assumed that the channel vector $\boldsymbol{h}_k$'s are independently distributed, without loss of generality we assume $a_k > 0, b_{ik} > 0, \forall i, k$.

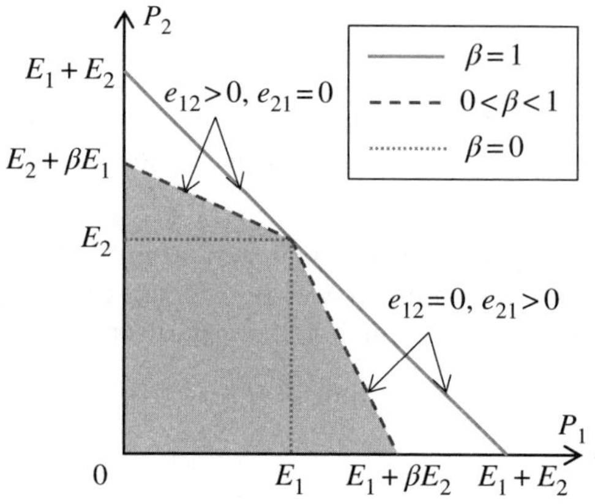

Figure 6.11 Feasible power region of BSs' transmit power under energy cooperation for a two-cell system.

implies that only unidirectional energy transfer between any pair of two BSs is needed to achieve the optimal solution of problem (6.38). It is also observed from Figure 6.11 that the power region in the case of $0 < \beta < 1$ is larger than that without energy sharing, i.e. $\beta = 0$, and at the same time smaller than that with ideal energy sharing, i.e. $\beta = 1$. Similarly, it is conjectured that for the general multicell scenario with $N > 2$, the feasible power region in the case of $0 < \beta_{ij} < 1, \forall i, j \in \mathcal{N}, i \neq j$ is also larger than that of $\beta_{ij} = 0, \forall i, j \in \mathcal{N}, i \neq j$, and at the same time smaller than that of $\beta_{ij} = 1, \forall i, j \in \mathcal{N}, i \neq j$. As a result, the optimal value of problem (6.38) under practical energy cooperation with $0 < \beta_{ij} < 1, \forall i, j \in \mathcal{N}, i \neq j$ should lie between those of the two extreme cases with the per BS power constraints ($\beta_{ij} = 0, \forall i, j \in \mathcal{N}, i \neq j$) and the BSs' sum power constraint ($\beta_{ij} = 1, \forall i, j \in \mathcal{N}, i \neq j$), as will be shown rigorously later in this paper.

6.7.2 Optimal Solution

In this section, we solve problem (6.38) for the general case of $0 \leq \beta_{ij} \leq 1, \forall i, j \in \mathcal{N}, i \neq j$ to obtain the optimal joint power allocation and energy transfer solution for BSs' joint communication and energy cooperation.

It can be verified that problem (6.38) is a convex problem, since the objective function is concave and all the constraints are affine. Thus, the Lagrange duality method can be applied to solve this problem optimally [18]. Let $\mu_i \geq 0, i \in \mathcal{N}$ be the dual variable associated with each of the N power constraints of problem (6.38) given in (6.39). Then the partial Lagrangian of problem (6.38) can be expressed as

$$\begin{aligned}
&\mathcal{L}(\{\mu_i\}, \{p_k\}, \{e_{ij}\}) \\
&= \sum_{k\in\mathcal{K}} \omega_k \log_2(1 + a_k p_k) \\
&\quad - \sum_{i\in\mathcal{N}} \mu_i \left(\sum_{k\in\mathcal{K}} b_{ik} p_k - E_i - \sum_{j\in\mathcal{N}, j\neq i} \beta_{ji} e_{ji} + \sum_{j\in\mathcal{N}, j\neq i} e_{ij} \right) \\
&= \sum_{k\in\mathcal{K}} \left(\omega_k \log_2(1 + a_k p_k) - \sum_{i\in\mathcal{N}} b_{ik} \mu_i p_k \right) \\
&\quad + \sum_{i,j\in\mathcal{N}, i\neq j} (\beta_{ij}\mu_j - \mu_i) e_{ij} + \sum_{i\in\mathcal{N}} \mu_i E_i.
\end{aligned} \tag{6.43}$$

Accordingly, the dual function is given by

$$f(\{\mu_i\}) = \max_{\{p_k\},\{e_{ij}\}} \mathcal{L}(\{\mu_i\}, \{p_k\}, \{e_{ij}\})$$
$$\text{s.t.} \quad p_k \geq 0, e_{ij} \geq 0, \qquad \forall k \in \mathcal{K}, i, j \in \mathcal{N}, i \neq j. \tag{6.44}$$

Thus, the dual problem is defined as

$$\min_{\{\mu_i \geq 0\}} \quad f(\{\mu_i\}). \tag{6.45}$$

Since problem (6.38) is convex and satisfies the Salter's condition, strong duality holds between problem (6.38) and its dual problem (6.45) [18]. Thus, we can solve problem (6.38) by solving its dual problem (6.45) equivalently. To solve problem (6.45), in the following we first solve problem (6.44) to obtain $f(\{\mu_i\})$ with a given set of $\mu_i \geq 0, i \in \mathcal{N}$, and then search over $\{\mu_i\}$ in $\mathbb{R}_N^+$ to minimize $f(\{\mu_i\})$ in (6.45).

We first give the following lemma.

Lemma 6.2 *In order for $f(\{\mu_i\})$ to be bounded from above, we have:*

(1) At least one of $\mu_i, \forall i \in \mathcal{N}$, must be strictly positive.
(2) $\beta_{ij}\mu_j - \mu_i \leq 0, \forall i, j \in \mathcal{N}, i \neq j$, must be true.

Proof: See [19]. □

According to Lemma 6.2, we only need to solve problem (6.44) with given $\{\mu_i\}$ satisfying $\mu_i \geq 0, \forall i \in \mathcal{N}$ (but not all equal to zero), and $\beta\mu_j - \mu_i \leq 0, \forall i, j \in \mathcal{N}, i \neq j$, since otherwise $f(\{\mu_i\})$ will be unbounded from above and thus need not to be considered for the minimization problem in (6.45). In this case, problem (6.44) can be decomposed into the following $K + N^2 - N$ number of subproblems (by removing the irrelevant constant term $\sum_{i \in \mathcal{N}} \mu_i E_i$ in (6.43)):

$$\max_{p_k \geq 0} \quad \omega_k \log_2(1 + a_k p_k) - \sum_{i \in \mathcal{N}} b_{ik}\mu_i p_k, \quad \forall k \in \mathcal{K}, \tag{6.46}$$

$$\max_{e_{ij} \geq 0} \quad (\beta_{ij}\mu_j - \mu_i)e_{ij}, \quad \forall i, j \in \mathcal{N}, i \neq j. \tag{6.47}$$

For the K subproblems in (6.46), note that $\sum_{i \in \mathcal{N}} b_{ik}\mu_i > 0$ always holds due to Lemma 6.2. Thus, it can be easily verified that the optimal solutions can be obtained as

$$p_k^{(\{\mu_i\})} = \left(\frac{\omega_k}{\ln 2 \sum_{i \in \mathcal{N}} b_{ik}\mu_i} - \frac{1}{a_k} \right)^+, \quad \forall k \in \mathcal{K}, \tag{6.48}$$

where $(x)^+ = \max(x, 0)$. Next, consider the remaining $N^2 - N$ subproblems in (6.47), for which it can be easily shown that with $\beta_{ij}\mu_j - \mu_i \leq 0, \forall i \neq j$ given in Lemma 6.2, the optimal solutions are

$$e_{ij}^{(\{\mu_i\})} = 0, \quad \forall i, j \in \mathcal{N}, i \neq j. \tag{6.49}$$

Notice that if $\beta_{ij}\mu_j - \mu_i = 0$, then the optimal solution of e_{ij} to problem (6.47) is not unique and can take any nonnegative value. In this case, we let $e_{ij}^{(\{\mu_i\})} = 0$ given in (6.49) only for solving the dual problem in (6.44), which may not be the optimal primary solution of e_{ij} to problem (6.38).

Combining the results in (6.48) and (6.49), we obtain $f(\{\mu_i\})$ with given $\{\mu_i\}$ satisfying $\mu_i \geq 0, \forall i \in \mathcal{N}$ (but not all equal to zero), and $\beta_{ij}\mu_j - \mu_i \leq 0, \forall i,j \in \mathcal{N}$. Then, we solve problem (6.45) by finding the optimal $\{\mu_i\}$ to minimize $f(\{\mu_i\})$. According to Lemma 6.2, the dual problem (6.45) can be equivalently reexpressed as

$$\begin{aligned} &\min_{\{\mu_i\}} f(\{\mu_i\}) \\ &\text{s.t.} \quad \mu_i \geq 0, \quad \forall i \in \mathcal{N} \\ &\qquad \beta_{ij}\mu_j - \mu_i \leq 0, \quad \forall i,j \in \mathcal{N}, i \neq j. \end{aligned} \tag{6.50}$$

Since the dual problem (6.45) is convex but not necessarily differentiable, a subgradient-based method such as the ellipsoid method [20] can be applied, for which it can be shown that the subgradients of $f(\{\mu_i\})$ for given μ_i are $E_i - \sum_{k\in\mathcal{K}} b_{ik} p_k^{(\{\mu_i\})} + \sum_{j\in\mathcal{N},j\neq i} \beta_{ji} e_{ji}^{(\{\mu_i\})} - \sum_{j\in\mathcal{N},j\neq i} e_{ij}^{(\{\mu_i\})} = E_i - \sum_{k\in\mathcal{K}} b_{ik} p_k^{(\{\mu_i\})}, \forall i \in \mathcal{N}$, where the equality follows from (6.49). Therefore, the optimal solution of problem (6.45) can be obtained as $\{\mu_i^*\}$.

With the optimal dual solution $\{\mu_i^*\}$ at hand, the corresponding $\{p_k^{(\{\mu_i^*\})}\}$ in (6.48) become the optimal solution for problem (6.38), denoted by $\{p_k^*\}$. Now, it remains to obtain the optimal solution of $\{e_{ij}\}$ for problem (6.38), denoted by $\{e_{ij}^*\}$. In general, $\{e_{ij}^*\}$ cannot be directly obtained from (6.49) with given $\{\mu_i^*\}$, since the solution of (6.49) is not unique if $\beta_{ij}\mu_j^* - \mu_i^* = 0$. Fortunately, it can be shown that given $\{p_k^*\}$, any $\{e_{ij}\}$ that satisfy the linear constraints in (6.39) and (6.40) are the optimal solution to problem (6.38). Thus, we can obtain $\{e_{ij}^*\}$ by solving the following feasibility problem:

$$\begin{aligned} &\text{find} \{e_{ij}\} \\ &\text{s.t.} \quad \sum_{k\in\mathcal{K}} b_{ik} p_k^* \leq E_i + \sum_{j\in\mathcal{N},j\neq i} \beta_{ji} e_{ji} - \sum_{j\in\mathcal{N},j\neq i} e_{ij}, \quad \forall i \in \mathcal{N} \\ &\qquad e_{ij} \geq 0, \qquad \forall i,j \in \mathcal{N}, i \neq j. \end{aligned} \tag{6.51}$$

Since problem (6.51) is a simple linear program (LP), it can be efficiently solved by standard convex optimization techniques such as the interior point method. As a result, we have finally obtained $\{e_{ij}^*\}$ and thus have solved problem (6.38) completely.

In summary, one algorithm to solve problem (6.38) for the general case of $0 \leq \beta_{ij} \leq 1, \forall i,j \in \mathcal{N}, i \neq j$, is given in Algorithm 6.3, in which $\mu_i = \mu > 0, \forall i \in \mathcal{N}$, are chosen as the initial point for the ellipsoid method in order to satisfy the constraints in (6.50). Note that Algorithm 6.2 needs to be implemented at a central unit, which is assumed to have all the CSI and energy information from all BSs in the same cluster.

6.7.3 Numerical Results

In this section, we provide simulation results to evaluate the performance of our proposed joint communication and energy cooperation algorithm, as compared with three benchmark schemes with communication cooperation only, energy cooperation only, and no cooperation. We set $\omega_k = 1, \forall k \in \mathcal{K}$ and thus consider the sum-rate throughput of all MTs as the performance metric. We also set $\beta_{ij} \triangleq \beta, \forall i,j \in \mathcal{N}, i \neq j$. We consider a

Algorithm 6.2: Solving the problem (6.38).

Input: $\mu_i = \mu > 0, \forall i \in \mathcal{N}$
Output: $\{e^*_{ij}\}$
1: **repeat**
2: Obtain $\{p_k^{(\{\mu_i\})}\}$ using (6.48) with given $\{\mu_i\}$
3: Compute the subgradients of $f(\{\mu_i\})$ as $E_i - \sum_{k\in\mathcal{K}} b_{ik} p_k^{(\{\mu_i\})}, \forall i \in \mathcal{N}$, then update $\{\mu_i\}$ accordingly based on the ellipsoid method [20], subject to the constraints of $\mu_i \geq 0, \forall i \in \mathcal{N}$, and $\beta\mu_j - \mu_i \leq 0, \forall i,j \in \mathcal{N}, i \neq j$
4: **until** $\{\mu_i\}$ all converge within a prescribed accuracy
5: $p_k^* \leftarrow p_k^{(\{\mu_i\})}, \forall k \in \mathcal{K}$
6: Compute $\{e^*_{ij}\}$ by solving the LP in (6.51)

practical three-cell cluster (with $N = 3$), where the cells are hexagonal with the inter-BS distance of one kilometer, and we set $M = 4$ and $K = 12$. We further consider the normalized wind and solar energy harvesting profiles over time as shown in Figures 6.6–6.8. We use a given constant $\bar{E}$ to denote the average energy harvesting rate. Moreover, since the channel coherence time is much shorter than the energy sample time of 15 minutes, for each sample of energy harvesting profiles, we randomly generate four MTs in the area covered by each hexagonal cell and run 100 independent channel realizations.

We assume that the channel is modeled by combining path loss and short-term fading. For each channel realization, it is assumed that the distance-dependent channel attenuation from BS i to MT k is fixed, which is determined by the path loss model of $\tilde{h}_{ik} = c_0\left(\frac{d_{ik}}{d_0}\right)^{-\varsigma}$, where $c_0 = -60$ dB is a constant equal to the path loss at a reference distance $d_0 = 10$ m, $\varsigma = 3.7$ is the path loss exponent, and d_{ik} denotes the distance from BS i to MT k. We also consider i.i.d. Rayleigh distributed short-term fading, i.e. the channels for each realization, denoted by $\boldsymbol{h}_{ik}$'s, are zero mean CSCG random vectors with covariance matrix specified by the corresponding path loss, i.e. $\tilde{h}_{ik}\boldsymbol{I}$. In addition, we assume that the background noise at each MT receiver is -85 dBm. As a result, supposing the transmit power at each BS is 10 dBW, the average SNR for each BS to an MT at each vertex of its covered hexagon is thus 0 dB.

In Figure 6.12, we show the average sum-rate performance over time, where we set $\bar{E} = 10$ dBW. It is observed that the performance gain of the "joint communication and energy cooperation" scheme against the "communication cooperation only" scheme is very significant in the nighttime (e.g. hours 0–7, 20–30), while this gain becomes much smaller during the daytime (e.g. hours 10–18). The reason is given as follows. In the nighttime, the sum-rate performance of the "communication cooperation only" scheme is limited by BS 3 that mainly relies on solar generation and thus has very low energy harvesting rates; hence, energy cooperation is most beneficial during the night period. In contrast, during the day time when the energy harvesting rates at three BSs are more evenly distributed, the energy cooperation gain becomes less notable. Furthermore, the performance gain of the "energy cooperation only" scheme over the "no cooperation" scheme is observed to behave similarly. The result shows the benefit of energy cooperation again when the energy harvesting rates are unevenly distributed, even in the case without communication cooperation.

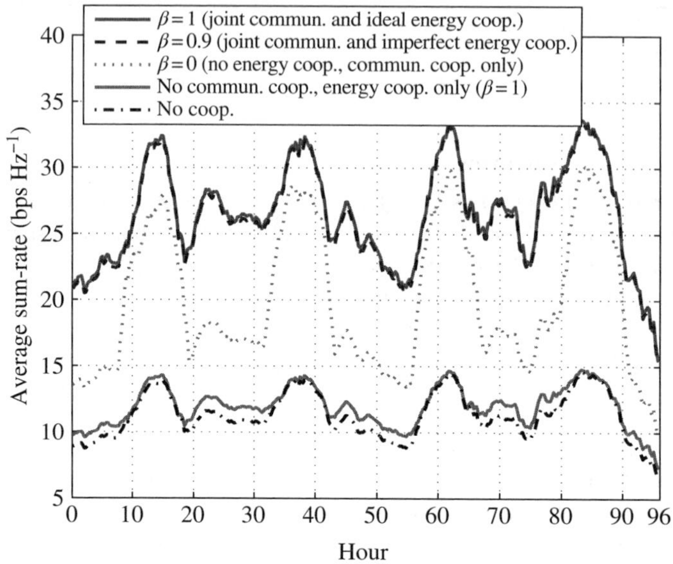

Figure 6.12 Sum-rate performance over time, where $\bar{E} = 10$ dBW.

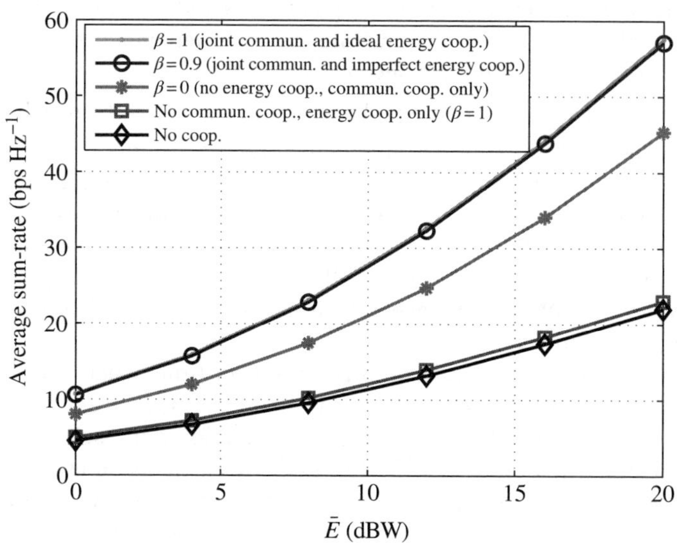

Figure 6.13 Sum-rate comparison with versus without communication and/or energy cooperation.

In Figure 6.13, we show the MTs' average sum-rate performance over four days versus $\bar{E}$. It is observed that the joint communication and energy cooperation with $\beta = 1$ or $\beta = 0.9$ considerably outperforms the other three suboptimal schemes at all values of $\bar{E}$, which shows the significant gains of joint communication and energy cooperation under this practical setup.

6.8 Extensions and Future Directions

Despite the aforementioned studies on energy and communication cooperation, there remain a lot of interesting topics unaddressed. We list several of them as follows for future study.

Practically, energy harvesting rates in general change slowly as compared with wireless channel and traffic load variations, and as a consequence, the time scale of implementing energy cooperation is normally longer than that of communication cooperation [23]. However, joint energy and communication cooperation in this chapter requires the energy cooperation to be realized in the same time scale as the communication cooperation, thus needing more frequent decision-making at BSs and higher implementation complexity at smart meters. To overcome this issue, it is promising to consider the multi-time-scale implementation of joint energy and communication cooperation, e.g. by employing a two-layer decision-making with energy cooperation in longer time scale and communications cooperation in shorter time scale, so as to balance the tradeoff between the cost-saving performance and the implementation complexity.

So far, we have focused on a single cellular system or multiple systems belonging to the same entity, aiming to minimize the total energy cost. In practice, however, multiple self-interested systems (owned by different operators) can coexist or colocate, and it is interesting to study their energy and/or communication cooperation. Unlike energy trading in Section 6.3, on the energy supply side, more than one aggregator may be needed to facilitate trading across different BS groups. As for the mutual energy sharing scheme, one selfish system may want to sell (buy) renewable energy to (from) the other system at a high (low) price. On the communication demand side, it is happening that some systems (e.g. Verizon and T-Mobile) are sharing spectrum in long term. Yet how to enable communication cooperation in short term (as in Section 6.4) requires intersystem communication compatibility and more coordination. Moreover, to establish joint energy and communication cooperation, cellular systems may seek for the advantage of resource complementarity. For example, in a preliminary study [14], it is shown that one system adequate in spectrum is willing to cooperate with another adequate in energy, since both systems can efficiently reduce their individual costs by exchanging spectrum and energy with each other. Overall, cooperation mechanism design is required to motivate or strengthen intersystem joint cooperation to realize a win-win situation for all systems involved.

Besides cellular networks, it is also appealing for heterogeneous communication networks (e.g. WiFi and small cells) to cooperate and reduce overall energy cost [24]. Offloading a mobile user from a macrocell to a small cell saves energy and better utilizes the wireline backhaul resource to expand the limited wireless spectrum. Yet these networks are different in service coverage, operated spectrum, and even energy harvesting availability (difficult indoor), and thus their joint energy and communication cooperation becomes more complicated than our design in Section 6.5. For example, scalability could be a problem, and one possible solution is to decompose the whole heterogeneous network into a number of micro-networks as in [6] with cooperation in each.

Up to now, we have focused on the case without the use of energy storage at BSs due to the cost consideration. With the advancement of battery technologies, we envision that energy storage may be employed in the future BSs and it is promising to study the energy and communication cooperation jointly with the storage management. In principle, the storage devices handle the renewable energy fluctuations at BSs to match the energy demand variations over time [25], while the energy and communication cooperation approaches introduced in this chapter do that over space. Therefore, the two approaches can be good complementarities. Nevertheless, such joint time and space domain optimization problems are very challenging to solve, since any present decisions made by BSs would affect their storage status and traffic loads served in the future. As an initial study, Chia et al. [26] has considered the joint energy cooperation and storage management problem for minimizing the total energy cost in a simplified cellular system with given energy demand at BSs.

Furthermore, it is worth noting that recently smart grids also enable time-varying energy prices to help stabilize the energy generation and transmission. For example, a hybrid electricity market has been successfully implemented in the United States and Norway [27, 28], which combines a day-ahead energy market and a real-time energy market. In the day-ahead energy market, electricity consumers can make commitment on tomorrow's energy purchase at low prices, whereas in the real-time energy market, they are free from commitment and can flexibly buy energy at higher prices or sell back the excessive energy commitment at prices lower than the day-ahead ones. In view of such a new hybrid electricity market, the energy purchased by cellular BSs in day ahead and in real time may lead to distinct unit costs at different time. This, together with the cost difference between renewable and on-grid energy, makes the cost-effective energy management of cellular networks more challenging. In this case, the cellular operator needs to jointly optimize many BSs' day-ahead and real-time energy cooperation (in energy purchase) and their communication cooperation, based on the time-varying wireless traffic load and energy prices, for the purpose of minimizing the energy costs [29].

6.9 Summary

This chapter investigates novel energy and communication cooperation approaches for energy cost saving in cellular networks with BSs powered by energy harvesting and smart grid. These approaches use both the two-way energy flow in smart grid and the communication cooperation in cellular networks to reshape the nonuniform energy supplies and energy demands over the cellular networks for cost saving. It is our hope that these new approaches can bring new insights on the energy demand management in smart grid by considering the unique properties of the cellular networks' communication demand and also on the wireless resource allocation in cellular networks by taking into account the new characteristics of the emerging renewable and smart grid energy supply.

References

1 Andrews, J.G., Buzzi, S., Choi, W. et al. (2014). What will 5G be. *IEEE Journal on Selected Areas in Communications* 32 (6): 1065–1082. https://doi.org/10.1109/JSAC.2014.2328098.

2 Hasan, Z., Boostanimehr, H., and Bhargava, V.K. (2011). Green cellular networks: a survey, some research issues and challenges. *IEEE Communication Surveys and Tutorials* 13 (4): 524–540. https://doi.org/10.1109/SURV.2011.092311.00031.

3 Wu, J., Rangan, S., and Zhang, H. (eds.) (2012). *Green Communications: Theoretical Fundamentals, Algorithms, and Applications*. CRC Press.

4 Bu, S., Yu, F.R., Cai, Y., and Liu, X.P. (2012). When the smart grid meets energy-efficient communications: green wireless cellular networks powered by the smart grid. *IEEE Transactions on Wireless Communications* 11 (8): 3014–3024. https://doi.org/10.1109/TWC.2012.052512.111766.

5 Han, T. and Ansari, N. (2014). Powering mobile networks with green energy. *IEEE Wireless Communications* 21 (1): 90–96. https://doi.org/10.1109/MWC.2014.6757901.

6 Fang, X., Misra, S., Xue, G., and Yang, D. (2012). Smart grid-the new and improved power grid: a survey. *IEEE Communication Surveys and Tutorials* 14 (4): 944–980. https://doi.org/10.1109/SURV.2011.101911.00087.

7 Saad, W., Han, Z., Poor, H.V., and Basar, T. (2012). Game-theoretic methods for the smart grid: an overview of microgrid systems, demand-side management, and smart grid communications. *IEEE Signal Processing Magazine* 29 (5): 86–105. https://doi.org/10.1109/MSP.2012.2186410.

8 Niu, Z., Wu, Y., Gong, J., and Yang, Z. (2010). Cell zooming for cost-efficient green cellular networks. *IEEE Communications Magazine* 48 (11): 74–78. https://doi.org/10.1109/MCOM.2010.5621970.

9 Zhang, R., Liang, Y.C., and Cui, S. (2010). Dynamic resource allocation in cognitive radio networks. *IEEE Signal Processing Magazine* 27 (3): 102–114. https://doi.org/10.1109/MSP.2010.936022.

10 Gesbert, D., Hanly, S., Huang, H. et al. (2010). Multi-cell MIMO cooperative networks: a new look at interference. *IEEE Journal on Selected Areas in Communications* 28 (9): 1380–1408. https://doi.org/10.1109/JSAC.2010.101202.

11 Xu, J., Duan, L., and Zhang, R. (2015). Cost-aware green cellular networks with energy and communication cooperation. *IEEE Communications Magazine* 53 (5): 257–263. https://doi.org/10.1109/MCOM.2015.7105673.

12 Gkatzikis, L., Koutsopoulos, I., and Salonidis, T. (2013). The role of aggregators in smart grid demand response markets. *IEEE Journal on Selected Areas in Communications* 31 (7): 1247–1257. https://doi.org/10.1109/JSAC.2013.130708.

13 Xu, J. and Zhang, R. (2015). CoMP meets smart grid: a new communication and energy cooperation paradigm. *IEEE Transactions on Vehicular Technology* 64 (6): 2476–2488. https://doi.org/10.1109/TVT.2014.2345415.

14 Guo, Y., Xu, J., Duan, L., and Zhang, R. (2014). Joint energy and spectrum cooperation for cellular communication systems. *IEEE Transactions on Communications* 62 (10): 3678–3691. https://doi.org/10.1109/TCOMM.2014.2353632.

15 Wiesel, A., Eldar, Y.C., and Shamai, S. (2006). Linear precoding via conic optimization for fixed MIMO receivers. *IEEE Transactions on Signal Processing* 54 (1): 161–176. https://doi.org/10.1109/TSP.2005.861073.

16 Bengtsson, M. and Ottersten, B. (2002). *Handbook of Antennas in Wireless Communications* (ed. L.C. Godara). CRC Press. https://doi.org/10.1201/9781420042146.

17 Yu, W. and Lan, T. (2007). Transmitter optimization for the multi-antenna downlink with per-antenna power constraints. *IEEE Transactions on Signal Processing* 55 (6): 2646–2660. https://doi.org/10.1109/TSP.2006.890905.

18 Boyd, S. and Vandenberghe, L. (2004). *Convex Optimization*. Cambridge: Cambridge University Press.

19 Xu, J. and Zhang, R. (2016). Cooperative energy trading in CoMP systems powered by smart grids. *IEEE Transactions on Vehicular Technology* 65 (4): 2142–2153. https://doi.org/10.1109/TVT.2015.2423317.

20 Boyd, S. Convex Optimization II. Stanford University. https://doi.org/http://www.stanford.edu/class/ee364b/lectures.html (accessed 2013).

21 Wiesel, A., Eldar, Y.C., and Shamai, S. (2008). Zero-forcing precoding and generalized inverses. *IEEE Transactions on Signal Processing* 56 (9): 4409–4418. https://doi.org/10.1109/TSP.2008.924638.

22 Zhang, R. (2010). Cooperative multi-cell block diagonalization with per-base-station power constraints. *IEEE Journal on Selected Areas in Communications* 28 (9): 1435–1445. https://doi.org/10.1109/JSAC.2010.101205.

23 Li, H., Xu, J., Zhang, R., and Cui, S. (2015). A general utility optimization framework for energy-harvesting-based wireless communications. *IEEE Communications Magazine* 53 (4): 79–85. https://doi.org/10.1109/MCOM.2015.7081079.

24 Cai, S., Che, Y., Duan, L. et al. (2016). Green 5G heterogeneous networks through dynamic small-cell operation. *IEEE Journal on Selected Areas in Communications* 34 (5): 1103–1115. https://doi.org/10.1109/JSAC.2016.2520217.

25 Rahbar, K., Xu, J., and Zhang, R. (2015). Real-time energy storage management for renewable integration in microgrid: an off-line optimization approach. *IEEE Transactions on Smart Grid* 6 (1): 124–134. https://doi.org/10.1109/TSG.2014.2359004.

26 Chia, Y.K., Sun, S., and Zhang, R. (2014). Energy cooperation in cellular networks with renewable powered base stations. *IEEE Transactions on Wireless Communications* 13 (12): 6996–7010. https://doi.org/10.1109/TWC.2014.2339845.

27 Ott, A.L. (2003). Experience with PJM market operation, system design, and implementation. *IEEE Transactions on Power Systems* 18 (2): 528–534. https://doi.org/10.1109/TPWRS.2003.810698.

28 Philpott, A.B. and Pettersen, E. (2006). Optimizing demand-side bids in day-ahead electricity markets. *IEEE Transactions on Power Systems* 21 (2): 488–498. https://doi.org/10.1109/TPWRS.2006.873119.

29 Xu, J., Duan, L., and Zhang, R. (2016). Energy group buying with loading sharing for green cellular networks. *IEEE Journal on Selected Areas in Communications* 34 (4): 786–799. https://doi.org/10.1109/JSAC.2016.2544603.

7

Energy Harvesting in Next-Generation Cellular Networks

7.1 Introduction

To handle the explosive growth of mobile traffic, next-generation cellular network will deploy more and more small-cell BSs (SBSs) in addition to the macro base stations (MBSs) [1]. The resultant network, namely, the heterogeneous network (HetNet), provides capacity boost on one hand but brings more energy consumption with the densely deployed SBSs on the other hand [2]. In fact, due to the dynamics of wireless traffic load, many BSs are lightly loaded but almost work at their peak power, due to the elements like power amplifiers and supporting circuits. Unfortunately, these BSs can hardly be turned off for the coverage guarantee. To solve this problem, a new separation architecture called hyper-cellular network (HCN) [3] is proposed, and the main idea is to decouple the function of control signaling from the function of data transmission, such that the data coverage can match the traffic dynamics in a more elastic way. Under HCN, SBSs are only utilized for high data rate transmission, whereas MBSs guarantee the network coverage and provide low data rate service. Therefore, SBSs can be turned off to save energy without worrying about the user coverage. To this end, its nature is to further power SBSs with renewable energy to save more grid power consumption [4].

However, as we have illustrated in Chapter 4, due to the randomness of renewable energy arrivals, it is challenging to manage wireless resource and the on–off states of energy harvesting (EH) BSs. It can be more challenging in HCN. First, diverse types of SBSs may be equipped with different kinds of energy sources, making the energy arrival statistically nonuniform over the space. In addition, the traffic load is nonevenly distributed across different base station (BS) tiers and also not in accordance with the energy arrivals over the spatial and temporal domains [5, 6]. To this end, on top of the techniques introduced in Chapter 4, in HCN the key to *match* the random energy arrival with the traffic load variation over *time and space* is to jointly optimize the working states of SBSs and the user traffic offloading. Although traffic offloading has been extensively studied in grid-powered cellular networks [7–11], the conventional offloading methods cannot be directly applied as they do not consider the energy states of BSs. Accordingly, energy-aware traffic offloading schemes are needed, and some energy-aware traffic offloading schemes have been proposed for single-tier homogeneous networks [12, 13] and two-tier HCN with one renewable energy-powered SBS [14], respectively. In the first part of the chapter, we will illustrate how to coordinate the on–off switching of SBSs with inter-tier traffic offloading, under the scenario with different types of SBSs, powered by various energy sources. The goal is to minimize the on-grid power consumption

Energy Harvesting Wireless Communications, First Edition.
Chuan Huang, Sheng Zhou, Jie Xu, Zhisheng Niu, Rui Zhang and Shuguang Cui.

of the whole HCN system while satisfying the quality of service (QoS) requirements of users [15].

Another emerging technology of next-generation cellular networks is to exploit edge caching [16–21] with proactive services, like push [22–25]. While the initial motivation of proactive caching and push is to reduce the duplicated content transmissions, and thus reduce the core network traffic load as well as the content delivery delay, it is also beneficial to address the mismatch between the energy and traffic in renewable energy-powered SBSs [26]. Specifically, the contents can be cached at the storage of SBSs and then pushed to users earlier than the actual demands when there is sufficient harvested energy. The users can successfully get the contents when they actually require it even if at that time the SBS does not have enough energy for transmission. Consequently, the energy waste due to the battery overflow can be avoided as the harvested energy can be used effectively and timely. It can be viewed as *transferring* the harvested energy along with the timeline to the *future* to *match* the random energy arrival with the traffic needs. In the second part of this chapter, we will demonstrate the concept of integrating proactive service provisioning with EH HCN and provide a detailed study on the optimal policy design for content push from an EH-based SBS [27].

This chapter is organized as follows. In Section 7.2, we demonstrate the use of traffic load offloading to better utilize the renewable energy in EH HCN, and this can be regarded as the matching of traffic and energy over the spatial domain. To further match the traffic and energy over the temporal domain, we use proactive content caching and push, as illustrated in Section 7.3.

7.2 Energy Harvesting Hyper-cellular Networks

7.2.1 System Model

In this section, the detailed model of HCN with hybrid energy supply is presented, together with the traffic model and power consumption model.

7.2.1.1 HCNs with Hybrid Energy Supply

A typical HCN is shown in Figure 7.1, where different types of SBSs, powered by various energy sources, are deployed in addition to the grid-powered MBS for capacity enhancement and elastic service. Depending on the energy sources being used to power the SBSs, SBSs can be classified into three types: (i) on-grid conventional SBSs (**CSBSs**); (ii) off-grid SBSs powered solely by renewable energy (**RSBSs**), for example, the solar and wind power; and (iii) hybrid SBSs (**HSBSs**) powered jointly by the renewable energy and the power grid. Denote by N_C, N_R, and N_H the number of CSBSs, RSBSs, and HSBSs in the network, respectively. Let $\mathfrak{B}_C = \{1, 2, \dots, N_C\}$, $\mathfrak{B}_R = \{1, 2, \dots, N_R\}$, and $\mathfrak{B}_H = \{1, 2, \dots, N_H\}$ be the set of CSBSs, RSBSs, and HSBSs, respectively. Let $\mathfrak{B} = \{\mathfrak{B}_C, \mathfrak{B}_R, \mathfrak{B}_H\}$ be the set of all SBSs. SBS$_n$ serves a circular area with radius $D_{s,n}$, and the small cells are assumed to have no overlap with each other. In HCN, the MBS is always active to guarantee the basic coverage, whereas SBSs can be dynamically turned on for traffic offloading or turned off for energy saving, depending on the traffic load and their energy status. For example, in Figure 7.1, the lightly loaded CSBS-1 is turned off to reduce the grid power consumption, while RSBS-1 is off due to the lack of harvested energy since it is purely powered by wind power.

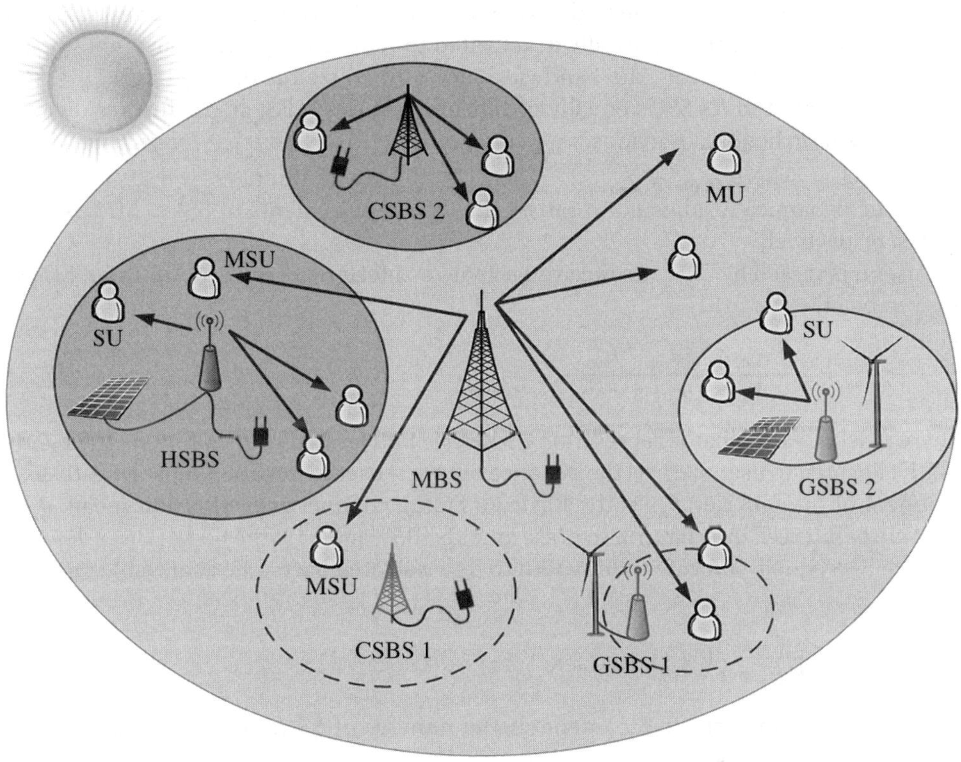

Figure 7.1 Illustration of a hyper-cellular network with SBSs powered by diverse energy sources.

7.2.1.2 Traffic and Channel Model

Two time scales are considered. In the large time scale, we divide the time of a day into T periods (e.g. $T = 24$ and the length of each time period is one hour) and assume the average EH rate and user density remain static in each time period t, but may change from period to period.

The user distribution in the spatial domain is modeled as a nonhomogeneous Poisson point process (PPP), whose density at time t is $\rho_n(t)$ in small cell n and $\rho_0(t)$ outside the coverage of any small cell. As shown in Figure 7.1, users not covered by small cells can only be served by the MBS, while users within small cells can be *partly* or fully served by the corresponding SBSs, according to the traffic and energy status. To this end, users can be classified into three types based on the serving BSs and locations: (i) macro–macro users (MMUs), users that are located outside small cells and served by the MBS; (ii) SBS–SBS users (SSUs), users located within small cells and served by SBSs; and (iii) macro–SBS users (MSUs), users located in small cells but served by the MBS. For theoretical analysis, random offloading scheme is adopted, where users in small cell n are served by SBS_n with probability $\varphi_n(t)$ and are served by the MBS with probability $1 - \varphi_n(t)$. As a result, the distributions of SSUs and MSUs in small cell n also follow PPP with density $\varphi_n(t)\rho_n(t)$ and $(1 - \varphi_n(t))\rho_n(t)$, respectively [28].

The spectrum allocated to the MBS tier and SBS tier is orthogonal to avoid cross-tier interference, whereas the intra-tier spectrum reuse factor is 1. Denote by W_{m} and W_{s}

the system bandwidth available to the MBS and each SBS, respectively. For each BS, its available bandwidth can be partially deactivated to reduce the power consumption, i.e. power control. At the SBS_n, the bandwidth actually utilized is denoted as $w_{\text{ss},n} \leq W_{\text{s}}$, which is allocated to its SSUs equally for fairness. At the MBS, W_{m} is further divided into different orthogonal portions: w_{mm} for serving MMUs and $w_{\text{ms},n}$ for serving MSUs in small cell n, where $w_{\text{mm}} + \sum_{n=1}^{N_{\text{C}}+N_{\text{R}}+N_{\text{H}}} w_{\text{ms},n} \leq W_{\text{m}}$. In addition, $w_{\text{ss},n}$, w_{mm}, and $w_{\text{ms},n}$ should be dynamically adjusted to satisfy the QoS requirements of SSUs, MMUs, and MSUs, respectively.

If user u is served by SBS_n, its received signal-to-interference-plus-noise-ratio (SINR) is given by [4]

$$\gamma_{\text{ss},nu} = I_n \frac{P_{\text{Ts},n} w_u}{W_{\text{s}}} \frac{d_{nu}{}^{-\alpha_{\text{s}}} h_{nu}}{(\theta_{\text{s}} + 1)\sigma^2 w_u}, \tag{7.1}$$

where I_n is a 0–1 indicator representing whether SBS_n is active or not, w_u is the bandwidth allocated to user u, d_{nu} is the distance between user u and SBS_n, α_{s} is the path-loss exponent of the SBS tier, h_{nu} is the Rayleigh fading channel gain with unit mean, θ_{s} is the ratio of intercell interference to noise among SBSs, and σ^2 is the noise power density. As each SBS equally allocates bandwidth to its associated users, the achievable rate of a user u is

$$r_{\text{ss},nu} = \frac{w_{\text{ss},n}}{K_{\text{ss},n} + 1} \log_2(1 + \gamma_{\text{ss},nu}), \tag{7.2}$$

where the random variable $K_{\text{ss},n}$ denotes the number of SSUs served by SBS_n except user u.

Similarly, if user u is served by the MBS as a MMU or MSU, its received SINR is given by

$$\gamma_{\text{m},u} = \frac{P_{\text{Tm}} w_u}{W_{\text{m}}} \frac{d_{0u}{}^{-\alpha_{\text{m}}} h_{0u}}{(\theta_{\text{m}} + 1)\sigma^2 w_u}, \tag{7.3}$$

where P_{Tm} is the transmit power of the MBS, d_{0u} is the distance between user u and the MBS, α_{m} is the path-loss exponent of the MBS tier, θ_{m} is the interference to noise ratio from other MBSs, and h_{0u} is the Rayleigh fading channel gain with the same probability distribution as h_{nu}. Then, the achievable rate of user u is given by

$$\begin{aligned} r_{\text{mm},u} &= \frac{w_{\text{mm}}}{K_{\text{mm}} + 1} \log_2(1 + \gamma_{\text{m},u}), \quad \text{for MMU,} \\ r_{\text{ms},nu} &= \frac{w_{\text{ms},n}}{K_{\text{ms},n} + 1} \log_2(1 + \gamma_{\text{m},u}), \quad \text{for MSU,} \end{aligned} \tag{7.4}$$

where K_{mm} and $K_{\text{ms},n}$ denote the number of MMUs and MSUs, respectively.

7.2.1.3 Power Consumption Model

BSs can work in either *active* mode or *sleep* mode, with different power consumption parameters. According to the EARTH project, the power consumption of a BS in *active* mode can be modeled as a constant term plus a radio frequency (RF)-related term [29]:

$$P_{\text{BS}} = P_{\text{C}} + \beta P_{\text{RF}}, \tag{7.5}$$

Table 7.1 Power model parameters for different types of BSs.

	Transmit power P_T (W)	Constant power P_C (W)	Coefficient β
Macro	20	130	4.7
Micro	6.3	56	2.6
Pico	0.13	6.8	4
Femto	0.05	4.8	8

where P_C includes the baseband processor, the cooling system, etc., coefficient β is the inverse of power amplifier efficiency factor, and P_{RF} is the RF power. The power-related parameters for different types of BSs are given in Table 7.1 [29].

The system bandwidth is further divided into orthogonal subcarriers, and a BS can decide to use how many subcarriers depending on the traffic demand. The RF power is proportional to the bandwidth of utilized subcarriers w, i.e.

$$P_{RF} = \frac{w}{W} P_T, \quad 0 < w \leq W, \tag{7.6}$$

where W is the available system bandwidth and P_T is the transmit power, which corresponds to P_{Tm} and $P_{Ts,n}$ for MBS and SBS, respectively. In this work, a constant power level is considered according to the LTE standard [30], i.e. P_T is treated as a system parameter, while we control the RF power by adjusting the utilized bandwidth W. Substituting P_{RF} in (7.5) into (7.6) leads to

$$P_{BS} = P_C + \frac{w}{W} \beta P_T. \tag{7.7}$$

If a BS is switched to the *sleep* mode, a small amount of power is still consumed so that the BS can be reactivated, which is negligible compared with P_C, and thus the power consumption in the sleep mode can be approximated as zero [30].

7.2.1.4 Green Energy Supply Model

Discrete energy model is adopted to describe the process of EH, and one unit of energy is denoted by E [31]. Denote by $\lambda_{E,n}(t)$ the arrival rate of per unit energy at SBS$_n$ and at time t. The harvested energy is saved in its battery, which is assumed to have sufficient capacity for realistic operation conditions, and thus we assume no battery overflow happens.

Figure 7.2 shows an example of the energy supply and consumption process at a typical SBS, whereby the harvested energy is used to power the SBS whenever the battery is not depleted. For RSBSs without grid power, they have to be shut down when the battery is depleted. Consequently, the corresponding users will be served by the upper-tier MBS for QoS guarantee. Note that handover procedure for users is conducted when the RSBS is shut down or reactivated, causing additional signaling overhead and power consumption. For HSBSs, they can use the grid power when there is no green energy, until renewable energy arrives.

The energy supply and consumption process of each SBS can be modeled by a queue, where the queue length denotes the amount of energy stored in the battery [32]. Based on the power consumption model of BSs (Eq. (7.7)), the equivalent service rate per unit

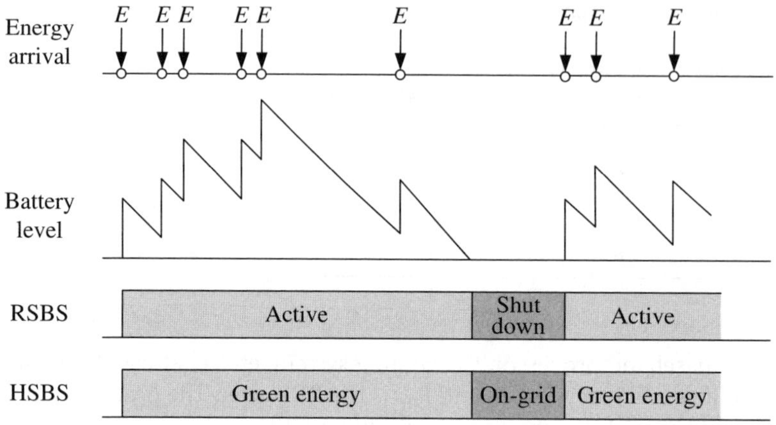

Figure 7.2 Illustrative process of renewable energy arrival and consumption.

energy for SBS_n is given by

$$\mu_{E,n}(t) = \frac{1}{E}\left(P_{Cs,n} + \frac{w_{ss,n}(t)}{W_s}\beta_{s,n}P_{Ts,n}\right), \tag{7.8}$$

where $P_{Ts,n}$, $P_{Cs,n}$, and $\beta_{s,n}$ are the transmit power, constant power, and power amplifier coefficient of SBS_n, respectively. We call μ_E as *energy consumption rate* for simplicity.

7.2.2 Analysis of Power Supply and Demand

We optimize the amount of offloaded traffic, the on–off state, and the RF power of each SBS in each time period t based on the stochastic information of traffic and energy (i.e. energy arrival rate and user density).[1]

To start with, we analyze the green power supply and demand. Specifically, the battery level is analyzed with queueing theory, and the QoS performance (i.e. outage probability) is derived with stochastic geometry, considering the *small-scale* randomness.

7.2.2.1 Energy Queue Analysis

For an SBS with EH, the variation of battery can be modeled as an M/D/1 queue with arrival rate λ_E and service rate μ_E given by (7.8). In what follows, we analyze the stable state of the energy queue. For the M/D/1 queue, the embedded Markov chain method is applied to obtain the stable state [33]. Denoted by L the queue length at any time and L^+ the queue length when a unit energy leaves the energy queuing. Due to the Poisson energy arrival, the state L^+ can be modeled as a Markov chain, with the transition probability matrix given by

$$M = \begin{pmatrix} m_0 & m_1 & m_2 & m_3 & \cdots \\ m_0 & m_1 & m_2 & m_3 & \cdots \\ 0 & m_0 & m_1 & m_2 & \cdots \\ 0 & 0 & m_0 & m_1 & \cdots \\ \vdots & \vdots & \vdots & \vdots & \ddots \end{pmatrix}, \tag{7.9}$$

1 The subscript t is omitted in the following to ease the presentation.

where

$$m_i = \frac{1}{i!}\left(\frac{\lambda_E}{\mu_E}\right)^{-i} e^{-\frac{\lambda_E}{\mu_E}}, \quad i = 0, 1, \ldots. \tag{7.10}$$

When $\frac{\lambda_E}{\mu_E} \geq 1$, the queue is not stable and the queue length goes to infinity, which means that the harvested energy is always sufficient. When $\frac{\lambda_E}{\mu_E} < 1$, the stationary probability distribution of L^+ is given by the Pollaczek–Khinchin formula [33], i.e.

$$\begin{aligned}
q_0 &= 1 - \frac{\lambda_E}{\mu_E},\\
q_1 &= \left(1 - \frac{\lambda_E}{\mu_E}\right)\left(e^{\frac{\lambda_E}{\mu_E}} - 1\right),\\
q_{L^+} &= \left(1 - \frac{\lambda_E}{\mu_E}\right)\left\{ e^{\frac{\lambda_E}{\mu_E}L^+} + \sum_{k=1}^{L^+-1} e^{k\frac{\lambda_E}{\mu_E}} (-1)^{\frac{\lambda_E}{\mu_E}-k} \right.\\
&\quad \left. \cdot \left[\frac{\left(k\frac{\lambda_E}{\mu_E}\right)^{L^+-k}}{(L^+-k)!} + \frac{\left(k\frac{\lambda_E}{\mu_E}\right)^{b-L^+-1}}{(b-L^+-1)!} \right] \right\} \quad (L^+ > 2),
\end{aligned} \tag{7.11}$$

For the M/D/1 queue, we have $q_L = q_{L^+}$ [33]. As a result, the stationary probability distribution of the amount of available green energy in the battery of an SBS is derived.

7.2.2.2 Outage Probability Analysis

Service outage occurs when the user's achievable data rate is less than its requirement R_Q, due to channel fading or bandwidth limitation. The outage probability should satisfy a certain threshold η. We are interested in analyzing the outage probability constraint for SSUs, MMUs, and MSUs, respectively, based on which the power demand can be obtained.

According to the traffic and channel model, the outage probability of a typical SSU u of SBS_n is given by

$$\begin{aligned}
G_{ss,n} &= \mathbb{E}_{\{K_{ss,n}, d_{ss,n}\}}[\mathbb{P}\{r_{ss,nu} < R_Q | K_{ss,n}, d_{ss,n}\}]\\
&= \int_0^{D_{s,n}} \sum_{k=0}^{\infty} \mathbb{P}\left\{ \gamma_{ss,n} < 2^{\frac{(k+1)R_Q}{w_{ss,n}}} - 1 | d \right\} Q_{K_{ss,n}}(k) f_{d_{ss,n}}(d) \mathrm{d}d,
\end{aligned} \tag{7.12}$$

where $Q_{K_{ss,n}}(k)$ is the probability that SBS_n serves k SSUs except SSU u and $f_{d_{ss,n}}(d) = \frac{2\pi}{D_n}d$ is the probability density function (PDF) of the distance between u and SBS_n. As the distribution of users follows PPP in each small cell, $K_{ss,n}$ follows the Poisson distribution of parameter $\pi\varphi\rho_n D_n^2$ according to the Slivnyak–Mecke theorem [34], where φ is the offloading ratio. Although the outage probability in (7.12) cannot be directly derived, the closed-form expression can be obtained in the high SINR and large system bandwidth regime, given as Theorem 7.1.

Theorem 7.1 As $\frac{P_{\mathrm{Ts},n}}{(\theta_s+1)\sigma^2 W_s} \to \infty$ (i.e. SBS_n provides high SINR) and $\frac{R_Q}{w_{\mathrm{ss},n}} \to 0$ (i.e. sufficient system bandwidth), the service outage probability is given as follows:

$$G_{\mathrm{ss},n} = \frac{2D_n{}^{\alpha_s}(\theta_s+1)\sigma^2}{(\alpha_s+2)P_{\mathrm{Ts},n}W_s{}^{-1}}\left(2^{\frac{R_Q}{w_{\mathrm{ss},n}}\left(1+\pi D_n{}^2\varphi_n\rho_n\frac{R_Q}{w_{\mathrm{ss},n}}\right)}-1\right). \tag{7.13}$$

Proof: Based on (7.3), we have

$$\begin{aligned}
&\mathbb{P}\left\{\gamma_{\mathrm{ss},n} \ge 2^{(K_{\mathrm{ss},n}+1)\frac{R_Q}{w_{\mathrm{ss},n}}}-1\right\}\\
&= \int_0^{D_n} \mathbb{P}\left\{h_n \ge \frac{(\theta_s+1)\sigma^2 W_s}{P_{\mathrm{Ts},n}d^{-\alpha_s}}\left(2^{(K_{\mathrm{ss},n}+1)\frac{R_Q}{w_{\mathrm{ss},n}}}-1\right)\right\}\frac{2d}{D_n^2}\mathrm{d}d\\
&= \int_0^{D_n} \exp\left(-\frac{(\theta_s+1)\sigma^2 W_s}{P_{\mathrm{Ts},n}d^{-\alpha_s}}\left(2^{(K_{\mathrm{ss},n}+1)\frac{R_Q}{w_{\mathrm{ss},n}}}-1\right)\right)\frac{2d}{D_n^2}\mathrm{d}d \qquad (7.14\mathrm{a})\\
&= \int_0^{D_n} \left(1-\frac{(\theta_s+1)\sigma^2 W_s}{P_{\mathrm{Ts},n}d^{-\alpha_s}}\left(2^{(K_{\mathrm{ss},n}+1)\frac{R_Q}{w_{\mathrm{ss},n}}}-1\right)\right)\frac{2d}{D_n^2}\mathrm{d}d \qquad (7.14\mathrm{b})\\
&= 1-\frac{2D_n^{\alpha_s}W_s}{\alpha_s+2}\frac{(\theta_s+1)\sigma^2}{P_{\mathrm{Ts},n}}\left(2^{(K_{\mathrm{ss},n}+1)\frac{R_Q}{w_{\mathrm{ss},n}}}-1\right),
\end{aligned}$$

where (7.14a) is based on the definition of Rayleigh fading and (7.14b) holds as $\frac{P_{\mathrm{Ts},n}}{(\theta_s+1)\sigma^2 W_s} \to \infty$, based on $\lim_{x\to 0} \mathrm{e}^{-x} = 1-x$.

Recall that the probability distribution of $K_{\mathrm{ss},n}$ follows Poisson distribution. By substituting (7.14) into (7.12), the outage probability of a typical SBS user is

$$\begin{aligned}
G_{\mathrm{ss},n} &= 1-\sum_{K=0}^{\infty}\mathbb{P}\left(\gamma_{\mathrm{ss},n} \ge 2^{(K+1)\frac{R_Q}{w_{\mathrm{ss},n}}}-1\right)P_{K_{\mathrm{ss},n}}(K)\\
&= 1-\sum_{K=0}^{\infty}\mathbb{P}\left\{\gamma_{\mathrm{ss},n} \ge 2^{(K+1)\frac{R_Q}{w_{\mathrm{ss},n}}}-1\right\}\frac{(\pi D_n^2\rho_n)^K}{K!}\mathrm{e}^{-\pi D_n^2\rho_n}\\
&= \frac{2D_n^{\alpha_s}(\theta_s+1)\sigma^2 W_s}{P_{\mathrm{Ts},n}(\alpha_s+2)}\left(2^{\frac{R_Q}{w_{\mathrm{ss},n}}}\exp\left(\pi D_n^2\rho_n\left(2^{\frac{R_Q}{w_{\mathrm{ss},n}}}-1\right)\right)-1\right).
\end{aligned} \tag{7.15}$$

As $\lim_{x\to 0}\frac{a^x-1}{x} = \ln a$ for $a > 1$, we have $\lim_{\frac{R_Q}{w_{\mathrm{ss}}}\to 0} 2^{\frac{R_Q}{w_{\mathrm{ss},n}}}-1 = \ln 2 \cdot \frac{R_Q}{w_{\mathrm{ss},n}}$. □

In fact, the assumptions in Theorem 7.1 are reasonable in practical cellular systems, where the SINR of users is generally high enough for reliable communications. Besides, each BS can support a large number of users simultaneously, which requires large amount of system bandwidth. Thus, the system bandwidth is much higher than the data rate requirement. Therefore, Eq. (7.13) can be applied to approximate the outage probability.

Notice that the outage probability (7.13) constrains the required bandwidth for given traffic density. Specifically, the service outage constraint of SSUs $G_{\mathrm{ss},n} \le \eta$ can be written as

$$\bar{w}_{\mathrm{ss},n}\tau_{\mathrm{ss},n} \ge R_Q, \tag{7.16}$$

where $\bar{w}_{\text{ss},n} = \frac{w_{\text{ss},n}}{1+\varphi_n \rho_n \pi D_n{}^2}$ is the expected bandwidth allocated to each SSU and $\tau_{\text{ss},n}$ denotes the spectrum efficiency of cell-edge users given by

$$\tau_{\text{ss},n} = \log_2\left(1 + \frac{P_{\text{Ts},n}}{(\theta_{\text{s}}+1)} \frac{\alpha_{\text{s}}+2}{2\sigma^2 W_{\text{s}}} \frac{\eta}{D_n{}^{\alpha_{\text{s}}}}\right). \tag{7.17}$$

The physical meaning of (7.16) is that the average data rate of the non-cell-edge users (with spectrum efficiency above $\tau_{\text{ss},n}$) should be no smaller than R_{Q}.

The outage probability of MMUs cannot be derived in closed form, which varies with the location and coverage of each SBS. We further assume that the MMUs to be uniformly distributed in the macro cell with density ρ_0' as follows:

$$\rho_0' = \frac{\rho_0}{D_0{}^2}\left(D_0{}^2 - \sum_{n\in\mathfrak{B}} D_n^2\right). \tag{7.18}$$

Thus approximation to the outage probability of MMUs can be derived in the same way as Theorem 7.1, i.e. the constraint $G_{\text{mm}} \leq \eta$ is equivalent to

$$\bar{w}_{\text{mm}} \tau_{\text{mm}} \geq R_{\text{Q}}, \tag{7.19}$$

where $\bar{w}_{\text{mm}} = \frac{w_{\text{mm}}}{1+\pi D_0{}^2 \rho_0'}$ and τ_{mm} are given by

$$\tau_{\text{mm}} = \log_2\left(1 + \frac{P_{\text{Tm}}}{(\theta_{\text{m}}+1)} \frac{\alpha_{\text{m}}+2}{2\sigma^2 W_{\text{m}}} \frac{\eta}{D_0{}^{\alpha_{\text{m}}}}\right). \tag{7.20}$$

Notice that the approximation is reasonable when the SBSs are uniformly distributed in the macro cell.

Next, we consider an MSU served by SBS_n. The approximation to the outage probability is given by Theorem 7.2.

Theorem 7.2 As $\frac{P_{\text{Tm}}}{(\theta_{\text{m}}+1)\sigma^2 W_{\text{m}}} \to \infty$ (i.e. the MBS provides high SINR) and $\frac{R_{\text{Q}}}{w_{\text{ms},n}} \to 0$ (i.e. sufficient system bandwidth), the outage probability constraint of the MSUs $G_{\text{ms},n} \leq \eta$ can be approximated as

$$\bar{w}_{\text{ms},n} \tau_{\text{ms},n} \geq R_{\text{Q}}, \tag{7.21}$$

where

$$\begin{aligned} \bar{w}_{\text{ms},n} &= \frac{w_{\text{ms},n}}{1+(1-\varphi_n)\rho_n \pi D_n{}^2}, \\ \tau_{\text{ms},n} &= \log_2\left(1 + \frac{\eta P_{\text{Tm}}}{\sigma^2 W_{\text{m}}(\theta_{\text{m}}+1) D_{\text{ms},n}{}^{\alpha_{\text{m}}}}\right), \end{aligned} \tag{7.22}$$

and $D_{\text{ms},n}$ denotes the distance between the MBS and SBS_n.

Theorem 7.2 can be proved in the similar way as Theorem 7.1. When SBS_n is in sleep mode, the service outage probability can be obtained based on Theorem 7.2, by setting $\varphi_n = 0$, i.e. no traffic is offloaded to SBS_n.

Notice that Eqs. (7.16), (7.19), and (7.21) determine the minimum bandwidth required for the given traffic demand (ρ_0, ρ_n) and the offloading scheme φ_n and the bandwidth requirement can reflect the power demand of each SBS and MBS according to the BS power consumption model (7.7).

7.2.3 Optimization in the Single-SBS Case

In this section, we optimize the traffic offloading for the single small-cell case, including the HSBS and RSBS. Note that a CSBS can be considered as a HSBS whose energy arrival rate is set to zero, i.e. $\lambda_E = 0$.

7.2.3.1 Single HSBS

For the single-HSBS case, the total on-grid power consumption P_{sum} consists of two parts:

$$P_{sum} = P_{MBS} + P_{HSBS}, \tag{7.23}$$

where P_{MBS} and P_{HSBS} denote the on-grid power consumptions of the MBS and HSBS, respectively, which can be derived from (7.7). Denoted by I_H a 0–1 variable indicating whether the HSBS is active or not, while $w_{ms}^{(a)}$ and $w_{ms}^{(o)}$ the corresponding bandwidth needed by the MSUs. Then we have

$$P_{MBS} = P_{Cm} + \frac{\beta_m P_{Tm}}{W_m} [w_{mm} + I_H w_{ms}^{(a)} + (1 - I_H)\, w_{ms}^{(o)}]. \tag{7.24}$$

In addition, as the HSBS consumes on-grid power only when the battery is empty, we have

$$P_{HSBS} = I_H q_0 \left(P_{Cs} + \frac{\beta_s P_{Ts}}{W_s} w_{ss} \right), \tag{7.25}$$

where q_0 is the probability of empty energy queue, obtained from (7.11).

Then, minimizing the on-grid power consumption of the single-HSBS case can be formulated as

$$\min_{I_H, \mu_E} \quad P_{sum} \tag{7.26a}$$

$$\text{s.t.} \quad G_{mm} \leq \eta, G_{ss} \leq \eta, G_{ms} \leq \eta \tag{7.26b}$$

$$0 \leq w_{mm} + I_H w_{ms}^{(a)} + (1 - I_H) w_{ms}^{(o)} \leq W_m \tag{7.26c}$$

$$0 \leq w_{ss} \leq W_s, \tag{7.26d}$$

where (7.26b) guarantees the QoS, and (7.26c) and (7.26d) are due to the bandwidth limitations of MBS and HSBS, respectively. μ_E can be derived based on the power consumption model of HSBS in (7.8).

Intuitively, there is a tradeoff between the power consumptions of the MBS and HSBS. By activating the HSBS for traffic offloading, the traffic load of the MBS decreases, leading to less RF power consumption of the MBS. Meanwhile, the activated HSBS introduces additional on-grid power consumption, especially when the renewable energy is insufficient. Furthermore, with higher energy consumption rate μ_E, more users are offloaded from the MBS to the HSBS, and thus RF power of the MBS decreases, but the power demand at the HSBS will increase accordingly. We therefore analyze this tradeoff relationship to solve problem (7.26).

Denoted by Δ_H the power saving gain by activating the HSBS:

$$\Delta_H = P_{sum}|_{I_H=0} - P_{sum}|_{I_H=1}. \tag{7.27}$$

Based on Theorems 7.1 and 7.2, the closed-form expression of Δ_H is given as the following theorem. Then, the optimal solution of problem (7.26) is derived as Theorem 7.4.

Theorem 7.3 The power saving gain by activating a HSBS for traffic offloading is given as follows:

$$\Delta_H = \begin{cases} \zeta_{EE}\mu_E E - \zeta_{EE} P_{Cs} - \frac{R_Q \beta_m P_{Tm}}{\tau_{ms} W_m} & \mu_E \leq \lambda_E \\ [\zeta_{EE} - 1]\mu_E E - \zeta_{EE} P_{Cs} - \frac{R_Q \beta_m P_{Tm}}{\tau_{ms} W_m} + \lambda_E E & \mu_E > \lambda_E, \end{cases} \tag{7.28}$$

where $\zeta_{EE} = \frac{W_s \tau_{ss} \beta_m P_{Tm}}{W_m \tau_{ms} \beta_s P_{Ts}}$.

Proof: With Eqs. (7.23)–(7.25), the power saving gain by activating the HSBS Δ_H is

$$\Delta_H = \frac{w_{ms}^{(o)} - w_{ms}^{(a)}}{W_m} \beta_m P_{Tm} - q_0 \left(P_{Cs} + \frac{w_{ss}}{W_s} \beta_s P_{Ts} \right). \tag{7.29}$$

To satisfy the outage probability constraints of (7.26b), we have

$$w_{ms}^{(o)} \geq \frac{R_Q}{\tau_{ms}} (1 + \rho_s \pi D_s^2), \tag{7.30a}$$

$$w_{ms}^{(a)} \geq \frac{R_Q}{\tau_{ms}} (1 + (1 - \varphi) \rho_s \pi D_s^2), \tag{7.30b}$$

based on Theorem 7.2. φ is offloading ratio constrained by the service capability of SBS (Eq. (7.16)):

$$\varphi \leq \frac{\frac{\tau_{ss} w_{ss}}{R_Q} - 1}{\rho_s \pi D_s^2}, \tag{7.31}$$

where w_{ss} depends on the energy consumption rate of the HSBS μ_E, which is given by

$$w_{ss} = \frac{1}{\beta_s W_s} (\mu_E E - P_{Cs}). \tag{7.32}$$

Note that $q_0 = 0$ when $\mu_E \leq \lambda_E$; otherwise, $q_0 = 1 - \frac{\lambda_E}{\mu_E}$. Substituting (7.30)–(7.32) into (7.29), the theorem is proved. □

Theorem 7.4 If the HSBS is active, the optimal energy consumption rate satisfies

$$\tilde{\mu}_E = \frac{1}{E} \left\{ P_{Cs} + \min \left\{ 1, \frac{R_Q}{\tau_{ss} W_s} (\rho_s \pi D_s^2 + 1) \right\} \beta_s P_{Ts} \right\}, \tag{7.33}$$

where ρ_s is the user density in the small cell and D_s and P_{Ts} denote the coverage radius and transmit power of the SBS, respectively. In addition, the HSBS should be activated (i.e. $\tilde{I}_H = 1$) if $\Delta_H|_{\mu_E = \tilde{\mu}_E} > 0$ or $w_{mm} + w_{ms}^{(o)} > W_m$; otherwise, $\tilde{I}_H = 0$.

Proof: According to (7.28), the power saving gain increases with μ_E if $\zeta_{EE} \geq 1$; otherwise, it achieves the maximum when $\mu_E = \lambda_E$. As $\zeta_{EE} > 1$ in real systems, the optimal solution

$\tilde{\mu}_E$ takes its maximal feasible value. Based on (7.26d), and $\varphi \leq 1$, we have

$$0 \leq w_{ss} \leq \min\left\{W_s, \frac{R_Q}{\tau_{ss}}(\rho_s \pi D_s^2 + 1)\right\}. \tag{7.34}$$

Recall the relationship between μ_E and w_{ss} in (7.26c), and the theorem is proved. □

Notice that Theorem 7.3 reflects the conversion rate of harvested energy (i.e. λ_E) into on-grid power (i.e. Δ_H), i.e. how much on-grid power can be saved with the utilization of a unit of harvested power. In addition, the physical meaning of $\frac{W_s \tau_{ss}}{\beta_s P_{Ts}}$ is the average energy efficiency of the SSUs in bits per joule, i.e. the amount of information transmitted per joule of transmit energy from the HSBS. Similarly, $\frac{W_m \tau_{ms}}{\beta_m P_{Tm}}$ is the average energy efficiency of MSUs at the MBS. Thus, ζ_{EE} compares the energy efficiency of the HSBS and MBS. Therefore, Theorem 7.3 indicates that users should be served by the BS (MBS or HSBS), which provides higher energy efficiency, if the harvested energy is insufficient to support the HSBS (i.e. $\mu_E > \lambda_E$). In practice, the HSBS usually provides higher energy efficiency compared with the MBS, due to shorter transmission distance and lower path loss. As a result, more subcarriers should be utilized to offload more users if the HSBS is active, which explains Theorem 7.4. As for cell activation, $\Delta_H|_{\mu_E=\tilde{\mu}_E} > 0$ indicates that activating the HSBS brings positive power saving gain. Besides, $w_{mm} + w_{ms}^{(o)} > W_m$ happens when the MBS is overloaded, and thus the HSBS should be activated to relieve the burden of the MBS. To get more insights, typical asymptotic results are illustrated in Corollaries 7.1 and 7.2.

Corollary 7.1 *When $\lambda_E \to 0$ and $\rho_s \to 0$, activating the HSBS does not reduce the RF power of MBS but consumes on-grid energy P_{Cs}. Thus, the HSBS should be deactivated.*

Corollary 7.2 *When $\lambda_E \geq \tilde{\mu}_E$ (i.e. sufficient green energy supply), offloading users to SBS reduces on-gird power consumption (i.e. $\Delta_H|_{\mu_E=\tilde{\mu}_E} \geq 0$), and therefore the HSBS should be active.*

7.2.3.2 Single-RSBS Case

Unlike the HSBS, the RSBS does not consume on-grid power. However, the SSUs have to be served by the MBS when the battery is empty, which causes handover, additional signaling cost, and on-grid power consumption. The total power consumption is given by

$$P_{sum} = P_{MBS} + P_{HO}, \tag{7.35}$$

where P_{MBS} is the power consumption of the MBS and P_{HO} reflects the additional power consumption of SSU handover.

Denoted by $I_R \in \{0, 1\}$ the on–off state of the RSBS. If the RSBS is active, handover happens in the following cases: (i) RSBS is shut down when the battery runs out. (ii) RSBS is reactivated when new energy arrives. According to the model of the energy queue, the first case corresponds to the event when L^+ transits from 1 to 0, with frequency of $q_1 A_{21} \mu_E$. Due to the duality between the two cases, the additional handover power consumption is given by

$$P_{\mathrm{HO}} = 2I_{\mathrm{R}}q_1A_{21}\mu_{\mathrm{E}}C_{\mathrm{HO}} = \begin{cases} 2I_{\mathrm{R}}\left(1-\dfrac{\lambda_{\mathrm{E}}}{\mu_{\mathrm{E}}}\right)\left(1-\mathrm{e}^{-\frac{\lambda_{\mathrm{E}}}{\mu_{\mathrm{E}}}}\right)\mu_{\mathrm{E}}C_{\mathrm{HO}} & \lambda_{\mathrm{E}} < \mu_{\mathrm{E}} \\ 0 & \lambda_{\mathrm{E}} \geq \mu_{\mathrm{E}}, \end{cases} \tag{7.36}$$

where C_{HO} denotes the energy consumed by one handover process in joule.

Note that the SBS may be shut down due to energy shortage even when it is in *on* state, in which the MBS has to utilize more bandwidth to serve the SSUs with additional bandwidth. Based on (7.7), the average on-grid power consumption of the MBS is given as follows:

$$P_{\mathrm{MBS}} = P_{\mathrm{Cm}} + \frac{\beta_{\mathrm{m}}P_{\mathrm{Tm}}}{W_{\mathrm{m}}}\left(w_{\mathrm{mm}} + I_{\mathrm{R}}((1-q_0)w_{\mathrm{ms}}^{(\mathrm{a})} + q_0 w_{\mathrm{ms}}^{(\mathrm{o})}) + (1-I_{\mathrm{R}})w_{\mathrm{ms}}^{(\mathrm{o})}\right), \tag{7.37}$$

where w_{mm} is constrained by (7.19) and $w_{\mathrm{ms}}^{(\mathrm{a})}$ and $w_{\mathrm{ms}}^{(\mathrm{o})}$ denote the bandwidth needed by the MBS when the RSBS is active and sleep, respectively.

Thus, the power consumption minimization problem can be formulated as follows:

$$\min_{I_{\mathrm{R}},\mu_{\mathrm{E}}} \quad P_{\mathrm{sum}} \tag{7.38a}$$

$$\text{s.t.} \quad G_{\mathrm{mm}} \leq \eta, G_{\mathrm{ss}} \leq \eta, G_{\mathrm{ms}} \leq \eta \tag{7.38b}$$

$$0 \leq w_{\mathrm{mm}} + w_{\mathrm{ms}}^{(\mathrm{o})} \leq W_{\mathrm{m}} \tag{7.38c}$$

$$0 \leq w_{\mathrm{ss}} \leq W_{\mathrm{s}}, \tag{7.38d}$$

where the objective is to minimize the total average on-grid power consumption, and the constraint (7.38b) guarantees the QoS, and constraints (7.38c) and (7.38d) account for the bandwidth limitation of MBS and RSBS, respectively.

Similar to the single-HSBS case, there exists a tradeoff between the power consumption of the MBS and the handover cost. By activating an RSBS for traffic offloading, the RF power of the MBS is reduced, but handover takes additional power if the renewable energy is insufficient. Besides, the energy consumption rate also has influences. Specifically, increasing energy consumption rate enables the RSBS to serve more users and thus reduces the RF power demand of the MBS. However, higher energy consumption rate may also result in higher handover cost, as the battery may become empty more frequently. In what follows, we analyze this tradeoff relationship to solve problem (7.38). Denoted by Δ_{R} the power saving gain through activating the RSBS, given by

$$\Delta_{\mathrm{R}} = P_{\mathrm{sum}}|_{I_{\mathrm{R}}=0} - P_{\mathrm{sum}}|_{I_{\mathrm{R}}=1}. \tag{7.39}$$

We summarize the relationship between Δ_{R} and μ_{E} in the following proposition.

Proposition 7.1 *Denoting* $\kappa = \zeta_{\mathrm{EE}}P_{\mathrm{Cs}} + \frac{\beta_{\mathrm{m}}P_{\mathrm{Tm}}R_{\mathrm{Q}}}{W_{\mathrm{m}}\tau_{\mathrm{ms}}}$, *the power saving gain* Δ_{R} *of an RSBS has the following properties:*

(1) Δ_{R} *increases linearly with* μ_{E} *for* $\mu_{\mathrm{E}} \leq \lambda_{\mathrm{E}}$.
(2) Δ_{R} *increases with* μ_{E} *if* $\kappa \geq 3\lambda_{\mathrm{E}}C_{\mathrm{HO}}$.

(3) Δ_R *decreases with* μ_E *for* $\mu_E > \lambda_E$ *if* $\kappa \le \left(1 - \frac{1}{e}\right)\lambda_E C_{HO}$.

(4) If $\left(1 - \frac{1}{e}\right)\lambda_E C_{HO} < \kappa < 3\lambda_E C_{HO}$, Δ_R *is a concave function of* $\frac{\lambda_E}{\mu_E}$ *for* $\mu_E > \lambda_E$, *and the optimal condition is*

$$\lambda_E C_{HO} \frac{e^{-\frac{\lambda_E}{\mu_E}}}{\left(\frac{\lambda_E}{\mu_E}\right)^2}\left(-e^{-\frac{\lambda_E}{\mu_E}} + 1 + \frac{\lambda_E}{\mu_E} - \left(\frac{\lambda_E}{\mu_E}\right)^2\right) = \kappa. \tag{7.40}$$

Proof: Parameter Δ_R can be written as

$$\Delta_R = (1 - q_0)\frac{W_{ms}^{(o)} - W_{ms}^{(a)}}{W_m}\beta_m P_{Tm} - P_{HO} \triangleq \Delta_{mbs} - P_{HO}, \tag{7.41}$$

where Δ_{mbs} represents the power saved at the MBS. Based on Eqs. (7.30)–(7.32), we have

$$\Delta_{mbs} = \begin{cases} \zeta_{EE}\mu_E E - \left(\zeta_{EE}P_{Cs} + \dfrac{\beta_m P_{Tm} R_Q}{W_m \tau_{ms}}\right) & \mu_E \le \lambda_E \\ \zeta_{EE}\lambda_E E - \dfrac{\lambda_E}{\mu_E}\left(\zeta_{EE}P_{Cs} + \dfrac{\beta_m P_{Tm} R_Q}{W_m \tau_{ms}}\right) & \mu_E > \lambda_E, \end{cases} \tag{7.42}$$

which reflects the equivalent conversion rate of the harvested energy into the on-grid power. Substituting Δ_{mbs} and P_{HO} into (7.41), the power saving gain Δ_R can be obtained.

When $\mu_E \le \lambda_E$, $P_{HO} = 0$, and $\Delta_R = \Delta_{mbs}$ increases linearly with μ_E. When $\mu_E > \lambda_E$, denote $x = \frac{\lambda_E}{\mu_E}$ where $x \in (0, 1)$. The derivative of Δ_R with respect to (w.r.t.) x is

$$\begin{aligned} \frac{d\Delta_R}{dx} &= -\kappa - \left[\left(1 - \frac{1}{x^2}\right)(1 - e^{-x}) + \left(\frac{1}{x} - 1\right)e^{-x}\right]\lambda_E C_{HO} \\ &= -\kappa - \lambda_E C_{HO}\frac{e^{-x}}{x^2}(-e^x + 1 + x - x^2) \triangleq -\kappa + \lambda_E C_{HO} f(x). \end{aligned} \tag{7.43}$$

Now we analyze the property of $f(x)$. As

$$\frac{df(x)}{dx} = \frac{2e^{-x}}{x^3}\left(-e^x + 1 + x + \frac{x^2}{2} - \frac{x^3}{2}\right) \tag{7.44a}$$

$$= \frac{2e^{-x}}{x^3}\left(-\frac{x^3}{2} - \sum_{i=3}^{\infty}\frac{x^i}{i!}\right) < 0, \tag{7.44b}$$

and $0 < x < 1$, we have $1 - e^{-1} < f(x) < \frac{3}{2}$. In addition, Δ_R is a concave function of x as $\frac{d^2\Delta_R}{dx^2} = f'(x) < 0$. Therefore, $\frac{d\Delta_R}{dx} < 0$ for $\kappa \ge 3\lambda_E C_{HO}$, and $\frac{d\Delta_R}{dx} > 0$ for $\kappa \le \left(1 - \frac{1}{e}\right)\lambda_E C_{HO}$. Otherwise, there exists $\tilde{x}$ satisfying $\frac{d\Delta_R}{dx}|_{\tilde{x}} = 0$, and the corresponding energy consumption rate $\mu_E = \frac{\lambda_E}{\tilde{x}}$ maximizes the power saving gain. Hence, the proposition is proved. □

Based on Eqs. (7.38b) and (7.38c) and $\varphi \le 1$, μ_E satisfies

$$\frac{P_{Cs}}{E} \le \mu_E \le \frac{1}{E}\left\{P_{Cs} + \min\left\{1, \frac{R_Q}{\tau_{ss}W_s}(\rho_s \pi D_s^2 + 1)\right\}\beta_s P_{Ts}\right\}. \tag{7.45}$$

By combining Propositions 7.1 and (7.45), the optimal energy consumption rate is obtained, while the details are not listed here for brevity. In addition, the RSBS should be activated for traffic offloading if the maximal power saving gain is positive; otherwise, it should stay in the *off* state. To get more insights, we further present following corollaries. Note that Corollary 7.3 corresponds to the case with high handover cost, while Corollary 7.3 corresponds to the case with sufficient energy supply.

Corollary 7.3 *If* $\kappa \leq (1 - \frac{1}{e})\lambda_E C_{HO}$ *and* $\lambda_E \leq P_{Cs}$, *the RSBS should be deactivated.*

Corollary 7.4 *If* $\lambda_E E > P_{Cs} + \beta_s P_{Ts}$, *the RSBS should be activated.*

7.2.4 Optimization in the Multi-SBS Case

In this section, we investigate the power consumption minimization problem for the case where multiple RSBSs, HSBSs, and CSBSs coexist.

7.2.4.1 Problem Formulation

When multiple small cells coexist, the power consumption of the MBS is given by

$$P_{MBS} = P_{Cm} + \frac{\beta_m P_{Tm}}{W_m} \left(w_{mm} + \sum_{n \in \mathfrak{B}} \tilde{w}_{ms,n} \right), \tag{7.46}$$

where $\tilde{w}_{ms,n}$ is the average bandwidth needed by the MBSs to serve users in cell n. Considering the characteristics of different SBSs, we have

$$\tilde{w}_{ms,n} = \begin{cases} w^{(a)}_{ms,n} I_n + w^{(o)}_{ms,n}(1 - I_n) & \text{for HSBSs or CSBSs} \\ I_n(w^{(a)}_{ms,n}(1 - q_{0,n}) + w^{(o)}_{ms,n} q_{0n}) + w^{(o)}_{ms,n}(1 - I_n) & \text{for RSBSs,} \end{cases} \tag{7.47}$$

where $w^{(a)}_{ms,n}$ (and $w^{(o)}_{ms,n}$) is the bandwidth needed by the MSUs in cell n when SBS_n is active (and in off state) and $q_{0,n}$ is the probability that the battery of $RSBS_n$ is empty.

The total on-grid power consumption is given by

$$P_{sum} = P_{MBS} + \sum_{n \in \mathfrak{B}_R} I_n P_{HO,n} + \sum_{i \in \mathfrak{B}_H} I_i P_{HSBS,i} + \sum_{j \in \mathfrak{B}_C} I_j P_{CSBS,j}, \tag{7.48}$$

where $P_{HO,n}$ reflects the handover cost of $RSBS_n$, while $P_{HSBS,i}$ and $P_{CSBS,j}$ denote the on-grid power consumption of the $HSBS_i$ and $CSBS_j$, respectively:

$$\begin{aligned} P_{HSBS,i} &= q_{0,i} \left(P_{Cs,i} + \frac{w_{ss,i}}{W_s} \beta_i P_{Ts,i} \right), \\ P_{CSBS,j} &= P_{Cs,j} + \frac{w_{ss,j}}{W_s} \beta_j P_{Ts,j}. \end{aligned} \tag{7.49}$$

Based on the analytical results of the single-SBS cases, the optimal energy saving gain through offloading traffic to each individual SBS can be obtained, and the corresponding amount of traffic offloaded to the SBSs $\tilde{\mu}_E$ is the optimal solution to minimize the grid power consumption. Then, the problem becomes deciding which SBSs should be activated. Firstly, the SBSs with positive power saving gain should be activated, and the on–off states is denoted by $\tilde{\mathcal{I}}$. If $(\tilde{\mathcal{I}}, \tilde{\mu}_{\mathrm{E}})$ is feasible under the bandwidth constraints of all SBSs and the MBS,

$$w_{\mathrm{m}}^{\max} = w_{\mathrm{mm}} + \sum_{n \in (\mathfrak{B}_{\mathrm{H}} \bigcup \mathfrak{B}_{\mathrm{C}})} \tilde{w}_{\mathrm{ms},n} + \sum_{n \in \mathfrak{B}_{\mathrm{R}}} w_{\mathrm{ms},n}^{(\mathrm{o})}, \tag{7.50}$$

where $w_{\mathrm{m}}^{\max}$ is the bandwidth needed from the MBS when all active RSBSs are in the *off* mode due to energy shortage, which should be no larger than W_m. Then $(\tilde{\mathcal{I}}, \tilde{\mu}_{\mathrm{E}})$ is optimal since all SBSs whose activation can reduce on-grid power consumption (i.e. the objective function) are turned on. Otherwise, the MBS is overloaded, and more SBSs should be turned off, with the price of increasing the on-grid power consumption. Therefore, the key problem is to determine which SBSs should be reactivated.

7.2.4.2 SBS Reactivation and TEATO Scheme

First, note that the on–off operation of RSBSs does not influence (7.50), i.e. the bandwidth constraint of the MBS. The reason is that bandwidth $w_{\mathrm{ms},n}^{(\mathrm{o})}$ should be reserved no matter RSBS$_n$ is activated or not, since the MBS needs to serve the SSUs when RSBS$_n$ is shut down due to energy shortage. Thus, activating new RSBSs only increases the total energy consumption without relieving the burden of MBS. In contrast, according to (7.50), offloading traffic to HSBSs and CSBSs can reduce the bandwidth requirement at the MBS. Therefore, only HSBSs and CSBSs should be considered for reactivation.

Denote by $\mathfrak{B}_{\mathrm{off}} = \{1, 2, \dots, N_{\mathrm{off}}\}$ the set of HSBSs and CSBSs in off state according to $\tilde{\mathcal{I}}$, where N_{off} is the number of SBSs in off state. Based on the analysis in Section 7.2.3, once a HSBS or CSBS is activated, it should serve as many users as possible, and the optimal energy consumption rate is given by (7.33). For each sleeping HSBS or CSBS i, activation increases the power consumption by $-\Delta_i$ (i.e. Eqs. (7.27) and (7.39)) but reduces the bandwidth demand at MBS by $\delta_i = w_{\mathrm{ms},i}^{(\mathrm{o})} - w_{\mathrm{ms},i}^{(\mathrm{a})}$. As a result, the SBS activating problem is a 0–1 knapsack problem as follows:

$$\max_{\mathcal{I}_{\mathrm{a}}} \quad \sum_{i=1}^{N_{\mathrm{off}}} \mathcal{I}_{\mathrm{a},i} \Delta_i \tag{7.51a}$$

$$\text{s.t.} \quad \sum_{i=1}^{N_{\mathrm{off}}} \mathcal{I}_{\mathrm{a},i} \delta_i \geq \tilde{w}_{\mathrm{m}}^{\max} - W_{\mathrm{m}} \tag{7.51b}$$

$$\mathcal{I}_{\mathrm{a},i} \in \{0, 1\}, \quad i \in \mathfrak{B}_{\mathrm{off}}, \tag{7.51c}$$

where $\tilde{w}_{\mathrm{m}}^{\max}$ is the bandwidth needed from the MBS under $(\tilde{\mathcal{I}}, \tilde{\mu}_{\mathrm{E}})$. The objective function is to minimize the increased power consumption caused by activating additional SBSs, and (7.51b) guarantees that the MBS is not overloaded.

The 0–1 knapsack problem is NP-hard, and the optimal solution cannot be obtained within polynomial time. Therefore, we relax $\mathcal{I}_{\mathrm{a},i}$ to a continuous variable $0 \leq \hat{\mathcal{I}}_{\mathrm{a},i} \leq 1$, which can represent the probability to activate SBS i. Then, Problem (7.51) becomes a linear programming function. Applying the Lagrange multiplier method, we obtain the necessary condition of the optimal solution, based on the Karush–Kuhn–Tucker conditions [35], given as the following proposition.

Proposition 7.2 *The optimal solution of the relaxed problem* $\hat{\mathcal{I}}_{\mathrm{a}}$ *satisfies*

(1) $\hat{\mathcal{I}}_{\mathrm{a},i} = 1$ *if* $\frac{-\Delta_i}{\delta_i} < \nu$,

(2) $\hat{I}_{\mathrm{a},i} = 0$ *if* $\frac{-\Delta_i}{\delta_i} > \nu$,

(3) $\sum_{i=1}^{N_{\mathrm{off}}} \hat{I}_{\mathrm{a},i}\delta_i = \tilde{w}_{\mathrm{m}}^{\max} - W_{\mathrm{m}}$, *where* $\nu \leq 0$ *is the Lagrange multiplier.*

In fact, Proposition 7.2 indicates that those SBSs that can offload traffic at lower extra power (i.e. smaller $\frac{-\Delta_i}{\delta_i}$) should be activated with higher priority. Therefore, a suboptimal solution of problem (7.51) is obtained.

With Proposition 7.2, we propose the two-stage energy-aware traffic offloading (TEATO) scheme. In the first stage, each SBS is analyzed independently to derive the maximal power saving gain Δ_i and the bandwidth relieved at the MBS δ_i, same as that in the single-SBS case. In the second stage, the on–off states of SBSs are optimized, where the SBSs with positive power saving gain are activated first. Then, additional HSBSs and CSBSs are reactivated if the MBS is still overloaded, in the order of increasing $\frac{-\Delta_i}{\delta_i}$.

7.2.5 Simulation Results

We first evaluate the energy saving gain of the optimal solution for the single-SBS case and then demonstrate the performance of the proposed TEATO scheme for an HCN consisting of one MBS, one RSBS, one HSBS, and three CSBSs. The SBSs are micro BSs, and the main simulation parameters can be found in Table 7.2. Solar power harvesting devices are equipped at RSBS and HSBS. Figure 7.3 shows typical daily traffic and energy profiles, where the energy profile is based on real solar power generation data provided by the Elia group,[2] and the adopted traffic profile proposed by the EARTH project has been widely used for performance evaluation [29].

7.2.5.1 Power Saving Gain of the Single-SBS Case

Figure 7.4 shows the maximal power saving gain achieved by activating the RSBS or the HSBS under different system parameters. Figure 7.4a shows the power saving gain for the RSBS. It can be seen that power saving gain decreases with the handover cost C_{HO}. Specifically, the power saving gain firstly decreases for $\lambda_{\mathrm{E}} < \lambda_{\mathrm{E}}^{\mathrm{th}}$ and then increases

Table 7.2 Simulation parameters.

Parameter	Value	Parameter	Value
D_{m}	1000 m	D_{s}	300 m
α_{m}	3.5	α_{s}	4
W_{m}	10 MHz	W_{s}	5 MHz
R_{Q}	300 kbps	η	0.05
σ^2	-105 dBm MHz^{-1}	θ_{m}	1000
θ_{s} (single SBS)	500	θ_{s} (multi-SBS)	2000

2 The power generation is sampled and averaged every 15 minutes, and the data was collected in Belgium on 1 August 2014. For details, please refer to http://www.elia.be/en/grid-data/power-generation/Solar-power-generation-data/Graph.

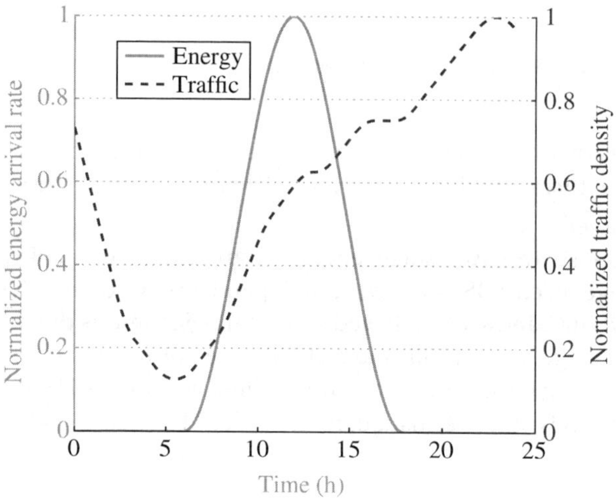

Figure 7.3 Daily traffic and energy profiles.

when $\lambda_E > \lambda_E^{th}$. Furthermore, λ_E^{th} increases with the handover cost C_{HO}. Theoretically, the power saving gain is a convex function of the energy arrival rate when the renewable energy supply is insufficient, i.e. $\lambda_E < \mu_E$, which can be proved by taking the second derivative of (7.41). Intuitively, when the energy arrival rate λ_E is small, the power saving gain of the MBS Δ_{mbs} is very low due to insufficient green energy, and thus the total power saving gain is dominated by the handover cost. As the energy queue is empty most of the time for $\lambda_E \to 0$, the handover frequency is approximately twice of the energy arrival rate, i.e. P_{HO} increases almost linearly as λ_E increases. Therefore, the power saving gain decreases for $\lambda_E < \lambda_E^{th}$. As λ_E becomes large, the probability that the battery runs out decreases, and the Δ_{mbs} balances P_{HO} out. Accordingly, the total power saving gain is dominated by the saved transmission power and thus increases with λ_E. Figure 7.4b shows the power saving gain for HSBSs, which increases with the density of users in the small cell. Moreover, it also increases with λ_E as the harvested energy reduces on-grid power consumption at the SBS. However, the power saving gain remains the same when the user density or λ_E is high, limited by the service capability of the HSBS.

Figure 7.5 shows the power saving performance of the proposed optimal solution and the conventional greedy scheme for the daily traffic and energy arrival profiles from Figure 7.3. For the conventional greedy scheme, the RSBS or HSBS is always active in order to offload users as many as possible, whose intuition is to make use of the harvested energy in greedy way to avoid battery overflow. Figure 7.5a shows the power saving gain w.r.t. the maximum energy arrival rate for the RSBS, when the maximal user density of the day is $\rho_s = 100$ km^{-2}. Under the greedy scheme, the power saving gain firstly decreases and then increases with the energy arrival rate due to the handover cost, same as the results in Figure 7.4a. However, the power saving gain of the optimal solution increases monotonously with the energy arrival rate, as the amount of offloading traffic and the on–off states of SBS are jointly optimized to avoid the frequent handover. In addition, the two schemes achieve the same performance if the handover cost can be ignored ($C_{HO} = 0$). Figure 7.5b shows the comparison of the power saving performance of the proposed optimal solution and the greedy scheme for HSBSs. It can be seen that

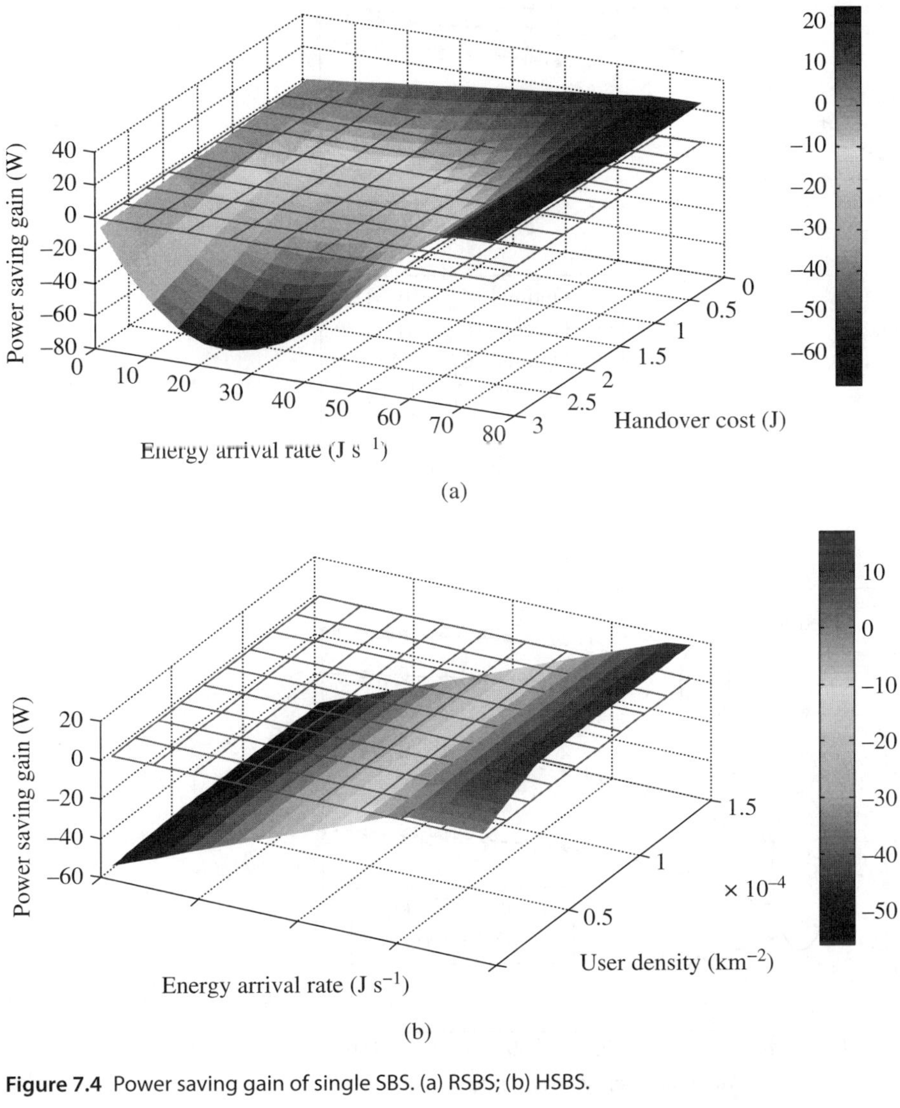

Figure 7.4 Power saving gain of single SBS. (a) RSBS; (b) HSBS.

the average power saving gain degrades as the traffic load decreases. The power saving gain of the greedy scheme even becomes negative for low energy arrival rate, whereas the proposed optimal solution always guarantees positive power saving gain.

7.2.5.2 Network Power Saving Gain

For the multi-SBS case, the maximal user density in macro cell is set to 20 km^{-2}, while the maximal energy arrival rate is set to 200 J s^{-1}. Suppose that the user density in small cells is twice as that covered by the macro cell. Figure 7.6 shows the normalized network power consumption when the handover cost per unit time is 2 J. The results are normalized by the power consumption of the HCN consisting of one MBS and four CSBSs, where no cell sleeping and power control are adopted. The difference between the two greedy schemes is whether CSBSs can go into sleep during low traffic hours

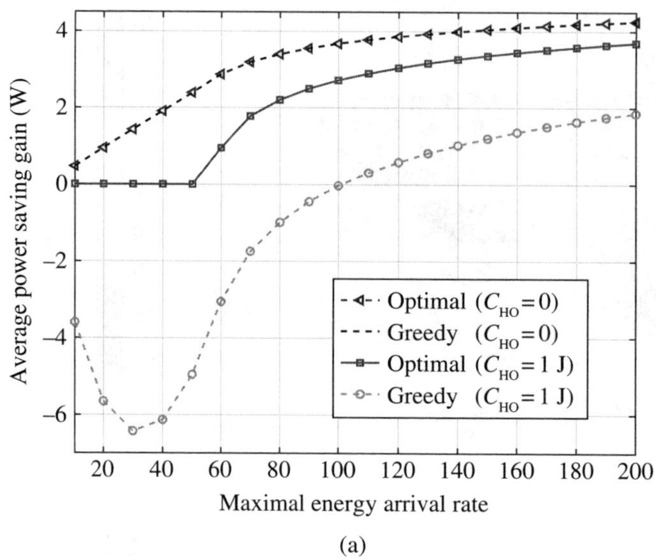

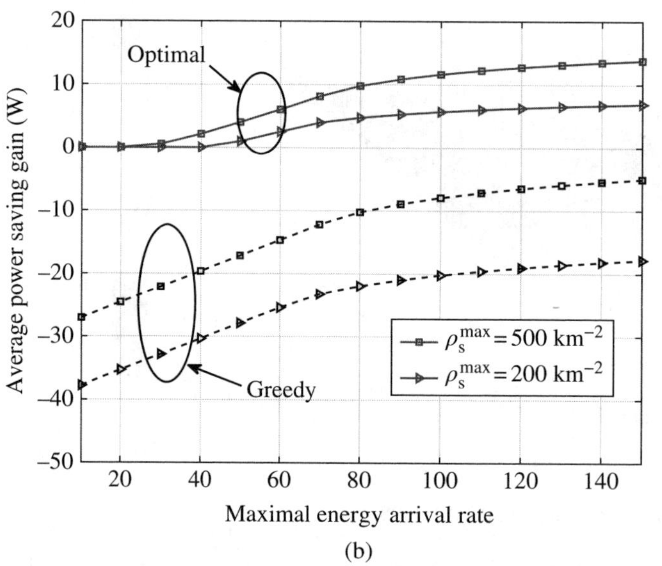

Figure 7.5 Average power saving gain of single SBS under daily traffic and energy profiles. (a) RSBS; (b) HSBS.

for energy saving. The TEATO scheme can achieve the best performance, as it adjusts the RF power and avoids activating SBSs, which may bring negative power saving gain. The performance with optimal solution obtained by exhaustive search is demonstrated as the green solid line. Notice that the performance of the proposed scheme TEATO is very close to the optimal solution, validating the 0–1 relaxation.

Figure 7.7 shows the average power saving gain of the three schemes under different weather conditions, where the maximal energy arrival rates for sunny and cloudy days are set to 500 and 50 W, respectively. It can be seen that the proposed TEATO can save around 50% energy for all scenarios compared with the greedy scheme without cell

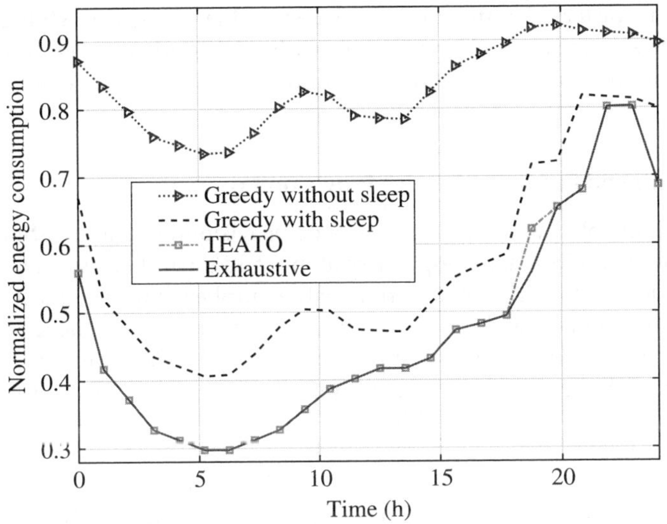

Figure 7.6 Daily power consumption.

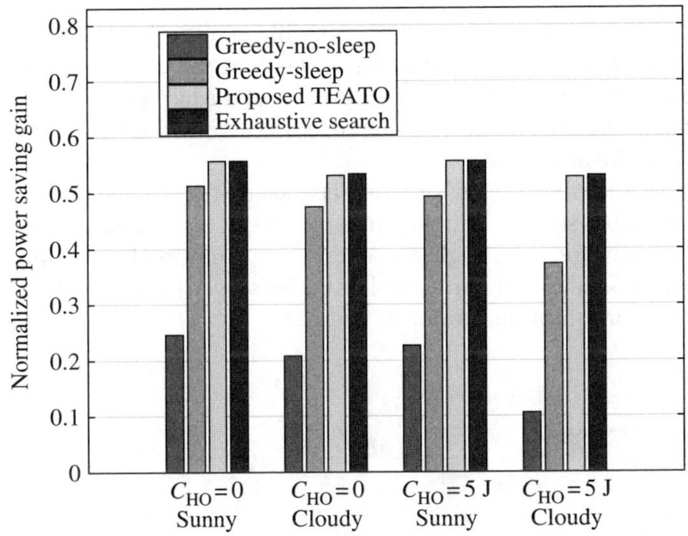

Figure 7.7 Average power saving gain.

sleeping. The results of Figures 7.6 and 7.7 demonstrate the benefits and necessity to conduct energy-aware power control and dynamic cell sleeping.

7.3 Proactive Content Caching and Push with Energy Harvesting-Based Small Cells

To match the unstable renewable energy supply with the dynamic traffic requirements, the proactive way of using the harvested energy can be effective, especially under the finite energy battery. In this section, we exploit the edge content caching and push technology at the BS side, in order to better utilize the renewable energy, satisfying the dynamic content requests from users. We will begin with a framework design for proactive content caching and push on BSs, and then dig into the optimal policy design for content push.

7.3.1 Network Architecture and Proactive Service Provisioning

The concept of an EH HCN enabled with proactive service provisioning is illustrated in Figure 7.8, where multiple EH SBSs cache popular contents and push them to users in a proactive manner. The design objective is to minimize the number of user requests handled by the MBS. The intuition of such metric is twofold: One is to save the grid power consumption. Because the grid power is mainly consumed by MBSs, minimizing the activity of the MBS reduces the grid power consumption, while the renewable energy used to power the SBSs can be regarded as free. The other is to enhance the users' QoS in terms of the file delivery time. For those contents already pushed to users, users can get the contents instantaneously without any delay. Even if for those requested contents without being pushed to users, as SBSs are closer to users, unicast from the SBS has higher transmission rate and thus guarantees shorter delay. The basic working flow of the EH SBS is described as follows:

- Fetch contents from the MBS via the backhaul link, possibly wireless backhaul link since these SBSs are EH based. As a result, the energy of fetching the contents is not negligible and accounts among major consumption portions of the harvested energy. Note that the wireline backhaul can also be considered as an option with less energy consumption but higher deployment cost.
- Cache the fetched content. The energy consumption depends on the storage method and the storage volume. For SBSs serving limited number of users, the contents can be stored locally in the SBS hardware, with negligible energy consumption for caching. When the cache size is large, additional hardware like cache server [36] is required, and in this case the energy consumption cannot be ignored.

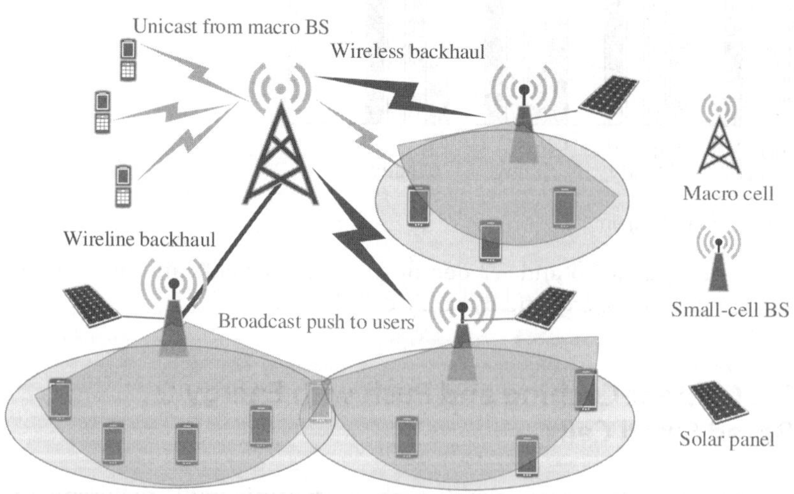

Figure 7.8 The concept an energy harvesting HCN enabled with proactive service provisioning. The small cells are energy harvesting based, while the macro BS is grid powered.

- Push the contents to users before the user actually requests it. Once contents have been fetched and cached at the SBS, they are selected to be pushed to users depending on their popularity and the battery status on the EH-based SBS. As shown in Figure 7.8, not only the users associated to the SBS but also those (as the handset in the middle of dark gray) in the overlapping coverage of multiple SBSs can enjoy the proactive push. In other words, the caching and push can be coordinated among multiple SBSs, and one example of caching coordination can be found in [17]. Note that for commonly interested contents, multicast/broadcast is performed, while for private contents, unicast push is adopted.
- Unicast the contents to users upon request. Users may request to its associated SBS for some content before it is pushed. If the SBS has the content fetched and the battery has enough energy, it will unicast the content to the user upon request. Pushing the requested content can also be considered, but if the requested content is private or not popular, it is not beneficial regarding the limited storage on user terminals.

On the user side, if the upper layer application requests some content, the user will check its local storage first to see if the content has already been pushed. If not, it will request over the air to its associated SBS. In this case, the request is still counted even if it is satisfied by the local storage. If the SBS is not able to handle the request, the MBS can unicast the content to the user.

The push mechanism can be realized by the existing broadcasting protocols without additional signaling overhead. Options include multimedia broadcast multicast services (MBMSs) proposed by 3GPP or broadcast and multicast service (BCMCS) [24]. The coordination of such broadcast channel falls into the category called integrated communication and broadcast networks (ICBNs) [37]. If specified broadcast channel does not exist, the SBS can reuse the unicast channel, but in this case the users besides the default unicast receiver should be notified to receive, which can be conveyed via the downlink control channel. Note that SBSs should have the popularity distribution of the contents, which can be updated by the MBS or the network provider periodically. This overhead is proportional to the update rate of the user interests, which is generally slower compared with the data transmission and EH.

7.3.1.1 Exploiting the Content and Energy Timeliness

The key benefit of proactive service provisioning is exploiting the timeliness of the contents and energy via intelligent caching and push, so as to match random energy arrivals and user requests over time and space. The timeliness of the contents corresponds to their popularity and life span. The contents are of interests to users for a finite period of time, and the popularity ranking of the contents may change over time. The timeliness of the harvested energy comes with the causality of energy usage and the limited battery capacity. The causality means that the harvested energy cannot be used before its arrival. Moreover, the limited battery capacity brings constraints on the delay of using such energy. In other words, if the arrived energy is not used timely, newly arrived energy will be wasted when the battery is full. To avoid this problem, the SBS can proactively fetches and caches the popular contents and then pushes them to users. In this way, the delay constraint of using the harvested energy is resolved, since the energy is used to provide stored contents to the users via push without waiting for the request. Or the harvested energy can be regarded as if it is transferred from the time when the content is pushed into the *future* when the user actually requests the content.

An illustrative example is shown in Figure 7.9. The time horizon is divided into equal-length slots, and user requests arrived during some slot are batched and responded at the beginning of the next slot. Energy is harvested and stored in the battery of the SBS at the beginning of each slot, and the corresponding battery state is shown in the figure. In this example the SBS can fetch or transmit (push or unicast) at most one content in each slot. The set of cached contents as of slot t is denoted by C_t'', and the set of pushed contents as of slot t is denoted by C_t'. Assume the length of the contents is the same and the height of the contents in the figure represents the energy used to fetch or push it, i.e. unit energy E for fetching and caching a content and $2E$ for pushing a content. As shown in Figure 7.9, the energy arrives in the second and fourth slots, and the SBS utilizes these energies (including the initial $3E$ energy in the battery) to fetch and cache two contents. The SBS then pushes the most popular content (the dark gray) in its cache to its users. Consequently, at the fourth and fifth slots, two requests for the dark gray content arrive, and since it has been pushed, the requests are instantaneously satisfied at user terminals. In the meantime, the SBS can push and fetch more contents, satisfying the requests in the seventh and eighth slots. Note that the request for the gray content is served with unicast from the SBS since it has not been pushed yet. In the last slot, since the SBS is running out of energy, the user request is responded by the MBS. In the example, without proactive caching and push, the four requests except the last one require $11E$ ($2E \times 4 = 8E$ for content unicast, and $E \times 3 = 3E$ for fetching the three contents), which cannot be satisfied since (i) the total energy budget is only $9E$ and (ii) the battery capacity is $6E$, which means that if the energy is not used proactively, $3E$ of the harvested energy will be wasted at the beginning of the fourth time slot.

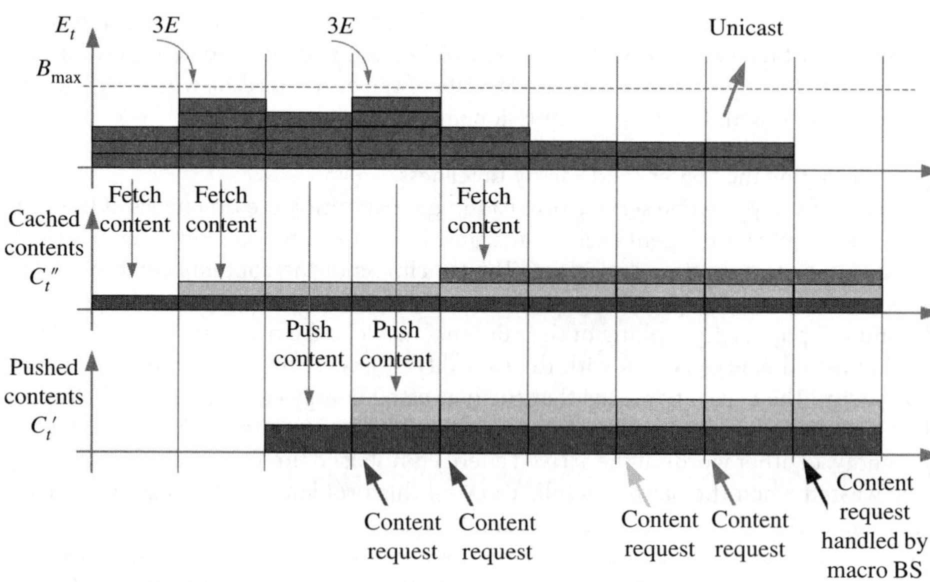

Figure 7.9 The behavior of an EH-based SBS enabled with proactive content caching and push: an example.

7.3.1.2 Energy Harvesting-Based Caching and Push: A Simple Policy Design Example

We illustrate the gain provided by proactive fetching and push, i.e. how proactive push can reduce the probability of handling user requests by the MBS.

Assume that the active content set C_t is time varying, i.e. new content comes into play over time and contents can also be outdated. This property is described as a birth–death process with birth rate λ_c and death rate μ_c, and the number of active contents at time slot t is given by $|C_t| = N$. The popularity of a content is defined as the probability that a user requests the content, denoted by f_i, which is assumed to follow the Zipf distribution [16]. By ranking the N contents with descending popularity, the popularity of the ith ranked content is

$$f_i = \frac{1/i^{\nu}}{\sum_{j=1}^{N} 1/j^{\nu}}, \tag{7.52}$$

where $\nu > 0$ is the Zipf parameter. Larger ν means that fewer contents account for most popular ones. The user request arrives at the beginning of a slot with probability p_r, and we count both kinds of requests: those satisfied by the user local storage thanks to the proactive push and the requests served over the air from the SBS. In addition, when a content gets outdated and departs from the active content set C_t, it will be removed from both C_t'' and C_t' (if it has already been fetched and pushed to users), and the corresponding storage space is freed.

At the beginning of each slot, the SBS harvests a random amount of energy and stores it in a battery with finite capacity E_{max}, and we denote the battery energy state by E_t. Assume one or more contents can be fetched in a slot, with energy consumption E_F for each content. Each fetched content can be transmitted, through either push or unicast, in a slot with E_P units of energy. For example, in Figure 7.9, $E_F = E$ and $E_P = 2E$. The channels from SBS to users are assumed to be static and identical, so that E_P amount of push energy can always guarantee successful delivery. Based on the current system state, including the active content set C_t, the pushed content set C_t', the cached content set C_t'', and the amount of energy units in the battery, the SBS makes the action decision for the current slot from fetching a content for caching, pushing a content, unicasting a required content in C_t'', and doing nothing. When the SBS decides to do nothing, the user request, if arrives, will be handled by the MBS. As explained, the policy design objective is to minimize the ratio of user requests handled by the MBS, denoted by η, subject to the energy causality constraint, i.e. the harvested energy cannot be used before its arrival.

First, consider push only by assuming $C_t'' = C_t$, which corresponds to the case that the SBS can get the content instantaneously via high-speed backhaul when it needs to push or unicast it. Therefore the energy consumption of fetch and caching is ignored. When the energy in the battery is sufficient and the SBS decides to push, the most popular content in C_t'' is pushed to users. We assume that the storage space of users is large enough to store the contents in set C_t'. A simple push policy works as follows. If there is no user request *over the air* in current slot and the battery energy is sufficient for pushing a content, the SBS will push the most popular content in C_t'', which has not been pushed. Otherwise, if a user requests a content, a unicast will be performed given the battery energy is sufficient. The user request is handled by the MBS when the energy in the SBS battery is not enough. Suppose the energy arrival follows Bernoulli distribution, i.e. at the beginning of each slot, the system can harvest E_H units of energy with probability p. Set $p = 0.5$, $\mu_c = 1 \times 10^{-3} s^{-1}$, $\nu = 1$, $E_P = 2$, and $E_H = 3$. The ratio of requests handled by the MBS is shown in Figure 7.10. To compare, the policy without proactive push serves as the baseline, where the SBS unicasts a required content as long as there is enough energy in the battery, or if no user request arrives, the SBS does nothing. It can be seen from Figure 7.10a that in the baseline scheme, the probability that the SBS cannot provide service increases as the request arrival rate p_r increases. But on the other hand, proactive push keeps such probability low and stable, i.e. almost irrelevant to p_r, hence substantially reducing the burden of the MBS. One can also note that when p_r is low and the content refreshing rate is high, i.e. $\lambda_c = 3$, proactive push does not bring any performance gain, because in this case user requests are diverse over different contents and a pushed content has low probability to be requested.

Figure 7.10b shows the influence of the battery capacity E_{max}, since one major benefit of proactive service is to address the limited battery. The ratio of requests handled by the MBS decreases as E_{max} increases in the figure. Compared with the baseline, the reduction of the ratio is significant. In other words, to achieve the same performance, the required battery capacity with push can be saved. In Figure 7.10c, the impact of the content popularity distribution is depicted, where the Zipf distribution parameter v varies from 0 to 2, i.e. from a uniform distribution to a more concentrated one. When the contents are uniformly distributed, it is better not to use proactive push, while the gain of proactive push increases with more concentrated content distribution and lower content refreshing rate.

We then take into account the cost of fetching the contents to the SBS. Initially the set of contents C_t is not available at the SBS, and the SBS has to firstly fetch the contents from the MBS via the backhaul. Remark we have $C'_t \subseteq C''_t \subseteq C_t$. It is reasonable to assume that the energy for fetching a content is less than pushing a content, and the SBS can possibly fetch multiple contents in one slot. A threshold-based fetch policy is proposed. When $C'_t = C''_t$ and there is no unicast request, and, in addition, the number of energy units in the battery is no less than a given threshold M_f, the SBS fetches at most K contents (the actual number of fetched contents is $\min\{K, |C_t \backslash C''_t|\}$) and therefore consumes E_F units of energy. There is another threshold M_p for push: the SBS will not push any content until the number of energy units in the battery is no less than M_p. Whether push or fetch depends on the relative quantity of sets C_t, C'_t, and C''_t. Larger threshold can help reduce the probability of battery outage when a unicast request arrives but on the other hand reduces the chances of push or fetch. To sum up, the threshold-based push and

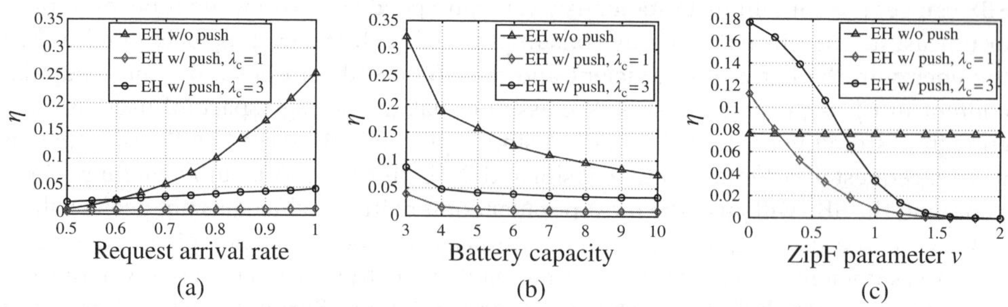

Figure 7.10 Ratio of the requests handled by the macro BS denoted by η, with proactive push, $E_P = 2, E_H = 3$. (a) $E_{max} = 10, v = 1$. (b) $p_r = 0.75, v = 1$. (c) $E_{max} = 10, p_r = 0.75$.

caching scheme is described as follows:

$$\text{Policy} = \begin{cases} \text{push} & \text{if } R_t \in \emptyset \bigcup C'_t \text{and } \dfrac{|C'_t|}{|C''_t|} < \dfrac{|C''_t|}{|C_t|} \text{ and } E_t \geq M_\text{p} \\ \text{fetch} & \text{if } R_t \in \emptyset \bigcup C'_t \text{ and } \dfrac{|C'_t|}{|C''_t|} \geq \dfrac{|C''_t|}{|C_t|} \text{ and } E_t \geq M_\text{f} \\ \text{unicast} & \text{if } R_t \in C''_t \backslash C'_t \text{ and } E_t \geq E_\text{P} \\ \text{idle} & \text{else.} \end{cases} \tag{7.53}$$

In this scheme, $\frac{|C'_t|}{|C''_t|} < \frac{|C''_t|}{|C_t|}$ means that the number of requests in C'_t relative to C''_t is less than that of C''_t relative to C_t. Hence, the cached contents in the SBS need to be pushed to reduce the unicast events. On the other hand, when $\frac{|C'_t|}{|C''_t|} \geq \frac{|C''_t|}{|C_t|}$, the number of cached contents is relatively small, which may result in more users served by the MBS.

We set $\lambda_\text{c} = 1, K = 3, E_\text{F} = 1, E_\text{max} = 10$, and other parameters are the same as those in the push-only case. Figure 7.11 compares the proposed algorithm with the case without proactive push. Similar to Figure 7.10, it shows that push can significantly improve the performance, and this also confirms the necessity of having push. The results also indicate that the thresholds (M_p and M_f) should be carefully selected regarding the system parameters, especially the battery capacity as shown in Figure 7.11b. From the figures, it is conjectured that having more aggressive fetch and caching (with smaller M_f) provides better performance, as contents to be pushed should already been cached. But one should also note that using too much energy for fetch and caching leaves less energy for push so that these two sides of activities should be balanced.

7.3.2 Policy Optimization for Content Push

To elaborate the proactive service provisioning with optimized push policy, we consider a two-tier HCN composed of an MBS and multiple SBSs as shown in Figure 7.8. Each SBS serves users in its coverage and treats the data transmissions of other SBSs as background interference. Since the SBSs are densely deployed, the sum interference is considered the same for every user. Hence, we only focus on the policy design of a

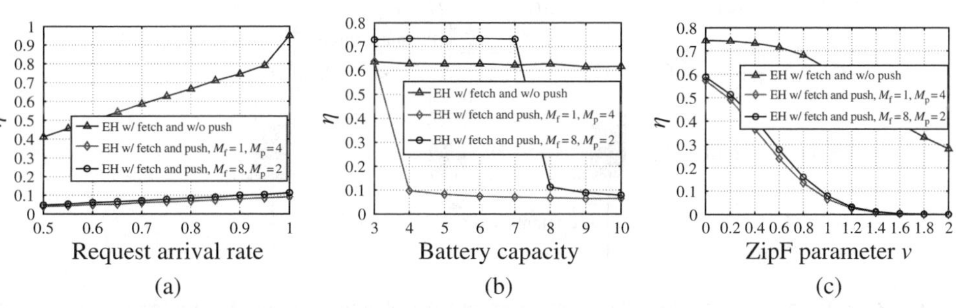

Figure 7.11 Evaluations of the ratio of the requests handled by the macro BS, with proactive fetch for caching and push, $E_\text{F} = 1, E_\text{P} = 2, E_\text{H} = 3$, where η denotes the ratio of user requests handled by the macro BS. (a) $E_\text{max} = 10, v = 1$. (b) $p_\text{r} = 0.75, v = 1$. (c) $E_\text{max} = 10, p_\text{r} = 0.75$.

single SBS, and all the SBSs can work similarly. The MBS is powered by the power grid, and the SBS is solely powered by renewable energy. The harvested energy is stored in a battery with finite capacity. Here we assume that the time and energy consumption to fetch contents is negligible and thus only focus on how to push contents from the SBS to users. We will provide detailed model for optimizing the push policy and then using Markov decision process (MDP) theory [38] to get the optimal policy. To reduce the complexity of policy design, we further introduce several threshold-based policies and evaluate their performance.

7.3.2.1 Model for Content Push at the Energy Harvesting-Based SBS

Figure 7.12 illustrates how the dynamic system evolves. Specifically, the system is slotted with slot length $T_{\rm p}$. In each slot t, a content request Q_t is generated from a user. Based on the request and the battery energy state, the SBS takes an action with energy usage U_t and then harvests a certain amount of energy A_t. At the user side, each user receives the contents and preserves a list of pushed contents C_t. The detailed description of the model is as follows.

Content Request and Channel Model The content request is assumed to arrive with the Bernoulli distribution, i.e. there is a content request with probability $p_u \in [0, 1]$ in each slot. The location of the request follows the uniform distribution in the cell coverage. Each content is transmitted in one time slot with average data rate r_0, which can be calculated as

$$r_0 = \mathbb{E}_h \left[W \log_2 \left(1 + \frac{P_t |h|^2 \beta d^{-\alpha}}{\sigma^2 + P_{\rm I}} \right) \right], \tag{7.54}$$

where W is the bandwidth of the SBS; P_t is the transmit power; h is the small-scale fast fading coefficient; β and α represent the path-loss constant and the path-loss exponent, respectively; d is the transmission distance; and $\sigma^2 + P_{\rm I}$ is the noise plus interference

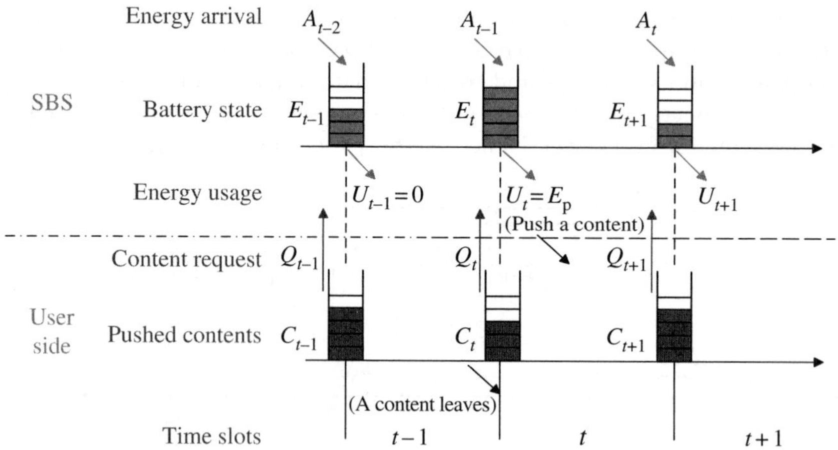

Figure 7.12 Timeline of the slotted system and examples of actions. At the beginning of slot $t-1$, the SBS takes sleep action ($U_{t-1} = 0$). By the end of slot $t-1$, a content leaves the system, and the number of pushed contents C_t is reduced by 1. At the beginning of slot t, the SBS takes push action ($U_{t-1} = E_{\rm p}$), which results in an increase of C_{t+1} by the end of the slot.

power. Assume the SBSs and the MBS are allocated with orthogonal frequency bands, and thus the interference is only caused by other SBSs. The interference power P_I is computed by the summation over all the interfering SBSs. $\mathbb{E}_h$ is the expectation operator w.r.t. h. That is, the average data rate r_0 is obtained by averaging the instantaneous channel rates of all fading blocks. In practice, as a content usually contains multiple data packets, each transmission slot contains multiple packet transmission frames, among which the small-scale fading h changes. Hence, the data rate averaging over h is a good approximation of the actual content delivery rate. We further assume that the user is of low mobility so that it remains in one cell during transmission.

Based on the above model, the content request state Q_t can be denoted as the amount of energy required for unicasting a content. As it is averaged over h, Q_t is a function of the transmission distance, i.e. $Q_t = E_u(d)$. Notice that $Q_t = 0$ indicates that there is no unicast request.

Action and Energy Model of the SBS The SBS has three possible actions: sleep, unicast, and push. When the SBS has sufficient energy, it can either unicast a content upon request or push a popular content to all the users in its coverage. When the battery energy is depleted, the SBS enters the sleep mode, and the content requests will be handled by the MBS. The action is denoted by u_t and is defined as

$$u_t = \begin{cases} 0 \text{ sleep} \\ 1 \text{ unicast a content} \\ 2 \text{ push a content.} \end{cases} \tag{7.55}$$

Notice that for the push action, the SBS always pushes the most popular content. This is most efficient as it minimizes the number of the duplicated data transmissions. The action to be taken is constrained by the available energy in the battery, i.e. the energy used in slot t satisfies $U_t \leq E_t$. In sleep mode, we assume the energy consumption is negligible, i.e. $U_t = 0$. In push mode, the energy $U_t = E_p$ is used to guarantee successful receipt for all the users in the cell coverage. Then the battery energy state is updated as $E_{t+1} = \min\{B_{\max}, E_t - U_t + A_t\}$, where $B_{\max}$ is the battery capacity and the harvested energy A_t is assumed ergodic and i.i.d. with mean λ_E (energy arrival rate).

Content Popularity Model Assume there are in total N active contents that the users are interested in, denoted by the set $C = (c_1, \ldots, c_N)^T$, where the ith ranked content c_i is requested with probability (or popularity) f_i. Statistical researches have shown that the content popularity distribution is well fitted by the Zipf distribution [16, 39]. The popularity of content c_i can be expressed as in (7.52).

In addition, as people's interest changes over time, old contents can become outdated, and new popular contents will emerge. We assume in each slot there is an old content replaced by a new one with probability $p_c \in [0, 1]$, and the old one is randomly picked over $\mathbf{C}$ with equal probability $\frac{1}{N}$. However the popularity changes as follows. If content c_i is replaced by the new content c', the active content set becomes $\mathbf{C}' = (c_1, \ldots, c_{i-1}, c_{i+1}, \ldots, c_N, c')^T$, i.e. the ranks of the contents $c_{i+1}, \ldots, c_N$ increase by one, and the new content c' is with rank N. Such a content updating model is used for two reasons. Firstly, it reflects the fact that the popularity of most contents increases gradually in real systems [39]. Secondly, it simplifies the problem formulation, i.e. we only need to consider the number of popular contents rather than the popularity of each individual content. Specifically, according to the push action and content model, it can be easily verified that the pushed content set is $C_t = (c_1, \dots, c_{C_t})$, where $C_t \leq N$ is the number of pushed contents. That is, the users always preserve the most popular contents. The push action is taken on the remaining content set $\widetilde{C}_t = (c_{C_t+1}, \dots, c_N)$, and the content c_{C_t+1} should be pushed at first. As a result, we can use the scalar C_k instead of the vector $\boldsymbol{C}_t$ to indicate the push state.

We summarize the randomness involved in the system as follows. Each content request is generated randomly according to Bernoulli distribution. Content requests follow the popularity distribution defined in (7.52), and its location follows uniform distribution. The random location of the user request indicates the randomness of channel. In addition, the energy arrives randomly, and the active content set is updated randomly. As the system state changes among slots, the action should also change accordingly.

Problem Formulation When a user requests a content that is not in its storage, and at the same time, the SBS chooses not to unicast the content, the request needs to be handled by the MBS (i.e. blocked by the SBS), which usually causes more energy consumption due to the larger transmission distance. We aim at optimizing the usage of harvested energy to minimize the ratio of content requests blocked by the SBS, termed as the *blocking probability* at the SBS. Mathematically, the objective is expressed as

$$\underset{u_1,u_2,\dots}{\text{minimize}} \lim_{T\to+\infty} \frac{\bar{K}}{T}, \tag{7.56}$$

where $\bar{K}$ is the number of blocked requests, T is the total number of slots, and the optimization variables are the SBS's actions $\{u_1, u_2, \dots\}$. We aim to find the optimal policy $\{\pi_1, \pi_2, \dots\}$, where the policy π_t is a mapping from the state space to the action space, i.e. the action in slot t is $u_t = \pi_t(x_t)$, where the system state x_t will be detailed in the next section. According to the statistics of the system including the energy arrival distribution, content request generating rate, content popularity, and updating process, we can find the optimal policy via dynamic programming, which will be detailed next with the MDP formulation.

7.3.2.2 Optimal Policy with Finite Battery Capacity

To find the optimal solution for the problem (7.56), we need to decide the SBS's action based on the system state at the beginning of each slot. MDP [38] is an effective mathematical framework to formulate this type of decision-making problems. A standard MDP problem contains the following elements: states, actions, cost function, and state transition. Notice that the energy states and the channel states are continuous in our problem. Therefore to make the problem tractable, we first discretize the system states and then reformulate the problem as an MDP optimization and use policy iteration algorithm to find the optimal policy.

System State The system state in slot t is denoted by $x_t = (E_t, Q_t, I_t, C_t)$, where the energy state E_t is the amount of energy in the battery, the request state Q_t is the energy required for unicast, the push state C_t is the number of contents that have been pushed, and $I_t \in \{0, 1\}$ indicates which content is requested. $I_t = 1$ if the required content is c_{C_t+1}, and $I_t = 0$ if the rank of required content is larger than $C_t + 1$. The reason to introduce I_t is that if the push action is chosen, $(C_t + 1)$th ranked content c_{C_t+1} will be pushed as it is the most popular content in the set $\widetilde{\boldsymbol{C}}_t = (c_{C_t+1}, \dots, c_N)^T$. Thus the user request for

c_{C_t+1} is simultaneously satisfied. Recall that Q_t represents the energy consumption for unicast, which is $Q_t = E_u(d) = P_t(d)T_p$, where $P_t(d)$ is the transmission power obtained by solving (7.54). And $Q_t = 0$, indicating that there is no content request. Finally, denote the state space by $\mathcal{S}$.

To make the DP algorithm tractable, we discretize the energy state E_t and the content request state Q_t into finite sets. Even though the optimal solution for the discretized finite state problem may not be optimal for the continuous state problem, the gap diminishes as the discretization becomes finer [38, Vol. I, Sec. 6.6.1]. Specifically, the energy is discretized with the unit energy E_{unit}. Then the energy state can be denoted by $E_t \in \{0, 1, \ldots, E_{\max}\}$ with $E_{\max}E_{\text{unit}} = B_{\max}$. $E_t = i$ corresponds to iE_{unit} amount of energy in the battery, and similarly, $A_t = i$ corresponds to iE_{unit} amount of energy arrived in slot t.

To discretize Q_t, select a series of distances $0 < d_1 < d_2 < \cdots < d_M = R$ so that $P_t(d_i)T_p = l_iE_{\text{unit}}$ where l_i is a positive integer for any $i = 1, 2, \ldots, M$. If the distance of a user to the SBS satisfies $d_{i-1} < d \leq d_i$, the SBS will unicast the requested content with energy $P_t(d_i)T_p$, guaranteeing the average data rate r_0 for all the users in this area. And $l_0 = 0$ denotes that the required energy for unicast is zero. Then the content request state is denoted by $Q_t \in \{0, 1, \ldots, M\}$, where $Q_t = i$ corresponds to the case that l_iE_{unit} amount of energy is required for unicast. With the discretization, the state space $\mathcal{S}$ is of finite dimension $(E_{\max} + 1) \times (2M + 1) \times (N + 1)$.

7.3.2.3 MDP Problem Formulation and Optimization

Besides the system state, the action, the cost function, and the state transition are further clarified to complete the MDP formulation. The action has been modeled as (7.55). Notice that in different states, the SBS may not be able to take all three actions. A simple example is that if $E_t = 0$, the SBS can do nothing but sleep, i.e. $u_t = 0$. Hence, the action space is state dependent, which can be expressed as $u_t \in \mathcal{U}_t(x_t)$. If $E_t \geq l_{Q_t} \geq 1$, i.e. the energy for unicast is available, we have $1 \in \mathcal{U}_t(x_t)$. On the other hand, if $E_t \geq E_p$ and $C_t < N$, i.e. the energy for push is available and there are contents to be pushed, we have $2 \in \mathcal{U}_t(x_t)$.

The per-slot cost function $g_t(x_t, u_t)$ denotes whether a content request is blocked or not. Mathematically, it can be expressed as

$$g_t(x_t, u_t) = \begin{cases} 1 & \text{if } Q_t > 0, u_t \neq 1, \text{and } I_t u_t = 0 \\ 0 & \text{otherwise.} \end{cases} \tag{7.57}$$

Notice that a content request is blocked if there is a request but the requested content is not transmitted by either unicast or push. When $g_t(x_t, u_t) = 0$, it refers to either there is no unicast request or the requested content is in the pushed list $\mathbf{C}_t$.

The state transition is expressed as the conditional probability

$$\begin{aligned} & p_{x_t \to x_{t+1}|u_t} \\ & = \Pr(E_{t+1}, Q_{t+1}, I_{t+1}, C_{t+1}|E_t, Q_t, I_t, C_t, u_t) \\ & = \Pr(E_{t+1}|E_t, Q_t, u_t)\Pr(C_{t+1}|C_t, u_t)\Pr(Q_{t+1}, I_{t+1}|C_{t+1}), \end{aligned} \tag{7.58}$$

where the second equality is derived by the definition of conditional probability. We calculate the state transition probability according to (7.58). Firstly, we denote $p_a(i), i = 0, 1, \ldots$ as the probability that iE_{unit} amount of energy arrives, which satisfies $p_a(i) \in [0, 1]$ and $\sum_i p_a(i) = 1$. To simplify the description, set

$p_a(i) = 0, \forall i = -1, -2, \ldots$, and then

$$
\Pr(E_{t+1}|E_t, Q_t, u_t) = \begin{cases} p_a(E_{t+1} - E_t) & \text{if } u_t = 0,\ E_{t+1} < E_{\max} \\ 1 - \sum_{i=0}^{E_{\max}-E_t-1} p_a(i) & \text{if } u_t = 0,\ E_{t+1} = E_{\max} \\ p_a(E_{t+1} - E_t + l_{Q_t}) & \text{if } u_t = 1, E_{t+1} < E_{\max} \\ 1 - \sum_{i=0}^{E_{\max}-E_t+l_{Q_t}-1} p_a(i) & \text{if } u_t = 1,\ E_{t+1} = E_{\max} \\ p_a(E_{t+1} - E_t + l_M) & \text{if } u_t = 2,\ E_{t+1} < E_{\max} \\ 1 - \sum_{i=0}^{E_{\max}-E_t+l_M-1} p_a(i) & \text{if } u_t = 2,\ E_{t+1} = E_{\max}. \end{cases} \tag{7.59}
$$

Notice that when $E_{t+1} = E_{\max}$, the battery may overflow, so the probability is calculated by summarizing all the possible energy arrival conditions.

Secondly, the push state C_t can only transit to its neighboring values $C_t + 1, C_t - 1$ or keep constant, i.e. $C_{t+1} \in \{\max\{0, C_t - 1\}, C_t, \min\{N, C_t + 1\}\}$. The transition probability of push state C_t is thus

$$
\Pr(C_{t+1}|C_t, u_t) = \begin{cases} p_c \frac{C_t}{N} & \text{if } u_t < 2, C_{t+1} = C_t - 1 \geq 0, \\ & \text{or } u_t = 2, C_{t+1} = C_t < N \\ 1 - p_c \frac{C_t}{N} & \text{if } u_t < 2, C_{t+1} = C_t, \\ & \text{or } u_t = 2, C_{t+1} = C_t + 1 \leq N \\ 0 & \text{else.} \end{cases} \tag{7.60}
$$

Finally, the content request state transition is

$$
\Pr(Q_{t+1}, I_{t+1}|C_{t+1}) = \begin{cases} (1 - p_u) + p_u \sum_{i=1}^{C_{t+1}} f_i & \text{if } Q_{t+1} = 0 \\ p_u f_{C_{t+1}+1} \frac{d_m^2 - d_{m-1}^2}{R^2} & \text{if } Q_{t+1} = m > 0, \\ & \text{and } I_{t+1} = 1 \\ p_u \left(1 - \sum_{i=1}^{C_{t+1}+1} f_i\right) \frac{d_m^2 - d_{m-1}^2}{R^2} & \text{if } Q_{t+1} = m > 0, \\ & \text{and } I_{t+1} = 0. \end{cases} \tag{7.61}
$$

where f_i is calculated according to (7.52) and $d_0 = 0$. The transition probability (7.61) is calculated based on our model, i.e. there is a content request with probability p_u, for content i with probability f_i, and the location of the request follows uniform distribution in the cell coverage. Notice that if $C_{t+1} = N$, we always have $Q_{t+1} = 0$. In summary, the state transition probability (7.58) is calculated based on the widely used models about energy arrival, content popularity, and user distribution. In fact, our framework is also applicable to some other models.

Based on the MDP framework, the original optimization problem (7.56) can be rewritten as

$$\min \lim_{T\to+\infty} \frac{1}{T}\mathbb{E}\left[\sum_{k=0}^{T-1} g(x_t, \pi_t(x_t))\right]. \tag{7.62}$$

The expectation is taken over all the random variables as previously described. The optimization is taken over all possible policies $\{\pi_1, \pi_2, \dots\}$. It can be proved that for any two states, there is a stationary policy $\boldsymbol{\pi} = \{\pi(x)\}_{x\in\mathcal{S}}$ so that one state can be accessed with nonzero probability from the other with finite steps. Consequently, the optimal value is independent of the initial state x_0, and there exists an optimal stationary policy $\boldsymbol{\pi}^*$ [38, Vol. II, Sec 4.2].

According to [38, Vol. II, Prop. 4.2.1], the optimal average cost λ^* together with some vector $\boldsymbol{h}^* = \{h^*(x)|x \in \mathcal{S}\}$ satisfies the Bellman equation:

$$\lambda^* + h^*(x) = \min_{u\in\mathcal{U}(x)}\left[g(x,u) + \sum_{y\in\mathcal{S}} p_{x\to y|u} h^*(y)\right]. \tag{7.63}$$

Furthermore, if $u = \pi^*(x)$ attains the minimum value of (7.63) for each x, the stationary policy $\boldsymbol{\pi}^*$ is optimal. Based on the Bellman equation, instead of the time average cost minimization, we only need to deal with (7.63), which only relates to per-slot cost $g(x, u)$ and state transition probability $p_{x\to y|u}$. The policy iteration algorithm [38, Vol. II, Sec. 4.4] can effectively solve the problem as described next.

Policy Iteration Algorithm The policy iteration algorithm starts with any feasible stationary policy and improves the objective iteratively. Suppose in the jth step we have a stationary policy denoted by $\boldsymbol{\pi}^{(j)}$. Based on this policy, we perform *policy evaluation* [38, Vol. II, Sec. 4.4], i.e. solving the following linear equations

$$\lambda^{(j)} + h^{(j)}(x) = g(x, \pi^{(j)}(x)) + \sum_{y\in\mathcal{S}} p_{x\to y|\pi^{(j)}(x)} h^{(j)}(y) \tag{7.64}$$

for $\forall x \in \mathcal{S}$ to get the average cost $\lambda^{(j)}$ and vector $h^{(j)}$. Notice that there are $(E_{\max} + 1) \times (2M + 1) \times (N + 1)$ equations but $(E_{\max} + 1) \times (2M + 1) \times (N + 1) + 1$ unknown parameters; hence multiple solutions exist, which are different with each other by a constant value for all $h^{(j)}(x)$. Without loss of generality, we can set, for example, $h^{(j)}(E_{\max} + 1, 2M + 1, N_S + 1) = 0$; then the solution for (7.64) is unique.

As $\boldsymbol{\pi}^{(j)}$ may not be the optimal policy, we subsequently perform *policy improvement* [38, Vol. II, Sec. 4.4] step to find the policy $\boldsymbol{\pi}^{(j+1)}$, which minimizes the right-hand side of the Bellman equation:

$$\pi^{(j+1)}(x) = \arg\min_{u\in\mathcal{U}(x)}\left[g(x,u) + \sum_{y\in\mathcal{S}} p_{x\to y|u} h^{(j)}(y)\right]. \tag{7.65}$$

If $\boldsymbol{\pi}^{(j+1)} = \boldsymbol{\pi}^{(j)}$, the algorithm terminates, and the optimal policy is obtained $\boldsymbol{\pi}^* = \boldsymbol{\pi}^{(j)}$. Otherwise, repeat the procedure by replacing $\boldsymbol{\pi}^{(j)}$ with $\boldsymbol{\pi}^{(j+1)}$. It is proved that the policy iteration algorithm terminates in finite number of iterations [38, Vol. II, Prop. 4.4.1] .

Although the DP-based solution can obtain the optimal policy, it has two drawbacks. Firstly, it usually encounters the *curse of dimensionality* [38] when the number of states is large. Secondly, it is difficult to obtain the closed-form expressions to reveal the solution structure. Therefore, we will present some heuristic algorithms in the following sections. Nevertheless, studying MDP formulation is meaningful as it provides a theoretical upper bound to evaluate other suboptimal algorithms.

7.3.2.4 Threshold-Based Policies

In this section, we propose several threshold-based policies. The intuition is that pushing the popular contents enhances the energy efficiency, while the push action itself consumes energy. As we always push the most popular contents, there should be a threshold on C_t, denoted by C_{thr}, so that it is favorable (in terms of energy consumption) to push contents in the list $\mathbf{C} = (c_1, \dots, c_{C_{\text{thr}}})$ rather than unicasting them with possible duplications. Also as will be shown later, threshold-based policies are analytically tractable and have some good asymptotic properties. Therefore, we consider that the SBS always takes push action as long as the number of pushed contents is smaller than the threshold C_{thr}. Notice that as the system is energy limited, the threshold should depend on the energy state E_t as well as the content request state Q_t. Note that it is difficult to analyze the state-dependent thresholds. Hence, we focus on the policies with constant threshold to obtain some closed-form results. Notice that unlike the MDP formulation, our analysis in this section is done in the continuous regime.

There are two steps to design a threshold-based policy. Step one is to determine the threshold C_{thr}. Step two is to determine the action to take when the threshold C_{thr} is achieved. With the objective of minimizing the number of requests blocked by the SBS, we first analyze the performance of push action with sufficient energy, which is detailed in the following lemmas.

Lemma 7.1 *If the required energy for push can always be satisfied, the stationary probability with which the SBS pushes a content in threshold-based policy is*

$$\Pr(u_t = 2) = \frac{p_c C_{\text{thr}}}{N + p_c}. \tag{7.66}$$

Proof: If the energy for push is always satisfied, C_t will not be smaller than $C_{\text{thr}} - 1$, since every time that $C_t = C_{\text{thr}} - 1$, a new content is pushed ($u_t = 2$). Thus, there are in total two stationary push states $C_t \in \{C_{\text{thr}}, C_{\text{thr}} - 1\}$, and the push action is always taken in state $C_t = C_{\text{thr}} - 1$. We have the transition probabilities $\Pr(C_{t+1} = C_{\text{thr}} - 1 | C_t = C_{\text{thr}}) = p_c \frac{C_{\text{thr}}}{N}$, $\Pr(C_{t+1} = C_{\text{thr}} - 1 | C_t = C_{\text{thr}} - 1) = p_c \frac{C_{\text{thr}}-1}{N}$. Therefore, the push state transition matrix can be expressed as

$$\mathbf{P}_C = \begin{pmatrix} p_c \frac{C_{\text{thr}}-1}{N} & 1 - p_c \frac{C_{\text{thr}}-1}{N} \\ p_c \frac{C_{\text{thr}}}{N} & 1 - p_c \frac{C_{\text{thr}}}{N} \end{pmatrix}. \tag{7.67}$$

By solving the equilibrium equation $q = q\mathbf{P}_C$, where $q = (q_{C_{thr}-1}, q_{C_{thr}})$, $q_i = \Pr(C_t = i)$, $i = C_{thr} - 1, C_{thr}$ are the stationary probabilities of push state, we can get

$$\Pr(u_t = 2) = q_{C_{thr}-1} = \frac{p_c C_{thr}}{N + p_c}. \tag{7.68}$$

□

The blocking probability of a threshold-based policy consists of two parts. The first happens because if the SBS is pushing a content while a user is requesting another content, the request is blocked. The second is due to energy shortage, i.e. the user request will be blocked if there is not enough energy for unicast. Based on the observation, we can derive the performance lower bound.

Lemma 7.2 *The blocking probability of the threshold-based policy is lower bounded by*

$$p_{blk,LB}(C_{thr}) = \frac{p_c p_u C_{thr}}{N + p_c}\left(1 - \sum_{i=1}^{C_{thr}} f_i\right). \tag{7.69}$$

Proof: We derive the lower bound by assuming that there is sufficient power supply for any actions. Thus, a request is blocked only when the SBS is pushing another content. Depending on (7.61) and (7.66), the stationary blocking probability satisfies

$$\begin{aligned} p_{blk} &\geq \Pr(Q_t > 0, I_t = 0, u_t = 2) \\ &= \Pr(u_t = 2)\Pr(Q_t > 0, I_t = 0 | u_t = 2) \\ &= \frac{p_c C_{thr}}{N + p_c} p_u \left(1 - \sum_{i=1}^{C_{thr}} f_i\right) \stackrel{\text{def}}{=} p_{blk,\,LB}(C_{thr}). \end{aligned} \tag{7.70}$$

□

The lower bound (7.69) is achievable when the required energy for unicast can always be satisfied. Nevertheless, with insufficient power supply, threshold-based policies can still provide substantial benefits, which will be shown with simulations. In the rest of this section, we will analyze several threshold-based policies based on different intuitions.

Push-Only Threshold-Based (POTB) Policy Notice that the more the contents are pushed to the user side, the fewer the unicast requests are triggered. Consequently, the number of blocking events due to energy shortage is reduced. Based on this intuition, we consider the push-only threshold-based (POTB) policy in which all the harvested energy is used to push contents. If the threshold is achieved, the SBS does not take unicast action but turns to sleep to save energy. According to Lemma 7.1, the threshold for the POTB policy can be chosen according to

$$\frac{p_c C_{thr,PO}}{N + p_c} E_p \leq \lambda_E, \tag{7.71}$$

which means that the average energy consumption for push does not exceed the available harvested energy. Hence,

$$C_{thr,PO} = \min\left\{N, \left\lfloor \frac{(N + p_c)\lambda_E}{p_c E_p} \right\rfloor\right\}, \tag{7.72}$$

where $\lfloor x \rfloor$ is the largest integer smaller than x. The performance of the POTB policy is described in the following theorem.

Theorem 7.5 Given the average energy arrival rate λ_E, if $B_{\max} \to +\infty$, the blocking probability of POTB policy is

$$p_{\mathrm{blk,PO}} = p_u \left(1 - \sum_{i=1}^{C_{\mathrm{thr,PO}}} f_i \right), \tag{7.73}$$

where $C_{\mathrm{thr,PO}}$ is expressed as in (7.72).

Proof: For simplicity, denote $C = C_{\mathrm{thr,PO}}$. Firstly, we assume that there is sufficient power supply for push. In this case, the request for a content in the list of contents $\widetilde{C}_t = (c_{C+1}, \dots, c_N)$ will be blocked. Obviously, the stationary blocking probability is expressed as (7.73).

Next, we prove that the blocking probability with EH and infinite battery capacity is also (7.73). It is equivalent to show that the stationary energy shortage probability, denoted by p_s, is zero. We prove it by contradiction. Specifically, we assume that

$$p_s = \Pr(E_t < E_p | C_t = C - 1) > 0 \tag{7.74}$$

and find the contradiction.

We consider the content updating process and the energy process separately. For the content updating process, we introduce the concept of *push request indicator* denoted by D_t. When a pushed content leaves, a new content needs to be pushed, i.e. a push request is generated. We use $D_t = 1$ to indicate that a push request is generated, and $D_t = 0$ otherwise. The push state achieves C only if all the push requests are satisfied. By definition, when a push request is generated, it is equivalent to that a pushed content leaves. Therefore, a push request will be generated with probability $\Pr(D_{t+1} = 1 | C_t = i) = \frac{p_c i}{N}, i \leq C$.

Denote the stationary distribution of C_t by $q_i = \Pr(C_t = i), i = 0, \dots, C$, which satisfies $q_i \geq 0$ and $\sum_{i=0}^{C} q_i = 1$. By the law of total probability, the stationary probability of push request can be calculated as

$$\begin{aligned}
\Pr(D_t = 1) &= \sum_{i=0}^{C} \Pr(C_{t-1} = i) \Pr(D_t = 1 | C_{t-1} = i) \\
&= \sum_{i=0}^{C} q_i \frac{p_c i}{N} \\
&\leq q_C \frac{p_c C}{N} + \sum_{i=0}^{C-1} q_i \frac{p_c (C-1)}{N} \\
&= q_C \frac{p_c}{N} + \frac{p_c (C-1)}{N},
\end{aligned} \tag{7.75}$$

where q_C can be calculated by solving the equilibrium $\boldsymbol{q} = \boldsymbol{q}\mathbf{P}_C$, where $\boldsymbol{q} = (q_0, \dots, q_C)$, and the elements (denoted by $p_{i,j} = \Pr(C_{t+1} = j | C_t = i), 0 \leq i \leq C, 0 \leq j \leq C$) of the state transition probability matrix $\mathbf{P}_C$ satisfy that (1) $\sum_j p_{i,j} = 1, p_{i,j} \geq 0, \forall i$ and (2) if $|i-j| > 1$, $p_{i,j} = 0$. The expression of q_C is

$$q_C = \left(1 + \frac{p_{C,C-1}}{p_{C-1,C}} + \cdots + \frac{\prod_{n=1}^{C} p_{n,n-1}}{\prod_{n=1}^{C} p_{n-1,n}}\right)^{-1}. \tag{7.76}$$

We compare with sufficient energy input case as in the proof of Lemma 7.1, in which only two stationary states $C_t \in \{C, C-1\}$ are accessible, and the push request generating probability can be calculated similarly:

$$\Pr'(D_t = 1) = q'_C \frac{p_c}{N} + \frac{p_c(C-1)}{N}, \tag{7.77}$$

where the stationary distribution

$$q'_C = \left(1 + \frac{p'_{C,C-1}}{p'_{C-1,C}}\right)^{-1} \tag{7.78}$$

is obtained by solving $\boldsymbol{q}' = \boldsymbol{q}'\mathbf{P}_C{}'$ where the elements of transition probability matrix $\mathbf{P}_C{}'$ satisfy $p'_{i,j} = 0, \forall i < C-1, j < C-1$. As

$$p'_{C,C-1} = p_{C,C-1} = p_c \frac{C}{N}, \tag{7.79}$$

based on our assumption $p_s > 0$, we have

$$\begin{aligned} p_{C-1,C} &= p_s \Pr(C_{t+1} = C | C_t = C-1, E_t < E_p) \\ &\quad + (1-p_s)\Pr(C_{t+1} = C | C_t = C-1, E_t > E_p) \\ &= p_s \cdot 0 + (1-p_s)p'_{C-1,C} < p'_{C-1,C}. \end{aligned} \tag{7.80}$$

Comparing (7.76) and (7.78), we have $q'_C > q_C$, which results in $\Pr(D_t = 1) < \Pr'(D_t = 1)$ according to (7.75) and (7.77). On the other hand, we have $q'_C = 1 - \frac{p_c C}{N + p_c}$ as derived in the proof of Lemma 7.1. Thus, we have

$$\Pr(D_t = 1) < \frac{p_c C}{N + p_c}. \tag{7.81}$$

For the energy process, each push request needs use energy E_p. The average energy consumption for push satisfies

$$\Pr(D_t = 1)E_p < \frac{p_c C}{N + p_c} E_p \leq \lambda_E. \tag{7.82}$$

Equation (7.82) means that the energy arrival is more than needed, and thus the energy shortage probability is zero, i.e. $p_s = 0$, which contradicts with the assumption (7.74). Hence, the theorem is proved. □

Notice that (7.73) is equivalent to the blocking probability with average power constraint. It is similar to the results in [40] that the AWGN channel capacity with infinite battery is equal to that with average power constraint. When λ_E is strictly larger than the required energy to push C_{thr} contents but is not enough to push $C_{thr} + 1$ contents, the remained energy in the battery can be used for unicast. In this way, the performance can be slightly improved. Based on Theorem 7.5, if λ_E is large enough so that $C_{\text{thr,PO}} = N$, we have $\sum_{i=1}^{C_{\text{thr,PO}}} f_i = 1$, and hence, the blocking probability is 0. Otherwise, $p_{\text{blk,PO}} > p_{\text{blk,LB}}$ holds.

As (7.73) decreases with $C_{\text{thr,PO}}$, the performance is better with larger threshold. It motivates us to design an improved version of POTB policy, named as always-push threshold-based (APTB) *policy*, which is equivalent to the POTB policy with threshold $C_{\text{thr}} = N$. However, if $\lambda_E < \frac{p_c N}{N+p_c} E_p$, there is not closed-form blocking probability expression for APTB policy as it depends on the distribution of energy arrival process.

Energy-Efficient Threshold-Based (EETB) Policy Apparently there is a tradeoff between the energy consumed for content push and the energy saved with less duplicated unicasts. From energy-efficiency point of view, the push action should be taken if the energy saved exceeds the energy consumed. The following lemma describes the condition under which content push is favorable in terms of energy efficiency.

Lemma 7.3 *If the index of content c_i satisfies*

$$i \leq \left(\frac{N p_u \bar{E}_u}{p_c E_p \sum_{j=1}^{N} 1/j^{\nu}} \right)^{\frac{1}{\nu}}, \tag{7.83}$$

where $\bar{E}_u = \int_0^R E_u(d) \frac{2d}{R^2} \text{d}d$ is the expected energy consumption of unicast, the energy consumption to push the content c_i is no more than the expected energy consumption for unicasting c_i upon requests.

Proof: Denote N_p as the number of slots in which a content stays in the content list C. According to the content updating process, N_p follows the geometric distribution, i.e.

$$\Pr(N_p = n) = \left(1 - p_c \frac{1}{N}\right)^{n-1} p_c \frac{1}{N}. \tag{7.84}$$

Denote N_r as the number of requests for the content c_i. In the case that the content c_i stays in the system for n slots, N_r follows the binomial distribution

$$\Pr(N_r = t | N_p = n) = \binom{n}{t} (p_u f_i)^t (1 - p_u f_i)^{n-t}. \tag{7.85}$$

By the law of total expectation, the average number of requests for the content with rank i is

$$\begin{aligned} \mathbb{E}[N_r] &= \mathbb{E}_{N_p}[\mathbb{E}[N_r | N_p]] \\ &= \sum_{n\geq 1} \Pr(N_p = n) n p_u f_i \\ &= \mathbb{E}[N_p] p_u f_i = \frac{N}{p_c} p_u f_i. \end{aligned} \tag{7.86}$$

Since the channel state is independent of the content request, the expected unicast energy consumption for all these requests is

$$\begin{aligned} \mathbb{E}[E_u] &= \mathbb{E}[N_r] \bar{E}_u \\ &= \frac{N}{p_c} p_u f_i \bar{E}_u. \end{aligned} \tag{7.87}$$

If the content of rank i is pushed to the users, the requests for this content are satisfied by the users' local storage. Hence, the unicast energy consumption (7.87) is avoided. In addition, pushing the content of rank i can reduce the energy consumption if

$$E_{\mathrm{p}} \le \frac{N}{p_{\mathrm{c}}} p_u f_i \bar{E}_u. \tag{7.88}$$

With some simple derivations, the condition (7.83) is obtained. □

Lemma 7.3 points out that if the popularity rank of a content satisfies (7.83), it is more energy efficient to push it to the user side in advance rather than unicasting it when required. Denote the threshold by

$$C_{\mathrm{thr,EE}} = \min\left\{N, \left\lfloor \left(\frac{N p_u \bar{E}_u}{p_{\mathrm{c}} E_{\mathrm{p}} \sum_{j=1}^{N} 1/j^{\nu}}\right)^{\frac{1}{\nu}} \right\rfloor\right\}. \tag{7.89}$$

Then pushing all the contents with rank $i \le C_{\mathrm{thr,EE}}$ is optimal from the energy-efficiency point of view.

Limited renewable energy requires to use the harvested energy in an efficient way. For the push action, it is more energy efficient to just push the contents with popularity rank $i \le C_{\mathrm{thr,EE}}$. And for the unicast action, sending a content to a user closer to the SBS is more energy efficient. Based on this intuition, the energy-efficient threshold-based (EETB) policy works as follows. The SBS firstly guarantees that the contents $c_1, \dots, c_{C_{\mathrm{thr,EE}}}$ are pushed to the user side and then responds to part of the unicast requests for the contents $c_{C_{\mathrm{thr,EE}}+1}, \dots, c_N$ according to the locations of users. It is detailed in Algorithm 7.1. In this algorithm, (7.92) is derived from the condition that the average energy consumption cannot exceed the average energy arrival, i.e.

$$\frac{p_{\mathrm{c}} C_{\mathrm{thr,EE}} E_{\mathrm{p}}}{N + p_{\mathrm{c}}} + \eta \int_0^{\tilde{d}} E_u(d) \frac{2d}{R^2} \mathrm{d}d \le \lambda_{\mathrm{E}}, \tag{7.90}$$

where

$$\eta = \left(1 - \frac{p_{\mathrm{c}} C_{\mathrm{thr,EE}}}{N + p_{\mathrm{c}}}\right) p_u \left(1 - \sum_{i=1}^{C_{\mathrm{thr,EE}}} f_i\right) \tag{7.91}$$

can be viewed as the probability of generating a unicast request for a content c_i, where $i \in \{C_{\mathrm{thr,EE}} + 1, \dots, N\}$ and the max and min operations in (7.92) are used so that if $\lambda_{\mathrm{E}} \le \frac{p_{\mathrm{c}} C_{\mathrm{thr,EE}} E_{\mathrm{p}}}{N+p_{\mathrm{c}}}$ or $\lambda_{\mathrm{E}} \ge \frac{p_{\mathrm{c}} C_{\mathrm{thr,EE}} E_{\mathrm{p}}}{N+p_{\mathrm{c}}} + \eta \bar{E}_u$, the equation still works. Also notice that we adopt the distance metric to decide whether to unicast or not, which can be replaced by the measured average SINR in practical systems.

The blocking probability of the EETB policy is summarized as follows.

Theorem 7.6 If $\lambda_{\mathrm{E}} \ge \frac{p_{\mathrm{c}} C_{\mathrm{thr,EE}} E_{\mathrm{p}}}{N+p_{\mathrm{c}}}$ and $B_{\max} \to +\infty$, the blocking probability of EETB policy is

$$p_{\mathrm{blk,EE}} = p_{\mathrm{blk,LB}}(C_{\mathrm{thr,EE}}) + \eta \frac{R^2 - \tilde{d}^2}{R^2}, \tag{7.93}$$

where $C_{\mathrm{thr,EE}}$ is expressed as (7.89).

Algorithm 7.1: EETB policy.

Input: Number of pushed contents C_t and the energy storage state E_t

Output: Action of the SBS u_t

1: Find the maximum $\tilde{d}$ satisfying

$$\int_0^{\tilde{d}} E_u(d)\frac{2d}{R^2}\mathrm{d}d \leq \min\left\{\bar{E}_u, \frac{1}{\eta}\max\left\{0, \lambda_\mathrm{E} - \frac{p_\mathrm{c} C_{\mathrm{thr,EE}} E_\mathrm{p}}{N + p_\mathrm{c}}\right\}\right\} \tag{7.92}$$

2: **if** $C_t < C_{\mathrm{thr,EE}}$ **then**
3: **if** $E_t \geq E_p$ **then** Push a content to all the users
4: **else** Keep sleep to store energy
5: **end if**
6: **else**
7: **if** $0 < Q_t \leq \min\{E_u(\tilde{d}), E_t\}$ **then** Unicast the requested content to the user
8: **else** Keep sleep to store energy
9: **end if**
10: **end if**

Proof: We prove the theorem by contradiction. Denote $C = C_{\mathrm{thr,\,EE}}$. If the required energy for EETB policy is always available, a request is blocked when either the SBS is pushing a content or the user's distance to the SBS is larger than $\tilde{d}$. Thus the blocking probability is

$$p_{\mathrm{blk}} = \Pr(Q_t > 0, I_t = 0, u_t = 2) + \Pr(u_t \neq 2)\Pr(Q_t > E_u(\tilde{d})|u_t \neq 2), \tag{7.94}$$

which results in (7.93). Then we assume the nonzero energy shortage probability and find the contradiction. The difference with the proof of Theorem 7.5 is that the energy shortage may happen in either push state or unicast state. Firstly, we assume that the battery energy is not enough for push, i.e. $p_\mathrm{s} = \Pr(E_t < E_\mathrm{p}|C_t = C-1) > 0$. Similar with the proof of Theorem 7.5, we can infer that $q_C < q'_C$ due to $p_\mathrm{s} > 0$, and the required energy satisfies

$$\begin{aligned}
&\Pr(D_t = 1)E_\mathrm{p} + q_C p_u\left(1 - \sum_{i=1}^{C} f_i\right)\int_0^{\tilde{d}} E_u(d)\frac{2d}{R^2}\mathrm{d}d \\
&< \frac{p_\mathrm{c} C}{N + p_\mathrm{c}}E_\mathrm{p} + q'_C p_u\left(1 - \sum_{i=1}^{C} f_i\right)\int_0^{\tilde{d}} E_u(d)\frac{2d}{R^2}\mathrm{d}d \leq \lambda_\mathrm{E}.
\end{aligned} \tag{7.95}$$

Secondly, assume the energy is insufficient for unicast, i.e. $\int_0^{\tilde{d}} p_d(d)\mathrm{d}d > 0$, where $p_d(d) = \Pr(E_t < E_u(d)|Q_t = E_u(d)), d > 0$. Then the required energy satisfies

$$\frac{p_\mathrm{c} C}{N + p_\mathrm{c}}E_\mathrm{p} + \eta\int_0^{\tilde{d}} (1 - p_d(d))E_u(d)\frac{2d}{R^2}\mathrm{d}d < \lambda_\mathrm{E}. \tag{7.96}$$

In both cases, the energy queue overflows. Hence, both (7.95) and (7.96) contradict the nonzero energy shortage probability assumption, and thus the theorem is proved. □

In this policy, the lower bound $p_{\text{blk,LB}}$ can be achieved if $\tilde{d} = R$, i.e. all the unicast requests are satisfied. Notice that if the condition $\lambda_{\text{E}} \geq \frac{p_c C_{\text{thr,EE}} E_{\text{p}}}{N+p_c}$ is not satisfied, the EETB policy degenerates to the APTB policy. Furthermore, with the closed-form expressions (7.73) and (7.93), we can easily find a better policy to minimize the blocking probability, i.e. we adopt the corresponding policy with smaller value calculated based on (7.73) and (7.93).

Greedy Optimal Threshold-Based (GOTB) Policy We have found the thresholds of the POTB policy and the EETB policy by minimizing the number of unicast requests and by maximizing the energy efficiency, respectively. We can also find the optimal threshold-based policy by searching over all the possible thresholds in a greedy way. Specifically, for each threshold, we calculate the average energy required to maintain the threshold. Secondly, we use the energy remained for energy-efficient unicast, i.e. determine a radius d_{thr} and serve the content requests generated inside the circle. The procedure is detailed in Algorithm 7.2.

Algorithm 7.2: GOTB policy.

Input: System parameters
Output: Optimal radius d_{thr}

```
1: Set C_thr = 0, p_blk = 1, d_thr = 0
2: for C = 0 to C_thr,PO do
3:   With threshold C, calculate d~ by (7.92), and then calculate p_blk,C by (7.93)
4:   if p_blk,C < p_blk then
5:     Set C_thr = C, p_blk = p_blk,C, d_thr = d~
6:   end if
7: end for
```

When the algorithm terminates, the threshold-based policy with parameters $C_{\text{thr}}, d_{\text{thr}}$ is optimized. Notice that the search ranges from 0 to $C_{\text{thr,PO}}$ since when $C > C_{\text{thr,PO}}$, there is no spare energy for unicast. The greedy optimal threshold-based (GOTB) policy can be used to evaluate the performance of the POTB policy and the EETB policy. Also notice that all the proposed threshold-based policies only depend on the statistic information of energy, contents, and user traffic, and thus they have general applications.

7.3.2.5 Numerical Results

We run some simulations to evaluate the performance of proposed algorithms. We set the cell radius $R = 50$ m, the required content delivery spectrum efficiency $r_0/W = 1$ bps Hz^{-1}, and the path-loss parameters $\beta = 10$ dB and $\alpha = 2$. The transmit power for cell-edge user is set $P_t(R) = 1$ W, which can be regarded as the maximum transmit power. The channel coefficient h follows Rayleigh fading, whose mean value and the interference plus noise power $\sigma^2 + I$ are set so that (7.54) holds for $r = r_0, d = R, P_t = P_t(R)$. The users are put into $M = 5$ classes, and the energy is quantized with unit $E_{\text{unit}} = \frac{P_t(R)T_{\text{p}}}{M}$. Hence, the number of energy units for push is $E_{\text{p}} = M$, and the content request in class m consumes m units of energy. Assume the energy arrival process follows Poisson distribution. We set $N = 20$ and the Zipf parameter $\nu = 1$.

Influence of Battery Capacity Firstly, we set $\lambda_E = 1.0, p_u = 0.9$ and evaluate the influence of the battery capacity, depicted in Figures 7.13 and 7.14. Comparing the two figures, the POTB policy is shown to be more tolerable to the limited battery capacity than the EETB policy. On the one hand, the performance degradation of the POTB policy with finite battery capacity is smaller than that of the EETB policy. For instance, with $p_c = 0.4$ and $E_{max} = 20$, the blocking probability of the POTB policy is about 10% more than the infinite battery limitation, while the EETB policy is more than 40%. On the other hand, the POTB policy converges to the infinite battery case much faster than the EETB policy. The reason is that if we view the battery as an energy queue, the service process of the

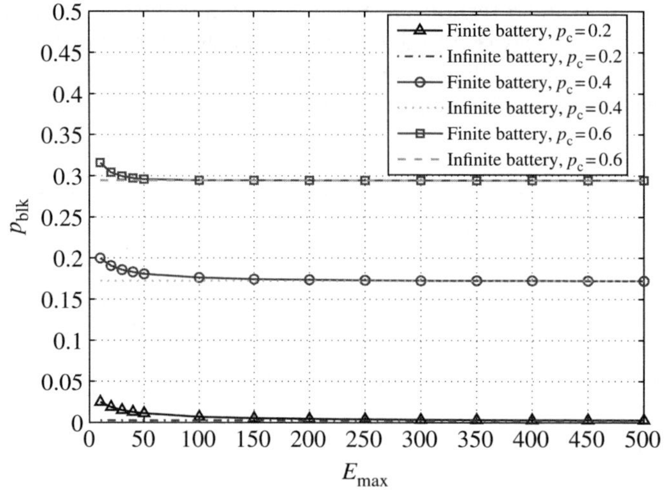

Figure 7.13 The influence of battery capacity on the POTB policy. $\lambda_E = 1.0, p_u = 0.9$.

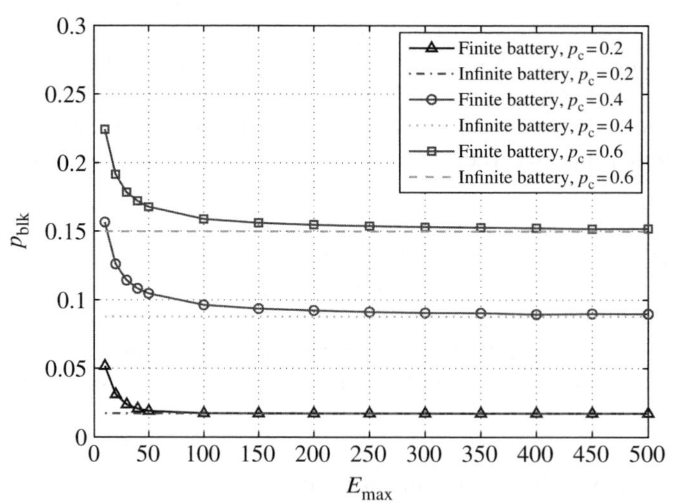

Figure 7.14 The influence of battery capacity on the EETB policy. $\lambda_E = 1.0, p_u = 0.9$.

POTB policy (only push) is less dynamic than that of the EETB policy (including both push and unicast with variable energy requirements).

Performance Comparison Then we evaluate the performance of the POTB policy, the EETB policy, and the APTB policy by comparing with the GOTB policy, the DP optimal policy, and the baseline policy without push, which is termed as *service-on-demand* policy, i.e. contents are delivered upon requests.

Figure 7.15 shows the performance comparison by varying the content updating rate p_c. We set the energy arrival rate $\lambda_E = 1.5$ and the content request generating rate $p_u = 0.9$. Compared with the service-on-demand policy, the proactive push policies perform better, and the gain increases as the content updating rate p_c decreases, which indicates that proactive push is more favorable with more stable content popularity. On the other hand, there is a cross point between the EETB policy and the POTB policy. The EETB policy achieves the same performance with the GOTB policy when $p_c \geq 0.4$, while the POTB policy performs the same when $p_c \leq 0.3$. But they converge to the same when $p_c \leq 0.1$ since $C_{\text{thr, PO}} = C_{\text{thr, EE}} = N$ in this situation. Also, the APTB policy shows a slight performance gain compared with the POTB policy. However, there is still a gap between the threshold-based policies and the DP optimal policy.

We further compare the policies by varying the energy arrival rate $\lambda_E = A$ with $p_c = 0.2$ and $p_u = 0.9$, as illustrated in Figure 7.16. As the content updating rate is relatively small, the performance gain of push-based policies is remarkable compared with the non-push policy. The APTB policy performs the same with the GOTB policy, while the POTB policy and the EETB policy perform the same as the GOTB policy with high energy arrival rate and low energy arrival rate, respectively. Notice that the curves corresponding to the threshold-based policies become flat as λ_E is large enough, which corresponds to the performance lower bound.

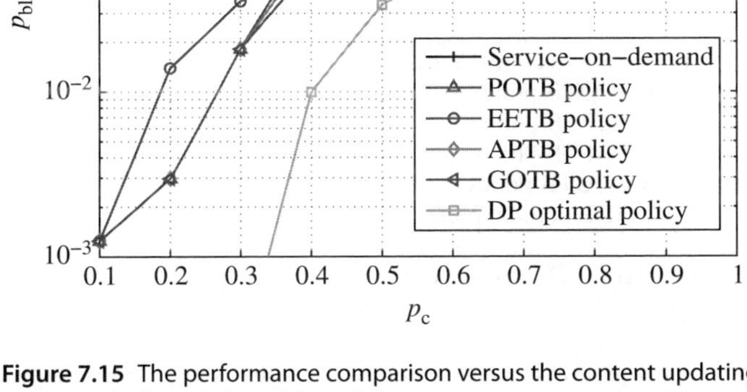

Figure 7.15 The performance comparison versus the content updating rate p_c with $N = 20$. The energy arrival rate is $\lambda_E = 1.5$, and the content request generating rate is $p_u = 0.9$.

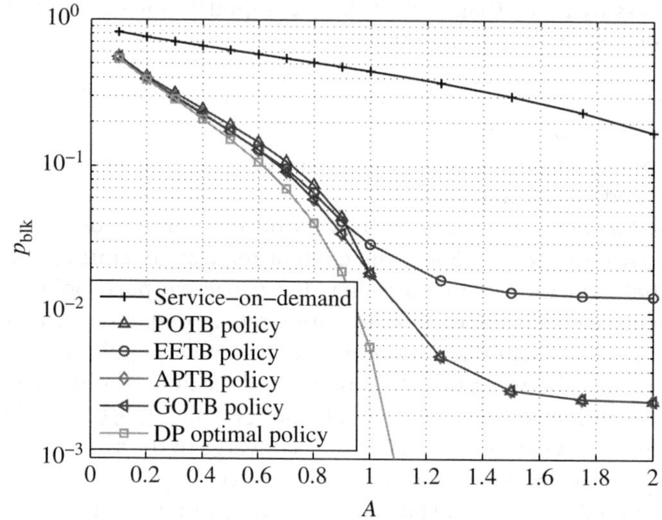

Figure 7.16 The performance comparison versus the energy arrival rate $\lambda_E = A$. The content updating rate is $p_c = 0.2$, and the content request generating rate is $p_u = 0.9$.

The performance variation w.r.t. the content request generating rate p_u is depicted in Figure 7.17. Notice that except the GOTB policy, our proposed threshold-based policies perform worse than the service-on-demand policy when the content request generating rate is very low. This is because if the content request arrives very slowly, it is not necessary to push contents in advance. However, as the search process in the GOTB policy contains the case $C = 0$, it will ultimately converge to the service-on-demand policy as

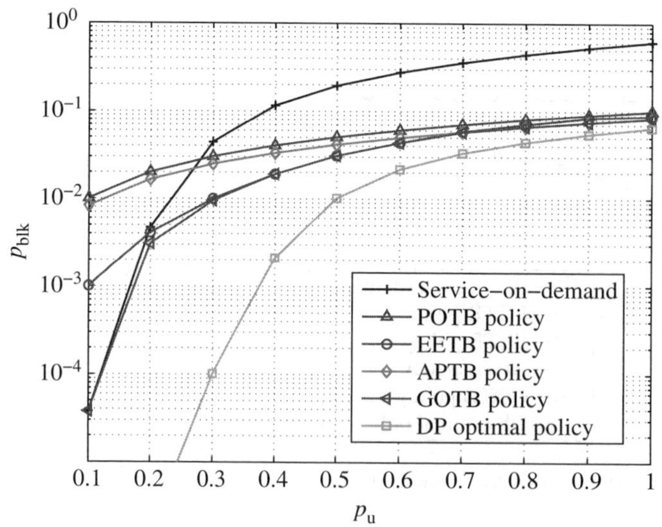

Figure 7.17 The performance comparison versus the content request generating rate p_u. The energy arrival rate is $\lambda_E = 0.75$, and the content updating rate is $p_c = 0.2$.

p_u decreases. In addition the performance gap between the threshold-based policies and the DP optimal policy becomes small as p_u increases.

7.4 Summary

In this chapter, we have addressed the problem of matching the randomly arrived renewable energy with user traffic variations over space and time in EH-based next-generation HCN. We explore two promising ways, namely, inter-tier traffic offloading and proactive caching and push, for spatial and temporal energy–traffic matching, respectively. First, we analyze the conversion rate of harvested energy into on-grid power through traffic offloading and then offer insights for practical green cellular network operations, e.g. whether an SBS should be activated and how much traffic should be offloaded to the SBS such that the on-grid power can be minimized. Based on the analysis, an energy-efficient traffic offloading scheme is proposed for HCN, and its performance is evaluated with extensive simulations. For HCN enabled with proactive service capability, we first provide an optimal policy via MDP formulation and then propose several low-complexity threshold-based push policies. Specifically, the push decision is made by comparing the number of un-pushed contents with a threshold, which is a function of the energy state of the SBS. Compared with the conventional service-on-demand policy, proactive push is shown to have tremendous performance gain in terms of service blocking probability, given the random energy arrivals and the limited battery at the SBS.

References

1 Wang, C.X., Haider, F., Gao, X. et al. (2014). Cellular architecture and key technologies for 5G wireless communication networks. *IEEE Communications Magazine* 52 (2): 122–130. https://doi.org/10.1109/MCOM.2014.6736752.

2 Ismail, M., Zhuang, W., Serpedin, E., and Qaraqe, K. (2014). A survey on green mobile networking: from the perspectives of network operators and mobile users. *IEEE Communication Surveys and Tutorials* 17 (3): 1535–1556. https://doi.org/10.1109/COMST.2014.2367592.

3 Zhou, S., Zhao, T., Niu, Z., and Zhou, S. (2016). Software-defined hyper-cellular architecture for green and elastic wireless access. *IEEE Communications Magazine* 54 (1): 12–19. https://doi.org/10.1109/MCOM.2016.7378420.

4 Dhillon, H., Li, Y., Nuggehalli, P. et al. (2014). Fundamentals of heterogeneous cellular networks with energy harvesting. *IEEE Transactions on Wireless Communications* 13 (5): 2782–2797. https://doi.org/10.1109/TWC.2014.040214.131201.

5 Han, T. and Ansari, N. (2014). Powering mobile networks with green energy. *IEEE Wireless Communications* 21 (1): 90–96. https://doi.org/\ignorespaces10.1109/MWC.2014.6757901.

6 Zheng, Z., Zhang, X., Cai, L. et al. (2014). Sustainable communication and networking in two-tier green cellular networks. *IEEE Wireless Communications* 21 (4): 47–53. https://doi.org/10.1109/MWC.2014.6882295.

7 Ye, Q., Rong, B., Chen, Y. et al. (2013). User association for load balancing in heterogeneous cellular networks. *IEEE Transactions on Wireless Communications* 12 (6): 2706–2716. https://doi.org/10.1109/TWC.2013.040413.120676.

8 Zhang, S., Wu, J., Gong, J. et al. (2014). Energy-optimal probabilistic base station sleeping under a separation network architecture. In: *2014 IEEE Global Communications Conference (GLOBECOM)*. Austin, TX, USA: IEEE. https://doi.org/10.1109/GLOCOM.2014.7037473.

9 Zhang, S., Gong, J., Zhou, S., and Niu, Z. (2015). How many small cells can be turned off via vertical offloading under a separation architecture? *IEEE Transactions on Wireless Communications* 14 (10): 5440–5453. https://doi.org/10.1109/TWC.2015.2438301.

10 Ismail, M. and Zhuang, W. (2014). Green radio communications in a heterogeneous wireless medium. *IEEE Wireless Communications* 21 (3): 128–135. https://doi.org/10.1109/MWC.2014.6845057.

11 Andrews, J., Singh, S., Ye, Q. et al. (2015). Energy-efficiency oriented traffic offloading in wireless networks: a brief survey and a learning approach for heterogeneous cellular networks. *IEEE Journal on Selected Areas in Communications* 33 (4): 627–640. https://doi.org/10.1109/JSAC.2015.2393496.

12 Han, T. and Ansari, N. (2013). On optimizing green energy utilization for cellular networks with hybrid energy supplies. *IEEE Transactions on Wireless Communications* 12 (8): 2872–2882. https://doi.org/10.1109/TCOMM.2013.051313.121249.

13 Gong, J., Thompson, J., Zhou, S., and Niu, Z. (2014). Base station sleeping and resource allocation in renewable energy powered cellular networks. *IEEE Transactions on Communications* 62 (11): 3801–3813. https://doi.org/10.1109/TCOMM.2014.2359883.

14 Zhang, S., Zhou, S., Gong, J. et al. (2015). Spatial traffic shaping in heterogeneous cellular networks with energy harvesting. In: *2015 IEEE Global Communications Conference (GLOBECOM)*. San Diego, CA: IEEE. https://doi.org/10.1109/GLOCOM.2015.7417251.

15 Zhang, S., Zhang, N., Zhou, S. et al. (2016). Energy-aware traffic offloading for green heterogeneous networks. *IEEE Journal on Selected Areas in Communications* 34 (5): 1116–1129. https://doi.org/10.1109/JSAC.2016.2520244.

16 Golrezaei, N., Molisch, A.F., Dimakis, A.G., and Caire, G. (2013). Femtocaching and device-to-device collaboration: a new architecture for wireless video distribution. *IEEE Communications Magazine* 51 (4): 142–149. https://doi.org/10.1109/MCOM.2013.6495773.

17 Shanmugam, K., Golrezaei, N., Dimakis, A.G. et al. (2013). FemtoCaching: wireless content delivery through distributed caching helpers. *IEEE Transactions on Information Theory* 59 (12): 8402–8413. https://doi.org/10.1109/TIT.2013.2281606.

18 Wang, X., Bao, Y., Liu, X., and Niu, Z. (2011). On the design of relay caching in cellular networks for energy efficiency. In: *2011 IEEE Conference on Computer Communications Workshops (INFOCOM WKSHPS)*. Shanghai, China: IEEE. https://doi.org/10.1109/INFCOMW.2011.5928820.

19 Bastug, E., Bennis, M., and Debbah, M. (2014). Living on the edge: the role of proactive caching in 5G wireless networks. *IEEE Communications Magazine* 52 (8): 82–89. https://doi.org/10.1109/MCOM.2014.6871674.

20 Blasco, P. and Gunduz, D. (2014). Learning-based optimization of cache content in a small cell base station. In: *2014 IEEE International Conference on Communications*. Sydney, Australia: IEEE. https://doi.org/10.1109/ICC.2014.6883600.

21 Zhou, B., Cui, Y., and Tao, M. (2016). Stochastic content-centric multicast scheduling for cache-enabled heterogeneous cellular networks. *IEEE Transactions on Wireless Communications* 15 (9): 6284–6297. https://doi.org/10.1109/TWC.2016.2582689.

22 Podnar, I., Hauswirth, M., and Jazayeri, M. (2002). Mobile push: delivering content to mobile users. In: *2002 IEEE 22nd International Conference on Distributed Computing Systems Workshops*. Vienna, Austria: IEEE Computer Society. https://doi.org/10.1109/ICDCSW.2002.1030826.

23 Nicopolitidis, P., Papadimitriou, G.I., and Pomportsis, A.S. (2002). Using learning automata for adaptive push-based data broadcasting in asymmetric wireless environments. *IEEE Transactions on Vehicular Technology* 51 (6): 1652–1660. https://doi.org/10.1109/TVT.2002.802978.

24 Wang, K., Chen, Z., and Liu, H. (2014). Push-based wireless converged networks for massive multimedia content delivery. *IEEE Transactions on Wireless Communications* 13 (5): 2894–2905. https://doi.org/10.1109/TWC.2014.040914.131422.

25 Chen, W. and Poor, H.V. (2017). Content pushing with request delay information. *IEEE Transactions on Communications* 63 (3): 1146–1161. https://doi.org/10.1109/TCOMM.2017.2648800.

26 Zhou, S., Gong, J., Zhou, Z. et al. (2015). GreenDelivery: proactive content caching and push with energy-harvesting-based small cells. *IEEE Communications Magazine* 53 (4): 142–149. https://doi.org/10.1109/MCOM.2015.7081087.

27 Gong, J., Zhou, S., Zhou, Z., and Niu, Z. (2017). Policy optimization for content push via energy harvesting small cells in heterogeneous networks. *IEEE Transactions on Wireless Communications* 16 (2): 717–729. https://doi.org/10.1109/TWC.2016.2628789.

28 Chiu, S.N., Stoyan, D., Kendall, W.S., and Mecke, J. (2013). *Stochastic Geometry and Its Applications*, 3e, 48–51. Chichester: Wiley.

29 Auer, G., Blume, O., Giannini, V. et al. (2010). D2. 3: energy efficiency analysis of the reference systems, areas of improvements and target breakdown. *Earth* 20 (10)1–68.

30 Sesia, S., Baker, M., and Toufik, I. (2009). *LTE - The UMTS Long Term Evolution: From Theory to Practice*. Wiley.

31 Zhang, X., Zheng, Z., Shen, Q. et al. (2014). Optimizing network sustainability and efficiency in green cellular networks. *IEEE Transactions on Wireless Communications* 13 (2): 1129–1139. https://doi.org/10.1109/TWC.2013.122713.131133.

32 Yang, J. and Ulukus, S. (2012). Optimal packet scheduling in an energy harvesting communication systems. *IEEE Transactions on Communications* 60 (1): 220–230. https://doi.org/10.1109/TCOMM.2011.112811.100349.

33 Nakagawa, K. (2005). On the series expansion for the stationary probabilities of an M/D/1 queue. *Journal of the Operations Research Society of Japan* 48 (2): 111–122. https://doi.org/10.15807/jorsj.48.111.

34 Heanggi, M., Andrews, J., Baccelli, F. et al. (2009). Stochastic geometry and random graphs for the analysis and design of wireless networks. *IEEE Journal on Selected Areas in Communications* 27 (7): 1029–1046. https://doi.org/10.1109/JSAC.2009.090902.

35 Boyd, S. and Vandenberghe, L. (2004). *Convex Optimization*, 243–247. Cambridge: Cambridge University press.

36 Sharma, N., Krishnappa, D.K., Irwin, D. et al. (2013). GreenCache: augmenting off-the-grid cellular towers with multimedia caches. In: *2013 4th ACM Multimedia*

Systems Conference. Oslo, Norway: ACM Press. https://doi.org/10.1145/2483977.2484009.

37 Niu, Z., Long, L., Song, J., and Pan, C. (2008). A new paradigm for mobile multimedia broadcasting based on integrated communication and broadcast networks. *IEEE Communications Magazine* 46 (7): 126–132. https://doi.org/10.1109/MCOM.2008.4557054.

38 Bertsekas, D.P. (2005). *Dynamic Programming and Optimal Control*, 3e. Belmont, MA: Athena Scientific.

39 Cha, M., Kwak, H., Rodriguez, P. et al. (2007). I tube, you tube, everybody tubes: analyzing the world's largest user generated content video system. In: *2007 the 7th ACM SIGCOMM Conference on Internet Measurement*. San Diego, CA: ACM Press. https://doi.org/10.1145/1298306.1298309.

40 Ozel, O. and Ulukus, S. (2012). Achieving AWGN capacity under stochastic energy harvesting. *IEEE Transactions on Information Theory* 58 (10): 6471–6483. https://doi.org/10.1109/TIT.2012.2204389.

Part III

Appendices

Appendix A

Convex Optimization

In this chapter, some basic concepts of convex optimization are briefly given, and some examples for basic convex optimization problems are also discussed.

A.1 Some Definitions

A.1.1 Convex Sets

To begin with, the definition of convex sets is introduced.

Definition A.1 Given a set C and any two points $x, y \in C$, if the following relationship holds

$$\theta x + (1-\theta)y \in C, \tag{A.1}$$

where $0 \le \theta \le 1$, then set C is convex.

The geometrical explanation for the above definition of convex sets is given as follows (see Figure A.1): Arbitrarily select two points x and y in set C, and draw a line segment to connect these two points. If all the points of this line segment fall into set C (see Figure A.1a), then C is convex; otherwise, C is non-convex, (see Figure. A.1b).

Example A.1 *To better understand Definition A.1, some examples are presented in the following:*

- The space with dimension n, *i.e.* $\mathbf{R}^n$[1]: *It is noticed that for* $\forall x, y \in \mathbf{R}^n$, $\theta x + (1-\theta) y \in \mathbf{R}^n$, *with* $0 \le \theta \le 1$.
- Norm balls, *i.e.* $\{\boldsymbol{x} : |\boldsymbol{x}| \le 1\}$, *with* $\boldsymbol{x} = [x_1, x_2, \dots, x_n]^T$ *and* $|\mathbf{x}| = \left(\sum_{i=1}^n x_i^2\right)^{1/2}$: *This conclusion is due to the fact that for* $\forall \boldsymbol{x}, \boldsymbol{y} \in \mathbf{R}^n$ *with* $|\boldsymbol{x}|, |\boldsymbol{y}| \le 1$ *and* $0 \le \theta \le 1$, *by using the triangle inequality and the positive homogeneity of norms, it follows*

$$\begin{aligned} |\theta \boldsymbol{x} + (1-\theta)\boldsymbol{y}| &\le |\theta \boldsymbol{x}| + |(1-\theta)\boldsymbol{y}| \\ &= \theta|\boldsymbol{x}| + (1-\theta)|\boldsymbol{y}| \le 1, \end{aligned} \tag{A.2}$$

which proves the conclusion.

1 $\mathbf{R}^n$ is the set of real numbers.

Energy Harvesting Wireless Communications, First Edition.
Chuan Huang, Sheng Zhou, Jie Xu, Zhisheng Niu, Rui Zhang and Shuguang Cui.

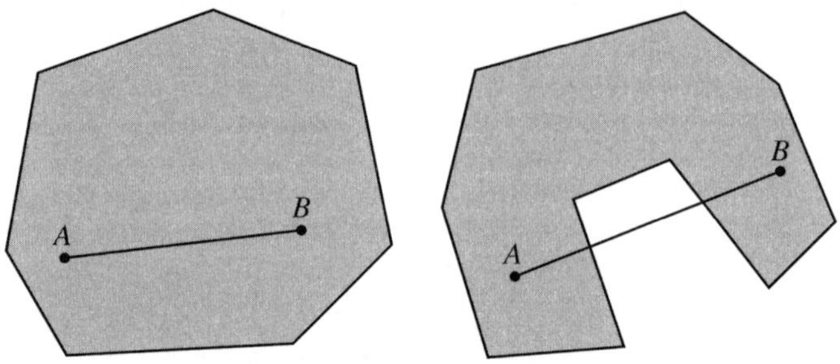

(a) (b)

Figure A.1 Examples of (a) convex and (b) non-convex sets.

- *Intersection of convex sets: If C_i $(i = 1, 2, \ldots, k)$ are all convex sets, then their intersection*

$$\bigcap_{i=1}^{k} C_i = \{x : x \in C_i, \forall i = 1, 2, \ldots, k\} \tag{A.3}$$

is also convex. To prove it, select two points $x, y \in \bigcap_{i=1}^{k} C_i$. Thus, for $0 \le \theta \le 1$, by the definition of convex set, it follows $\theta x + (1-\theta)y \in C_i$, $\forall i = 1, 2, \ldots, k$, which implies $\theta x + (1-\theta)y \in \bigcap_{i=1}^{k} C_i$. We should also point out that the union *of convex sets is not convex in general.*

- *Positive semidefinite matrices: Denote the set of all symmetric positive semidefinite matrices*[2] *as $\mathbf{S}_+^n$, which is a convex set. To prove it, choose two symmetric positive semidefinite matrices $A, B \in \mathbf{S}_+^n$. Then, for $\forall x \in \mathbf{R}^n$ and $0 \le \theta \le 1$, it follows*

$$\boldsymbol{x}^T[\theta A + (1-\theta)B]\boldsymbol{x} = \theta \boldsymbol{x}^T A \boldsymbol{x} + (1-\theta)\boldsymbol{x}^T B \boldsymbol{x} \ge 0. \tag{A.4}$$

Similarly, it is not difficult to prove that the sets of all positive definite, negative semidefinite, and negative definite matrices are also convex.

A.1.2 Convex Functions

Definition A.2 A function $f : \mathbf{R}^n \to \mathbf{R}$ is convex, if $\forall x, y \in \mathcal{D}(f)$ and $0 \le \theta \le 1$, we have

$$f(\theta x + (1-\theta)y) \le \theta f(x) + (1-\theta)f(y), \tag{A.5}$$

with $\mathcal{D}(f)$ being a convex set and the domain of function f.

A geometrical explanation of convex function is shown in Figure A.2: arbitrarily select two points on the curve determined by the convex function f, denoted as $(x, f(x))$ and $(y, f(y))$, respectively, and draw a line segment to connect them. If the part of the function between these two points locates under this line segment, function f is convex.

If Definition A.2 stands with strict inequality for $x \neq y$ and $0 < \theta < 1$, then function f is said to be strictly convex. If $-f$ is convex, then f is concave. If $-f$ is strictly convex, then f is strictly concave.

---

2 A symmetric matrix A is positive semidefinite, if and only if $x^T A x \ge 0$ for $\forall x \in \mathbf{R}^n$.

A.1.2.1 First-Order Condition for Convexity

Consider the case that a function $f : \mathbf{R}^n \to \mathbf{R}$ is differentiable, i.e. for any $x \in \mathcal{D}(f)$, its gradient always exists. We claim that function f is convex, if and only if for any $x, y \in \mathcal{D}(f)$, we have

$$f(y) \geq f(x) + \nabla_x f(x)^T (y - x), \tag{A.6}$$

where $\mathcal{D}(f)$ is also a convex set and $\nabla_x f(x)$ is the first-order derivative of function f. Function $f(x) + \nabla_x f(x)^T (y - x)$ is the *first-order approximation* to function f at point x.

The first-order condition given in (A.6) is a necessary and sufficient condition for the convex function f, i.e. function f is convex if and only if the tangent line defined in the right hand side of (A.6) is a global underestimator of function f, see Figure A.3.

The same as the definition of convex function, if the strict inequality in (A.6) holds, function f is said to be strictly convex; if the inequality in (A.6) is reversed, function f is concave; and if the reverse inequality is strict, function f is strictly concave.

A.1.2.2 Second-Order Condition for Convexity

Consider the case that a function $f : \mathbf{R}^n \to \mathbf{R}$ is twice differentiable, i.e. for any $x \in \mathcal{D}(f)$, the corresponding Hessian[3] $\nabla_x^2 f(x)$ always exists. Under this condition, a necessary and sufficient condition for the convex function f is that the corresponding Hessian is positive semidefinite, i.e. for $\forall x \in \mathcal{D}(f)$, we have

$$\nabla_x^2 f(x) \succeq 0, \tag{A.7}$$

and $\mathcal{D}(f)$ is convex. Here, the symbol $\succeq$ means positive semidefiniteness, i.e. $A \succeq 0$ implies $x^T A x \geq 0$ for any vector x. For the one-dimensional case, the second order condition given in (A.7) means that the second order derivative $f''(x)$ is nonnegative.

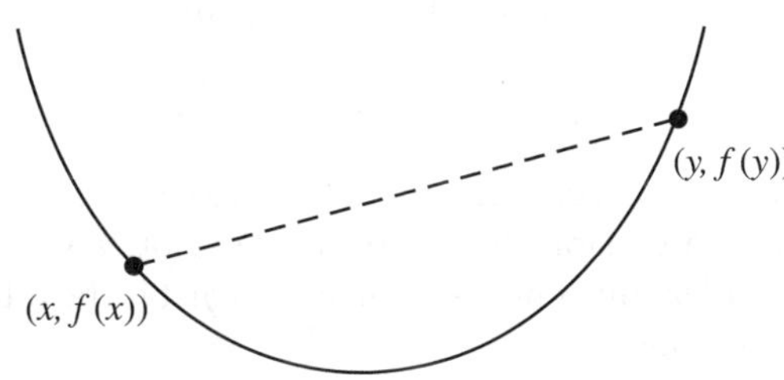

Figure A.2 An example of convex function. According to the definition of convex function, the line segment between the two points $(x, f(x))$ and $(y, f(y))$ should be above the curve determined by function f.

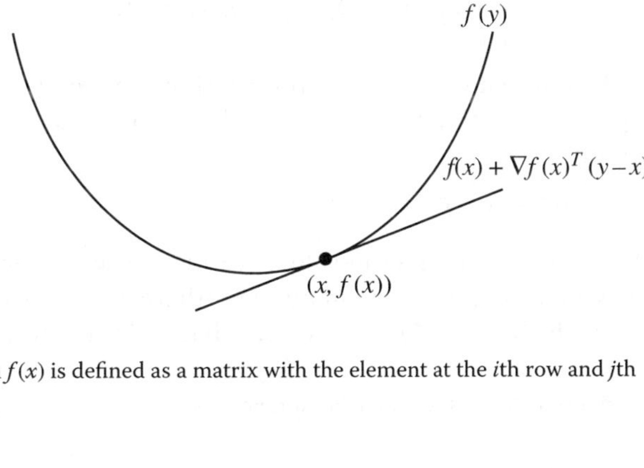

Figure A.3 First-order condition of convex function.

3 The Hessian $\nabla_x^2 f(x) \in \mathbf{R}^{n \times n}$ of function $f(x)$ is defined as a matrix with the element at the ith row and jth column being given as $(\nabla_x^2 f(x))_{ij} = \frac{\partial^2 f(x)}{\partial x_i \partial x_j}$.

If the Hessian of function f is positive definite, function f is strictly convex. If its Hessian is negative semidefinite, function f is concave. If its Hessian is negative definite, function f is strictly concave.

A.1.2.3 Examples of Convex Functions

Some examples of convex functions are listed as follows:

- *Exponential*: $f(x) = e^{ax}$, for $\forall a \in \mathbf{R}$. To prove that function f is convex, its second order derivative is given as $f''(x) = a^2 e^{ax}$, and obviously, it follows $f''(x) \geq 0$ for $\forall x$.
- *Negative logarithm*: $f(x) = -\log x$ with $\mathcal{D}(f) = \mathbf{R}^{++}$, where $\mathbf{R}^{++}$ is the set $\{x : x > 0\}$. It is easy to obtain that $f''(x) = \frac{1}{x^2} > 0$ for all x.
- *Quadratic functions*: $f(x) = \frac{1}{2}x^T Ax + b^T x + c$, where $A \in \mathbb{S}^n$ is a symmetric matrix, $b \in \mathbf{R}^n$ and $c \in \mathbf{R}$. Its Hessian is given as $\nabla_x^2 f(x) = A$. If A is a positive semidefinite matrix, function f is convex (similarly for strictly convex, strictly concave, concave).
- *Nonnegative weighted sums of convex functions*: Choose convex functions f_i and nonnegative real numbers w_i, $i = 1, 2, \ldots, k$. Then, the function $f(x) = \sum_{i=1}^k w_i f_i(x)$ is convex.

A.2 Convex Optimization Problems

Now, we are ready to introduce the convex optimization problems. A convex optimization problem is with the following forms:

$$\min \quad f_0(x) \tag{A.8}$$

$$\text{s.t.} \quad f_i(x) \leq 0, \quad i = 1, \ldots, m, \tag{A.9}$$

$$h_i(x) = 0, \quad i = 1, \ldots, p, \tag{A.10}$$

where f_0 is a convex function, f_i are convex functions, h_i are affine functions,[4] and x is the design variable. It is important to notice that the feasible set of a convex optimization problem must be also convex, since it is the intersection of the domains of the constraint functions, i.e.

$$\mathcal{D} = \bigcap_{i=1}^{m} \{x | f_i(x) \leq 0\} \bigcap_{i=1}^{p} \{x | h_i(x) = 0\}. \tag{A.11}$$

The optimal value of the optimization problem (A.8)–(A.10) is denoted as $p^\star$ and is equal to the minimum possible value of the objective function in the convex feasible region:

$$p^\star = \inf\{f_0(x) | x \in \mathcal{D}\}. \tag{A.12}$$

We allow $p^\star$ to take the value $+\infty$ or $-\infty$ when the problem is infeasible (the feasible region is empty) or unbounded (there exist feasible points such that $f_0(x) \to -\infty$), respectively. If $f_0(x^\star) = p^\star$, we say that $x^\star$ is an optimal point.

4 Affine function is the sum of linear functions.

A.2.1 Optimality in Convex Problems

Before stating the conclusion about global optimality for convex optimization problems, let us first define the local and global optimal points, respectively.

A feasible point is called local optimal if there are no "nearby" feasible points leading to a lower objective value. A more rigorous definition is given as follows.

Definition A.3 A point x is local optimal, if it is feasible (i.e. it satisfies all the constraints of the optimization problem (A.8)–(A.10)) and if there exists $\epsilon > 0$ such that all feasible points z with $|x - z|_2 \leq \epsilon$ satisfy $f_0(x) \leq f_0(z)$.

A feasible point is called *global optimal* if there is no feasible point with a lower objective value. A more rigorous definition is given as follows.

Definition A.4 A point x is global optimal, if it is feasible, and for all feasible points $z \in \mathcal{D}$, it follows $f_0(x) \leq f_0(z)$.

Now we come to the core conclusion of this subsection.

Theorem A.1 For a convex optimization problem, all local optimal points are global optimal.

Proof: We prove this property by contradiction. Suppose that x is a local optimal point while not global optimal, i.e. there exists a feasible point y such that $f_0(x) > f_0(y)$. By the definition of local optimality, there exists no feasible point z such that $|x - z|_2 \leq \epsilon$ and $f_0(z) < f_0(x)$. Now we choose a point $z = \theta y + (1 - \theta)x$ with $\theta = \frac{\epsilon}{2|x-y|_2}$, and we obtain

$$\begin{aligned}
|x - z|_2 &= \left| x - \left(\frac{\epsilon}{2|x - y|_2} y \left(1 - \frac{\epsilon}{2|x - y|_2} x \right) \right) \right|_2 \\
&= \left| \frac{\epsilon}{2|x - y|_2} (x - y) \right|_2 \\
&= \epsilon/2 \leq \epsilon.
\end{aligned}$$

In addition, due to the convexity of function f_0, we have

$$f_0(z) = f_0(\theta y + (1 - \theta)x) \leq \theta f_0(y) + (1 - \theta) f_0(x) < f_0(x).$$

Furthermore, since the corresponding feasible set is convex and x and y are both feasible, it follows that $z = \theta y + (1 - \theta)x$ is also feasible. Thus, z is a feasible point with $|x - z|_2 < \epsilon$ and $f_0(z) < f_0(x)$, which contradicts to the assumption. Therefore, this theorem is proved. □

A.2.2 Examples

Consider a least-squares problem, where we aim to minimize $|Ax - b|_2^2$ for matrix $A \in \mathbf{R}^{m \times n}$ and vector $b \in \mathbf{R}^m$. This problem can be analytically solved by the normal equation[5]. However, when we want to constrain the entries of x to lie within a

5 Any of a set of simultaneous equations involving experimental unknowns and derived from a larger number of observation equations in the course of least-squares adjustment of observations.

predefined range, this method will not work. In other words, consider the following new optimization problem:

$$\min \quad \frac{1}{2}|Ax - b|_2^2 \tag{A.13}$$

$$\text{s.t.} \quad l \preceq x \preceq u, \tag{A.14}$$

where $l \in \mathbf{R}^n$ and $u \in \mathbf{R}^n$. With the extra constraint (A.14), there will no longer exist an analytical solution of problem (A.13)–(A.14) in general, and algorithms should be adopted to compute its optimal solution.

A.3 Duality

A.3.1 The Lagrange Dual Function

Consider an optimization problem in a similar form as problem (A.8)–(A.10), whose domain is given in (A.11) and optimal value is denoted as $p^\star$. Here, we do not need to assume that problem (A.8)–(A.10) is convex.

The key idea in Lagrangian duality is to take the constraints (A.9)–(A.10) into account by augmenting the objective function (A.8) with a weighted sum of the constraint functions (A.9)–(A.10). Then, problem (A.8)–(A.10) becomes an unconstrained optimization problem. Following the above idea, we define the **R**$\mathcal{L}$: $\mathbf{R}^n \times \mathbf{R}^m \times \mathbf{R}^p \to \mathbf{R}$ associated with problem (A.8)–(A.10) as

$$\mathcal{L}(x, \lambda, \nu) = f_0(x) + \sum_{i=1}^{m} \lambda_i f_i(x) + \sum_{i=1}^{p} \nu_i h_i(x), \tag{A.15}$$

where the domain of (A.15) is given as $\mathcal{D}(\mathcal{L}) = \mathcal{D} \times \mathbf{R}^m \times \mathbf{R}^p$, and $\lambda = [\lambda_1, \lambda_2, \dots, \lambda_m]^T$ and $\nu = [\nu_1, \nu_2, \dots, \nu_p]^T$ denote the Lagrange multipliers associated with the inequality constraints in (A.9) and the equality constraints in (A.10), respectively.

Then, we define the Lagrange dual function (or dual function) for problem (A.8)–(A.10) as follows:

$$g(\lambda, \nu) = \inf_{x \in D} \left(f_0(x) + \sum_{i=1}^{m} \lambda_i f_i(x) + \sum_{i=1}^{p} \nu_i h_i(x) \right), \tag{A.16}$$

where $g : \mathbf{R}^m \times \mathbf{R}^p \to \mathbf{R}$ is the minimum value of the Lagrangian $\mathcal{L}(x, \lambda, \nu)$ over x for fixed $\lambda \in \mathbf{R}^m$, $\nu \in \mathbf{R}^p$. Obviously, if the Lagrangian is unbounded in x, the dual function $g(\lambda, \nu)$ takes the value $-\infty$. The dual function $g(\lambda, \nu)$ is always concave, since it is the pointwise infimum of a family of affine functions of (λ, ν), even when the original problem (A.8)–(A.10) is not convex.

Theorem A.2 The dual function $g(\lambda, \nu)$ defined in (A.16) is a lower bound on the optimal value $p^\star$ of problem (A.8)–(A.10), i.e. for any $\lambda \succeq 0$ and ν, we have

$$g(\lambda, \nu) \leq p^\star. \tag{A.17}$$

Proof: This important property can be easily verified. Suppose that $\tilde{x}$ is a feasible point of problem (A.8)–(A.10), i.e. it satisfies $f_i(\tilde{x}) \leq 0$, $h_i(\tilde{x}) = 0$, and $\lambda > 0$. Then, we can derive

$$\sum_{i=1}^{m} \lambda_i f_i(\tilde{x}) + \sum_{i=1}^{p} v_i h_i(\tilde{x}) \leq 0. \tag{A.18}$$

From the constraints (A.9) and (A.10), we observe that the first term of (A.18) is non-positive and the second term is zero, and it follows

$$\mathcal{L}(\tilde{x}, \lambda, v) = f_0(\tilde{x}) + \sum_{i=1}^{m} \lambda_i f_i(\tilde{x}) + \sum_{i=1}^{p} v_i h_i(\tilde{x}) \leq f_0(\tilde{x}), \tag{A.19}$$

which implies

$$g(\lambda, v) = \inf_{x \in D} \mathcal{L}(x, \lambda, v) \leq \mathcal{L}(\tilde{x}, \lambda, v) \leq f_0(\tilde{x}). \tag{A.20}$$

Thus, we proved that $g(\lambda, v) \leq f_0(\tilde{x})$ holds for all feasible points, and it follows that the inequality in (A.17) is valid. □

A.3.2 The Lagrange Dual Problem

Based on Theorem A.2, for each pair (λ, μ) with $\lambda \succeq 0$, the Lagrange dual function defined in (A.16) is a lower bound on the optimal value $p^\star$ of optimization problem (A.8)–(A.10). A natural question is: what is the best lower bound that we can achieve from the Lagrange dual function (A.16)? The solution of this question is given by the following optimization problem:

$$\max \quad g(\lambda, v) \tag{A.21}$$

$$\text{s.t.} \quad \lambda \succeq 0. \tag{A.22}$$

We call this problem as the Lagrange dual problem associated with problem (A.8)–(A.10). In this context, the original problem (A.8)–(A.10) is called the primal problem. If $(\lambda^\star, v^\star)$ are optimal for problem (A.21)–(A.22), we call $(\lambda^\star, v^\star)$ as the dual optimal Lagrange multipliers.

Since the objective function in (A.21) is concave and the constraint in (A.22) is convex, the Lagrange dual problem (A.21)–(A.22) is a convex optimization problem. Moreover, this conclusion always holds, even though the primal problem (A.8)–(A.10) is non-convex.

A.3.3 Strong Duality

Denote $d^\star$ as the optimal value of the Lagrange dual problem (A.21)–(A.22), and it is the best lower bound on the optimal value of $p^\star$ for the primal problem (A.8)–(A.10), i.e.

$$d^\star \leq p^\star. \tag{A.23}$$

This inequality holds even when the original problem (A.8)–(A.10) is non-convex. This property is called as weak duality. If the equality in (A.23) is achieved, i.e. $d^\star = p^\star$, the optimal duality gap between the prime and dual problems is zero. This property is called as strong duality.

A.3.4 KKT Conditions

We now assume that functions $f_0, \dots, f_m, h_1, \dots, h_p$ are differentiable. Thus, they have open domains. However, we do not need assume that they are convex.

Let $x^\star$ and $(\lambda^\star, \nu^\star)$ be any dual optimal points with zero duality gap. Since $x^\star$ minimizes $L(x, \lambda^\star, \nu^\star)$, it is obvious that its gradient must vanish at $x^\star$, i.e.

$$\nabla f(x^\star) + \sum_{i=1}^{m} \lambda_i^\star \nabla f_i(x^\star) + \sum_{i=1}^{p} \nu_i^\star \nabla h_i(x^\star) = 0. \tag{A.24}$$

Therefore, we have

$$\begin{cases} f_i(x^\star) \le 0, & i = 1, \dots, m & \text{(A.25)} \\ h_i(x^\star) = 0, & i = 1, \dots, p & \text{(A.26)} \\ \lambda_i^\star \ge 0, & i = 1, \dots, m & \text{(A.27)} \\ \lambda_i^\star f_i(x^\star) = 0, & i = 1, \dots, m & \text{(A.28)} \\ \nabla f(x^\star) + \sum_{i=1}^{m} \lambda_i^\star \nabla f_i(x^\star) + \sum_{i=1}^{p} \nu^{\star_i} \nabla h_i(x^\star) = 0, & & \text{(A.29)} \end{cases}$$

which are the Karush-Kuhn-Tucker (KKT) conditions. The KKT conditions play a significant role in optimization. In some cases, we can solve the KKT conditions analytically. Generally, there are many algorithms for convex optimization to solve the KKT conditions.

Appendix B

Markov Decision Process

In this chapter, some basic concepts associated with the Markov decision process are briefly introduced, and the solutions to the Markov decision problems under different scenarios are also discussed.

B.1 Definition and Notation

Five elements of the Markov decision process, i.e. decision moments, states, actions, transition probabilities, and rewards, are defined in this section.

B.1.1 Decision Moments

The time point that a decision is made is called a *decision moment*. The set of all decision moments is denoted as T, which in general is a subset of the nonnegative real number. There are three cases for set T:

(1) *T is a finite set*: Without loss of generality, T can be denoted as $T = \{0, 1, \dots, N\}$, and the corresponding Markov decision problem is called a *finite-horizon problem*.
(2) *T is a countable set*: Similarly, T can be defined as $T = \{0, 1, \dots, N, \dots\}$, and the corresponding Markov decision problem is called an *infinite-horizon problem*.
(3) *T is a continuum*: The decisions are made at all continuous decision moments, random points of time, or the opportune times chosen by decision maker.

B.1.2 State and Action Sets

At every decision moment, the considered system is described by *state*. The set of all possible system states are denoted as S. Set S is with the following cases: (i) finite set, (ii) countable infinite set, or (iii) compact subset of finite-dimensional Euclidean space. In general, set S may not be identical at different decision moments. However, in most applications, we assume that S is independent of t.

At any decision moment, if the decision maker observes the state $i \in S$, it can choose an action a from a set A_i, where A_i is the action set corresponding to state i. Set A_i can be with the following cases: (i) finite set, (ii) countable infinite set, (iii) compact subset of finite-dimensional Euclidean space; or (iv) non-empty Borel subset of complete and separable metric spaces. Here, A_i is also assumed to be not dependent on the

Energy Harvesting Wireless Communications, First Edition.
Chuan Huang, Sheng Zhou, Jie Xu, Zhisheng Niu, Rui Zhang and Shuguang Cui.

decision moment t, and denote $A = \bigcup_{i \in S} A_i$ as the action set of the Markov decision process. We should point out that the choice of action can be either deterministic, i.e. the corresponding action a is determined by a given state i, or random among multiple available actions, i.e. the corresponding action a is a random variable with the probability distribution $q(a)$. In this case, we denote $\text{Dis}(A_i)$ as all probability distributions on the Borel subsets of A_i, with $\text{Dis}(A)$ being all probability distributions on the Borel subsets of A.

B.1.3 Rewards and Transition Probabilities

At any decision moment t, after choosing action $a \in A_i$ for state i, there are two consequences:

(1) The decision maker gets the reward $r(i, a)$.
(2) The probability distribution $p(j|i, a)$ determines the state of system at the next decision moment, i.e. $p(j|i, a)$ is the probability that the system transfers under action a from state i to state j at the next decision moment. $p(j|i, a)$ is also known as the *transfer probability function*. Obviously, it follows

$$\sum_{j \in S} p(j|i, a) = 1. \tag{B.1}$$

The *reward* $r(i, a)$ is a real-valued function and defined over $i \in S$, $a \in A_i$. In general, the reward also depends on the state j at the next decision moment, and it is denoted as $r(i, a, j)$. Thus, by utilizing the theory of total probability, the expected reward of action a in state i is given as

$$r(i, a) = \sum_{j \in S} r(i, a, j) p(j|i, a). \tag{B.2}$$

Based on the definitions in Section B.1, the reorganization

$$\{T, S, A_i, p(j|i, a), r(i, a)\} \tag{B.3}$$

is referred as a *Markov decision process*.

B.1.4 Decision Rules

A *decision rule* describes how the decision maker chooses an action for the Markov decision process at a decision moment for different states. There are two cases:

(1) *Markovian decision rules*: For each $i \in S$, there exists $d_i \in A_i$. Then, we call the function $d : S \to A_i$ as a Markovian decision rule. Obviously, for this case, the decision rule d only depends on the current state i and action a.
(2) *History-dependent decision rules*: If a decision rule is not Markovian, it is history dependent. From the process of decision-making, a track of the Markov decision process is made up of successive states and actions, i.e. the track h_t of the Markov decision process at decision moment t is given as

$$h_t = (i_0, a_0, i_1, a_1, \dots, i_{t-1}, a_{t-1}, i_t), \quad t \geq 0. \tag{B.4}$$

Denote the set of all possible h_t as H_t, which is defined as

$$H_t = H_{t-1} \times A \times S = (S \times A)^t \times S, \tag{B.5}$$

with $A \times S$ being the *Cartesian product* of A and S. When $t = \infty$, H_∞ is the collection of all tracks.

The action at each decision moment can be either deterministic or randomized. The action is said to be deterministic because it chooses an action with certainty, and randomized is the opposite. Thus, based on the history dependency and action choice of the Markov decision process, the decision rules can be classified as four cases, as shown in Table B.1. Here, $q_{d_t}(\cdot)$ is the probability distribution on the set of actions for a randomized decision rule d_t.

Here, the decision rules are classified as history dependent and randomized (HR), history dependent and deterministic (HD), Markovian and randomized (MR), and Markovian and deterministic (MD), depending on their degree of dependence on the past information and the method of action choices. Denote the set of all possible decision rules at time t as $D_t^K = \{d_t\}$, where K designates a class of decision rules, i.e. $K = \{\text{HR}, \text{HD}, \text{MR}, \text{MD}\}$.

For D_t^{MD}, the reward and the transition probability are $r_t(i, d_t(i))$ and $p_t(j|i, d_t(i))$, respectively. For D_t^{HD}, these are given as $r_t(i, d_t(h_t))$ and $p_t(j|i, d_t(h_t))$. For D_t^{MR}, we obtain its reward as

$$r_t(i, d_t(i)) = \sum_{a \in A_i} r_t(i, a) q_{d_t(i)}(a), \tag{B.6}$$

and its transition probability as

$$p_t(j|i, d_t(i)) = \sum_{a \in A_i} p_t(j|i, a) q_{d_t(i)}(a). \tag{B.7}$$

For D_t^{HR}, its reward and transition probability are obtained in a similar way as those of D_t^{MR}.

B.1.5 Policies

The decision rules at all decision moments are called as *policy*, which provides a prescription for the decision maker with action selections under any possible future system state. Mathematically, a policy π is a sequence of decision rules, i.e. $\pi = (d_1, d_2, \dots, d_N)$ with $d_t \in D_t^K, t = 1, 2, \dots, N$, and $N \leq \infty$. Furthermore, we can define the set of all possible policy as $\prod^K = D_1^K \times D_2^K \times \dots \times D_N^K$, $N \leq \infty$. If $d_t = d$ for all $t \in T$, the corresponding policy $\pi = (d, d, \dots)$ is called a stationary policy, and it is denoted as d^∞.

Table B.1 Classification of decision rules.

	Action choice	
History dependency	**Deterministic**	**Randomized**
Markovian	$d_t(i_t) \in A_{i_t}$	$q_{d_t(i_t)}(\cdot) \in \text{Dis}(A_{s_t})$
History dependent	$d_t(h_t) \in A_{i_t}$	$q_{d_t(h_t)}(\cdot) \in \text{Dis}(A_{s_t})$

B.2 Finite-Horizon Case

When the set of decision moments is $T = \{0, 1, 2, \dots, N\}$, the Markov decision problem is called a *finite-horizon problem*. In this section, we will show a general method to solve the finite-horizon Markov decision problem.

B.2.1 Expected Reward

For a finite-horizon Markov decision process, denote its expected total reward at decision moment N for policy π in state i as $v_N^\pi(i)$, and it is defined as

$$v_N^\pi(i) = \mathbb{E}_i^\pi \left\{ \sum_{t=1}^{N-1} r_t(X_t, Y_t) + r_N(X_N) \right\}, \tag{B.8}$$

where the random variables X_t and Y_t are the state and action at decision moment t and take values in S and A, respectively. For example, if $\pi \in \prod^{\text{HD}}$, Table B.1 shows $Y_t = d_t(h_t)$.

To account for the time value of rewards, we refer a discount parameter λ, $0 \leq \lambda \leq 1$, and add a weight λ^{t-1} to the reward $r_t(X_t, Y_t)$ obtained at decision moment t. Thus, the total reward defined in (B.8) can be modified as the expected total discounted reward:

$$v_{N,\lambda}^\pi(i) = \mathbb{E}_i^\pi \left\{ \sum_{t=1}^{N-1} \lambda^{t-1} r_t(X_t, Y_t) + \lambda^{N-1} r_N(X_N) \right\}. \tag{B.9}$$

Obviously, when $\lambda = 1$, (B.9) is equal to (B.8).

B.2.2 Optimal Policies

A policy $\pi^* \in \prod$ of a Markov decision process at state i is said to be *optimal*, if it achieves the largest expected reward defined in (B.8), i.e.

$$v_N^{\pi^*}(i) \geq v_N^\pi(i), i \in S, \tag{B.10}$$

where $\pi \in \prod$ is an arbitrary policy.

The *value* of the Markov decision problem at state i is denoted as $v_N^*(i)$ and is defined as

$$v_N^*(i) = v_N^{\pi^*}(i) = \sup_{\pi \in \prod} v_N^\pi(i). \tag{B.11}$$

B.2.3 Policy Evaluation

In this section, we show how to utilize *dynamic programming* to evaluate the expected total reward of a given policy for the considered Markov decision process. Without loss of generality, we consider the randomized history-dependent policy, and the corresponding except total reward u_t^π starting from decision time t with the given track h_t is given as

$$u_t^\pi(h_t) = \mathbb{E}_{h_t}^\pi \left\{ \sum_{n=t}^{N-1} r_n(X_n, Y_n) + r_N(X_N) \right\}. \tag{B.12}$$

Obviously, when $h_1 = i$, it follows $u_1^\pi(i) = v_N^\pi(i)$.

Thus, $u_t^\pi(h_t)$ for a given policy $\pi \in \prod^{\text{HR}}$ can be computed from the last to the first decision moment: at the last decision moment, $u_N^\pi(h_N)$ can be directly obtained by $r_N(X_N)$ for all the realizations of state X_N; then, $u_t^\pi(h_t)$ can be recursively computed by $u_{t+1}^\pi(h_{t+1})$ with the law of total probability as

$$u_t^\pi(h_t) = r_t(i_t, q_{d_t(h_t)}) + \sum_{j\in S} p_t(j|i_t, q_{d_t(h_t)}) u_{t+1}^\pi(h_t, q_{d_t(h_t)}, j), \tag{B.13}$$

where $u_{t+1}^\pi(h_t, q_{d_t(h_t)}, j)$ is the except total reward at next decision moment. (B.13) is the well-known Bellman equation. Thus, we summarize in Algorithm B.1.

Algorithm B.1: Compute $u_t^\pi(h_t)$ for a given policy $\pi \in \prod^{\text{HR}}$.

Input: $t = N$
Output: $u_t^\pi(h_t)$
1: Set $u_N^\pi(h_N) = r_N(i_N)$ for all $h_N = (h_{N-1}, a_{N-1}, i_N) \in H_N$
2: **repeat**
3:     Substitute $t-1$ for t
4:     Compute $u_t^\pi(h_t)$ for each $h_t = (h_{t-1}, a_{t-1}, i_t) \in H_t$ by (B.13)
5: **until** $t = 1$

B.2.4 Optimality Equation

Inspired by (B.8) and (B.11), the optimal policy π^* should maximize (B.13) at each decision moment t. Thus, the following equation, termed as *optimality equations*, plays a key role for a Markov decision process, i.e.

$$u_t^*(h_t) = \sup \left\{ r_t(i_t, q_{d_t(h_t)}) + \sum_{j\in S} p_t(j|i_t, q_{d_t(h_t)}) u_{t+1}(h_t, q_{d_t(h_t)}, j) \right\}, \tag{B.14}$$

where $t = 1, \ldots, N-1$ and $h_t = (h_{t-1}, a_{t-1}, i_t) \in H_t$. For $t = N$, the optimality equation is given as

$$u_N(h_N) = r_N(i_N), \tag{B.15}$$

where $h_N = (h_{N-1}, a_{N-1}, i_N) \in H_N$.

In Markov decision theory, the optimality equations (B.14) are fundamental tools and have some important properties: (i) solutions to the optimality equations (B.14) are the optimal returns from decision moment t onward. (ii) The optimality equations provide a method to determine whether a policy is optimal. If the expected total reward induced by policy π from decision moment t onward satisfies these equations, then it is optimal.

B.2.5 Backward Induction

Based on the optimality equations (B.14), a backward induction (dynamic programming) algorithm, summarized as Algorithm B.2, is proposed to efficiently solve the finite-horizon Markov decision problem, and it can find the optimal policy and optimal value. The main idea of this algorithm is the same as that in Section B.2.3, and in each iteration, we compute the optimality equation to obtain the optimal decision at each decision moment, i.e.

$$u_t^*(i_t) = \max_{a \in A_{i_t}} \left\{ r_t(i_t, a) + \sum_{j \in S} p_t(j|i_t, a) u_{t+1}^*(j) \right\}, \tag{B.16}$$

and the corresponding optimal action set $A_{i_t,t}^*$ is

$$A_{i_t,t}^* = \arg\max_{a \in A_{i_t}} \left\{ r_t(i_t, a) + \sum_{j \in S} p_t(j|i_t, a) u_{t+1}^*(j) \right\}. \tag{B.17}$$

Algorithm B.2: Backward induction for computing the optimal value and optimal policy for finite-horizon Markov decision problem.

Input: $t = N$
Output: $u_t^*(i_t)$
1: Set $u_N^*(i_N) = r_N(i_N)$ for all $i_N \in S$
2: **repeat**
3: Substitute $t - 1$ for t
4: Compute the maximum reward $u_t^*(i_t)$ for each $i_t \in S$ by (B.16)
5: **until** $t = 1$

B.3 Infinite-Horizon Case

When the set T of decision moments is $T = \{0, 1, 2, \dots, N, \dots\}$, the corresponding Markov decision problem is called an infinite-horizon problem. Consider the case that the Markov decision problem is stationary (time homogeneous), i.e. the rewards $r(i, a)$, the transition probabilities $p(j|i, a)$, and the decision sets $D^K (K = \text{HR, HD, MR, and MD})$ do not vary from one decision moment to another.

B.3.1 The Expected Average Reward Criterion

For a stationary infinite-horizon Markov decision process, each policy $\pi = (d_1, d_2, \dots)$ induces a bivariate discrete time reward process $\{(X_t, r(X_t, Y_t)), t = 1, 2, \dots\}$, where X_t is the state of the considered system at decision moment t and $r(X_t, Y_t)$ is the received reward with action Y_t in state X_t.

For a given policy π, the expected average reward v^π for the infinite-horizon Markov decision problem is defined as the long-term average of the reward $r(X_t, Y_t)$, i.e.

$$v^\pi(i) = \lim_{N \to \infty} \frac{1}{N} \mathbb{E}_i^\pi \left\{ \sum_{t=1}^{N} r(X_t, Y_t) \right\} = \lim_{N \to \infty} \frac{1}{N} v_{N+1}^\pi(i), \tag{B.18}$$

where $v_{N+1}^\pi(i)$ denotes the expected total reward in with N decision moments and is similarly defined as (B.8).

For some cases, the limit in (B.18) does not exist. Therefore, a condition is imposed to ensure the existence of this limit. We define the lim sup average reward v_+ and the lim sup average reward v_- by

$$v_+^\pi(i) \equiv \lim_{N\to\infty} \sup \frac{1}{N} v_{N+1}^\pi(i), \tag{B.19}$$

and

$$v_-^\pi(i) \equiv \lim_{N\to\infty} \inf \frac{1}{N} v_{N+1}^\pi(i). \tag{B.20}$$

The reward functions (B.19) and (B.20) provide upper and lower bounds on the average reward per period attainable by policy π. It is noticed that the limit exists if and only if $v_+^\pi = v_-^\pi$.

Similarly, the expected total discounted reward for the infinite-horizon Markov decision problem is defined as

$$v_\lambda^\pi(i) = \lim_{N\to\infty} \mathbb{E}_i^\pi \left\{ \sum_{t=1}^{N} \lambda^{t-1} r(X_t, Y_t) \right\}, \tag{B.21}$$

where λ is a discount parameter with $0 \leq \lambda \leq 1$. The limit in (B.21) exists if

$$\sup_{i\in S} \sup_{a\in A_i} |r(i,a)| = M, \tag{B.22}$$

with M being finite.

B.3.2 Optimal Policies

For the infinite-horizon Markov decision process, we aim to find a policy π^* that maximizes the expected total reward $v^\pi(i)$, i.e.

$$\pi^* = \arg\sup v^\pi(i), \tag{B.23}$$

where $v^\pi(i)$ is defined in (B.18). Furthermore, we define the *value* of the infinite-horizon Markov decision problem at state i as $v^*(i)$, i.e.

$$v^*(i) \equiv \sup_{\pi\in\prod} v^\pi(i). \tag{B.24}$$

According to the above analysis, a policy $\pi^* \in \prod$ is optimal in the sense of achieving the maximum expected total reward, if it satisfies

$$v^{\pi^*}(i) \geq v^\pi(i). \tag{B.25}$$

B.3.3 Policy Evaluation and Optimality Equations

For a policy $\pi = (d_1, d_2, d_3, \ldots)$ of the considered Markov decision problem, we observe from (B.18) that its corresponding reward can be computed as

$$\begin{aligned} v^\pi &= \lim_{N\to\infty} \frac{1}{N} \sum_{t=1}^{N} P_\pi^{t-1} r_{d_t} \\ &= r_{d_1} + P_{d_1} r_{d_2} + P_{d_1} P_{d_2} r_{d_3} + \cdots \\ &= r_{d_1} + P_{d_1}(r_{d_2} + P_{d_2} r_{d_3} + P_{d_2} P_{d_3} r_{d_4} + \cdots) \\ &= r_{d_1} + P_{d_1} v^{\pi'}, \end{aligned} \tag{B.26}$$

where $\pi' = (d_2, d_3, \ldots)$ and P is a vector that consists of the probability of rewards, i.e. $P_{d_1} v^{\pi'} = \sum_{j \in S} p_{d_1}(j|i,a) v^{\pi'}(j)$. Consider the case that π is stationary, and it follows $\pi' = \pi$. Therefore, we use $d^{\infty} \equiv (d, d, \ldots)$ to denote the stationary policy for the considered infinite-horizon Markov decision problems, and (B.26) becomes

$$v^{d^{\infty}} = r_d + P_d v^{d^{\infty}}. \tag{B.27}$$

Therefore, we solve (B.27) and can obtain the reward corresponding to the given policy π. For most of the practical cases, we can adopt an iterative algorithm to efficiently compute $v^{d^{\infty}}$ given in (B.27), and we will discuss it in the next subsection.

Based on (B.27), we can obtain the optimality equation, which is the well-known Bellman equation, for the infinite-horizon Markov decision problems as

$$v(i) = \sup_{a \in A_i} \left\{ r(i,a) + \sum_{j \in S} p(j|i,a) v(j) \right\}. \tag{B.28}$$

B.3.4 Iteration Algorithms

Now, we show how to utilize the policy evaluation (B.27) and the optimality equations (B.28) to iteratively compute the optimal policy and optimal value, i.e. *policy iteration* and *value iteration*, for the infinite-horizon Markov decision problems. The two algorithms are equivalent while may be applied to different scenarios.

Algorithm B.3: Policy iteration algorithm.

Input: $d_0 \in D$ with $r_{d_0} \geq 0$
Output: v^n, d^*

1: Set $n = 0$
2: (Policy evaluation) Obtain v^n by (B.29)
3: (Policy improvement) Choose d_{n+1} to satisfy (B.30)
4: **if** $d_{n+1} = d_n$ **then**
5:     Set $d^* = d_n$
6: **else**
7:     Set $n = n + 1$ and return step 2
8: **end if**

The policy iteration algorithm is summarized in Algorithm B.3. Its main idea is as follows: in the nth iteration, update the reward v^n and the decision rule d_{n+1} with an old version decision rule d_n as follows:

$$v^n = r_{d_n} + P_{d_n} v^n, \tag{B.29}$$

$$d_{n+1} = \arg\max_{d \in D} \{ r_d + P_d v^n \}, \tag{B.30}$$

where (B.29) and (B.30) are based on (B.26). When the value of state converges, the algorithm ends. It is worth mentioning that the policy evaluation in (B.29) can also be computed iteratively.

Value iteration is an alternative algorithm to obtain the close-to-optimal policy and value for the infinite-horizon Markov decision problems. For this iterative algorithm, we update the value and the policy by utilizing the old version of the value obtained in the previous iteration, i.e.

$$v^{n+1}(i) = \max_{a \in A_i} \left\{ r(i,a) + \sum_{j \in S} p(j|i,a) v^n(j) \right\}, \tag{B.31}$$

$$d_\varepsilon(i) = \arg\max_{a \in A_i} \left\{ r(i,a) + \sum_{j \in S} p(j|i,a) v^{n+1}(j) \right\}. \tag{B.32}$$

Compared with the policy iteration algorithm in Algorithm B.4, value iteration takes less time at each iteration while the total number of iterations is large.

Algorithm B.4: Value iteration algorithm.

Input: $v^0 \in V, \varepsilon > 0$
Output: $v^{n+1}(i), d_\varepsilon(i)$
1: Set $n = 0$
2: For each $i \in S$, compute $v^{n+1}(i)$ by (B.31)
3: **if** $\|v^{n+1} - v^n\| < \varepsilon$, **then**
4:     For each $i \in S$, choose $d_\varepsilon(i)$ by (B.32), and stop
5: **else**
6:     Set $n = n + 1$ and return step 2
7: **end if**

Appendix C

Optimal Stopping Theory

In this chapter, we briefly introduce the optimal stopping theory. The definition of the optimal stopping problems, the existence of the optimal stopping rules for general optimal stopping problems, and the structure of the optimal stopping rules for some classes of optimal stopping problems are summarized.

C.1 Optimal Stopping Problems

C.1.1 Definition

Optimal stopping problems are defined by the following two objects:

(1) A sequence of random variables, $X_1, X_2, \ldots$, whose joint distribution is known.
(2) A sequence of real-valued reward functions

$$y_0, y_1(x_1), y_2(x_1, x_2), \ldots, y_\infty(x_1, x_2, \ldots), \tag{C.1}$$

where x_i is the realization of random variable X_i, $i = 1, 2, \ldots$.

The decision process of a stopping problem can be described as follows. If you select not to take any observation, you receive the constant reward y_0. At the nth decision moment, $n = 1, 2, \ldots$, you observe $X_1 = x_1$, $X_2 = x_2$, … , $X_n = x_n$: if you choose to stop at this moment, you obtain the reward $y_n(x_1, \ldots, x_n)$; otherwise, you continue the next round of observation and observe $X_{n+1} = x_{n+1}$. If you never stop, reward $y_\infty(x_1, x_2, \ldots)$ is received. The optimal stopping problem is to select an appropriate decision moment to stop to maximize the expected reward, and thus randomized decisions are permitted.

With the observations $x_1, \ldots, x_n$, we denote the probability that you stop after the nth observation as $\phi_n(x_1, \ldots, x_n)$, satisfying $0 \le \phi_n(x_1, \ldots, x_n) \le 1$. Thus, a *stopping rule* ϕ is a sequence of ϕ_n functions, i.e.

$$\phi = (\phi_0, \phi_1(x_1), \phi_2(x_1, x_2), \ldots). \tag{C.2}$$

The stopping rule is deterministic, if each $\phi_n(x_1, \ldots, x_n)$ is either 0 or 1; otherwise, the stopping rule is randomized.

Based on the definition of stopping rule, the *stopping time* N, $0 \le N \le \infty$, is defined as the first decision moment that the stopping rule ϕ chooses to stop. Obviously, N is also a random variable and determined by the stopping rule ϕ and the sequence of observations $x_1, \ldots, x_N$. If stopping never occurs, N turns to ∞.

Energy Harvesting Wireless Communications, First Edition.
Chuan Huang, Sheng Zhou, Jie Xu, Zhisheng Niu, Rui Zhang and Shuguang Cui.

The probability density function of N for given $\boldsymbol{X} = \boldsymbol{x}$ with $\boldsymbol{X} = (X_1, X_2, \ldots)$ and $\boldsymbol{x} = (x_1, x_2, \ldots)$ is denoted as $\psi = (\psi_0, \psi_1, \psi_2, \ldots, \psi_\infty)$, where

$$\psi_n(x_1, \ldots, x_n) = P(N = n | \boldsymbol{X} = \boldsymbol{x}), \quad n = 0, 1, 2, \ldots, \tag{C.3}$$

$$\psi_\infty(x_1, x_2, \ldots) = P(N = \infty | \boldsymbol{X} = \boldsymbol{x}). \tag{C.4}$$

The relationship between the stopping rule ϕ and the probability density function ψ for stopping time N is given as

$$\begin{aligned}
&\psi_0 = \phi_0,\\
&\psi_1(x_1) = (1 - \phi_0)\phi_1(x_1),\\
&\vdots\\
&\psi_n(x_1, \ldots, x_n) = \left[\prod_{j=0}^{n-1}(1 - \phi_j(x_1, \ldots, x_j))\right] \phi_n(x_1, \ldots, x_n), \qquad \text{(C.5)}\\
&\vdots\\
&\psi_\infty(x_1, x_2, \ldots) = 1 - \sum_{j=0}^{\infty} \phi_j(x_1, \ldots, x_j).
\end{aligned}$$

Now, we are ready to rigorously define the optimal stopping problem. Our goal is to select a stopping rule ϕ to maximize the expected reward $V(\phi)$, i.e.

$$\phi^* = \arg\max_{\phi} V(\phi), \tag{C.6}$$

where

$$\begin{aligned}
V(\phi) &= \mathbb{E}[y_N(X_1, \ldots, X_N)]\\
&= \mathbb{E}\left[\sum_{j=0}^{\infty} \psi_j(X_1, \ldots, X_j) y_j(X_1, \ldots, X_j)\right].
\end{aligned} \tag{C.7}$$

C.1.2 Existence of Optimal Stopping Rules

In this subsection, we consider the existence of the optimal stopping rules for the general optimal stopping problems with observations $X_1, X_2, \ldots$ and rewards $Y_0, Y_1, \ldots, Y_\infty$, where $Y_n = y_n(X_1, \ldots, X_n)$. We will rigorously show that when the following two conditions are satisfied, i.e.

$$\text{A1.} \quad \mathbb{E}\{\sup_n Y_n\} < \infty, \tag{C.8}$$

$$\text{A2.} \quad \limsup_{n \to \infty} Y_n \leq Y_\infty \quad \text{a.s.}, \tag{C.9}$$

the optimal stopping rule of Problem (C.6) exists. If either one of the conditions in (C.8) and (C.9) is not satisfied, an optimal stopping rule of Problem (C.6) will not exist. We precede the main theorem about the existence of the optimal stopping rules by firstly introducing the following definitions and lemmas:

Definition C.1 A stopping rule N is regular, if for each n, we have

$$\mathbb{E}\{Y_N | \mathcal{F}_n\} > Y_n, \quad \text{for} \quad N > n, \tag{C.10}$$

where $\mathcal{F}_n$ denotes the sub-σ-field of example space generated by random variables $X_1, X_2, \dots X_n$ and the conditional expectation means $\mathbb{E}\{Y_N|\mathcal{F}_n\} = \mathbb{E}\{Y_N|X_1, X_2, \dots X_n\}$.

In another meaning, for all $N > n$, N is regular if

$$\mathbb{E}\{Y_N|X_1 = x_1, \dots, X_n = x_n\} > y_n(x_1, \dots, x_n). \tag{C.11}$$

Then, the first lemma shows how to replace a given stopping rule by a regular one without hurting the expected reward.

Lemma C.1 *Under A1, given any stopping rule N, there is a regular stopping rule N′ such that* $\mathbb{E}[Y_{N'}] \geq \mathbb{E}[Y_N]$.

The second lemma shows that we can possibly improve the stopping rule by two regular stopping rules.

Lemma C.2 *Under A1, if N and N′ are regular stopping rules,* $N'' = \max(N, N')$ *is also a regular stopping rule and satisfies*

$$\mathbb{E}[Y_{N''}] \geq \max\{\mathbb{E}[Y_N], \mathbb{E}[Y_{N'}]\}. \tag{C.12}$$

With the two lemmas above, we can apply the Fatou–Lebesgue lemma to prove the existence of optimal stopping rule, which states: If $X_1, X_2, \dots$ is a sequence of real-valued random variables, satisfying $\mathbb{E}[X_n] < \infty$ for all n, then it follows $\limsup_n \mathbb{E}[X_n] \leq \mathbb{E}[\limsup_n X_n]$.

Theorem C.1 Under A1 and A2, there exists a stopping rule N^* such that $\mathbb{E}[Y_{N^*}] = V^*$, with $V^* = \sup_N \mathbb{E}[Y_N]$.

Proof: In this proof, we first show how to construct this N^*. Suppose that $N_1, N_2, \dots$ are a sequence of stopping rules, satisfying $\mathbb{E}[Y_{N_j}] \to V^*$. Then, by Lemma C.1, we can obtain that there exist a sequence of regular stopping rules $N_1', N_2', \dots$ such that $\mathbb{E}[Y_{N_j'}] \to V^*$. Define $N_j'' = \max\{N_1', N_2', \dots\}$, and by Lemma C.2, it follows that $\mathbb{E}[Y_{N_j''}] \geq \mathbb{E}[Y_{N_j'}]$ and thus $\mathbb{E}[Y_{N_j''}] \to V^*$. We claim that this N_j'' is the desired N^*.

Next, we prove that $\mathbb{E}[Y_{N^*}] = V^*$. Since $\mathbb{E}[Y_{N^*}] \leq V^*$, we only need to prove $\mathbb{E}[Y_{N^*}] \geq V^*$. We can obtain

$$V^* = \limsup \mathbb{E}[Y_{N_j''}] \tag{C.13}$$
$$\leq \mathbb{E}[\limsup Y_{N_j''}] \tag{C.14}$$
$$\leq \mathbb{E}[Y_{N^*}], \tag{C.15}$$

where (C.13) is due to the definition of N_j'' and Lemma C.2 and (C.14) is due to the well-known Fatou–Lebesgue lemma. Here, (C.15) is due to the fact that $\{N_j''\}$ is a monotonically non-decreasing sequence (and thus it follows $\limsup_{j\to\infty} Y_{N_j''} \leq Y_{N^*}$ by condition A2) and $\{Y_{N_j''}\}$ is upper bounded by $\sup_N Y_N$, which is integrable by condition A1. Therefore, the proof is completed. □

C.2 Optimal Conditions

C.2.1 Optimality Equation

In order to select a stopping rule to maximize the expected reward, the optimality equation will be given in this subsection.

Definition C.2 Let X_t, $t \in T$, be a collection of random variables, and T is a countable or uncountable set. It is said that a random variable Z is an essential supremum of $\{X_t\}_{t\in T}$, and write $Z = \text{ess sup}_{t\in T} X_t$, if

(1) $P(Z \geq X_t) = 1$ for all $t \in T$.
(2) If Z' is any other random variable such that $P(Z' \geq X_t) = 1$ for all $t \in T$, then $P(Z' \geq Z) = 1$.

Lemma C.3 *There always exists an essential supremum $Z = \text{ess sup}_{t\in T} X_t$ and a countable subset $C \subset T$, such that $Z = \sup_{t\in C} X_t$.*

According to the above definitions and lemmas, define the optimal expected reward as

$$V_n^* = \operatorname*{ess\,sup}_{N\geq n} \mathbb{E}\{Y_N|\mathcal{F}_n\}, \tag{C.16}$$

and then, we derive the theorem about the optimality equation as follows.

Theorem C.2 Under A1, the optimality equation of Problem (C.6) is given as

$$V_n^* = \max(Y_n, \mathbb{E}\{V_{n+1}^*|\mathcal{F}_n\}). \tag{C.17}$$

Proof: First, we prove the following inequality:

$$V_n^* = \operatorname*{ess\,sup}_{N\geq n} \mathbb{E}\{Y_N|\mathcal{F}_n\} \leq \max(Y_n, \mathbb{E}\{V_{n+1}^*|\mathcal{F}_n\}). \tag{C.18}$$

Considering the case of $N > n$, we can obtain

$$\begin{aligned}
\mathbb{E}\{Y_N|\mathcal{F}_n\} &= \mathbb{E}\{\mathbb{E}[Y_N|\mathcal{F}_n, X_{n+1}]\} && \text{(C.19)}\\
&= \mathbb{E}\{\mathbb{E}[Y_N|\mathcal{F}_n, X_{n+1}]|\mathcal{F}_n\} && \text{(C.20)}\\
&= \mathbb{E}\{\mathbb{E}[Y_N|\mathcal{F}_{n+1}]|\mathcal{F}_n\} && \text{(C.21)}\\
&\leq \mathbb{E}\{V_{n+1}^*|\mathcal{F}_n\}, && \text{(C.22)}
\end{aligned}$$

where (C.19) is due to the independence of $X_1, X_2, \ldots$, (C.20) is due to the law of total expectation $\mathbb{E}[X] = \mathbb{E}\{\mathbb{E}[X|Y]\}$, and (C.22) is due to (C.16).

Then, based on the above analysis, we obtain the following result for the case of $N \geq n$:

$$\begin{aligned}
\mathbb{E}\{Y_n|\mathcal{F}_n\} &= I_{\{N=n\}} Y_n + I_{\{N>n\}} \mathbb{E}\{Y_N|\mathcal{F}_n\}\\
&= I_{\{N=n\}} Y_n + I_{\{N>n\}} \mathbb{E}\{V_{n+1}^*|\mathcal{F}_n\}\\
&\leq \max(Y_n, \mathbb{E}\{V_{n+1}^*|\mathcal{F}_n\}), && \text{(C.23)}
\end{aligned}$$

where $I_{\{N=n\}}$ is the index function. If the condition $N = n$ is true, it follows $I_{\{N=n\}} = 1$; otherwise, it follows $I_{\{N=n\}} = 0$. Therefore, with (C.23), we prove (C.18).

Next, we prove the reverse inequality for (C.18). Considering the case of $N > n$, we construct a sequence of regular stopping rules $N_1', N_2', \ldots$ and define $N_k'' = \max(N_1', N_2', \ldots)$ similar as the proof of Theorem C.1. Then, we can obtain

$$V_n^* \geq \mathbb{E}\{Y_{N_k''} | \mathcal{F}_n\} \tag{C.24}$$

$$= \mathbb{E}\{\mathbb{E}[Y_{N_k''} | \mathcal{F}_{n+1}] | \mathcal{F}_n\} \tag{C.25}$$

$$\geq \mathbb{E}(\max_{1 \leq j \leq k} \mathbb{E}\{Y_{N_j'} | \mathcal{F}_{n+1}\} | \mathcal{F}_n) \tag{C.26}$$

$$\geq \mathbb{E}(\max_{1 \leq j \leq k} \mathbb{E}\{Y_{N_j} | \mathcal{F}_{n+1}\} | \mathcal{F}_n), \tag{C.27}$$

$$\to \mathbb{E}\{V_{n+1}^* | \mathcal{F}_n\}. \tag{C.28}$$

where (C.24) is obtained by (C.16) and Lemmas C.1 and C.2, (C.25) is obtained similar to (C.20), (C.26) and (C.27) are from the proof of Theorem C.1, and (C.28) is obtained from Lemmas C.3 and (C.16). Therefore, for all $N \geq n$, the reverse inequality of (C.18) is obtained, and Theorem C.2 is proved. □

It is noticed that the optimality equation (C.6) holds under condition A1 alone, even if an optimal rule problem (C.6) does not exist.

C.2.2 Principle of Optimality

Under condition A1, define a stopping rule N^* as

$$N^* = \min_n \{n \geq 0 : Y_n = V_n^*\}. \tag{C.29}$$

In the following theorem, we will show that the stopping rule given in (C.29) is optimal.

Theorem C.3 Under A1 and A2, the stopping rule N^* given in (C.29) is optimal for the considered optimal stopping problem. It means that for an arbitrarily optimal rule N_0, we have $N^* \leq N_0$.

Proof: We consider an arbitrarily optimal rule N_0 and define $N = \min\{N_0, N^*\}$. It is easy to show that N is also optimal and $N \leq N^*$. If $N = N^*$, the proof is completed.

Then, the case of $N < N^*$ is considered. For some n, it follows $N = n < N^*$, and then $Y_n < V_n^*$. Next, we will show that we can improve N to let Y_n be close to V_n^*. Similar to the previous subsection, we construct a regular stopping rule $N_k'' = \max(N_1', N_2', \ldots)$ by Lemmas C.1 and C.2. Thus, there exists N_k'' such that

$$\mathbb{E}\{I_{\{N=n<N^*\}} Y_{N_k''}\} > \mathbb{E}\{I_{\{N=n<N^*\}} Y_n\}. \tag{C.30}$$

Similar to the proof of (C.24) in Theorem C.2, we obtain

$$\mathbb{E}\{I_{\{N=n<N^*\}} Y_{N_k''}\} \to \mathbb{E}\{I_{\{N=n<N^*\}} V_n^*\}. \tag{C.31}$$

In order to improve N, we set $N_k'' = N$ to let Y_n be close to V_n^*. If $N = N^0 < N^*$, we obtain $\mathbb{E}[Y_{N^*}] > \mathbb{E}[Y_N]$, and it contradicts the optimality of N. Therefore, $N = N^*$ and Theorem C.3 is proved. □

C.3 Example: House-Selling Problem

Here, we offer a specific problem as an example to show how the optimal stopping theorem works. Consider a house-selling problem, for which offers come in daily for a house desired to be sold. X_n denotes the amount of the offer on day n. The values of offers are unknown before they come in. However, it can be assumed that the offers are independent and all have the same distribution $F(x)$. Every offer costs an amount $c > 0$ to observe, which can be thought as a cost of living. When the offer X_n is received, one must decide to whether accept it or wait for a better one. Before deciding the eventual offer, one has to pay the observational cost up to the current day.

We distinguish two problems with different payoffs, relying on whether or not you can recall and accept a past offer. If you are allowed to recall the past offers, then the reward at each day is given as

$$\begin{aligned} y_0 &= 0, \\ y_n(x_1, \dots, x_n) &= \max\{x_1, \dots, x_n\} - nc, \quad \text{for } n = 1, 2, \dots, \\ y_\infty(x_1, \dots, x_n) &= -\infty. \end{aligned} \tag{C.32}$$

In this case, you can receive the largest outstanding offer. If you cannot recall a past offer, then the reward is given as

$$\begin{aligned} y_0 &= 0, \\ y_n(x_1, \dots, x_n) &= x_n - nc, \quad \text{for } n = 1, 2, \dots, \\ y_\infty(x_1, \dots, x_n) &= -\infty. \end{aligned} \tag{C.33}$$

In the sequel, we mainly focus on the case without recall. From (C.33), we define

$$Y_0 = -\infty, Y_1 = X_1 - c, \dots, Y_n = X_n - nc, \dots, Y_\infty = -\infty, \tag{C.34}$$

where Y_0 is set as $-\infty$ to force to accept at least one observation, and let $Y_\infty = -\infty$ since the cost of an infinite number of observation is infinite. First, we check the existence of the optimal stopping rule, and we can obtain that conditions A1 and A2 are satisfied for the house-selling problem without recall:

A1: If $\mathbb{E}[X^+ < \infty]$, then $\sup Y_n < \infty$ a.s. and $Y_n \to -\infty$ a.s.;

A2: If $\mathbb{E}[(X^+)^2 < \infty]$, then $\mathbb{E}[\sup Y_n < \infty]$.

Since the two assumptions are satisfied, an optimal stopping rule exists and is given by the principle of optimality (C.29).

Suppose that you pay the cost c and observe $X_1 = x_1$ and V^* denotes the expected return. It is noticed that if you continue to observe from this time point, x_1 will be lost, and the cost has already been paid. It seems that at the next time point, the problem is started over again, i.e. the problem is invariant time.

Hence, you can get an expected return V^* but no more if you continue from this point. The principle of optimality states that if $x_1 < V^*$, you should continue, and if $x_1 \geq V^*$, you should stop. The optimal stopping rule given by the principle of optimality is

$$N^* = \min\{n \geq 1 : X_n \geq V^*\}. \tag{C.35}$$

Then, the problem is to calculate V^* via the following optimality equation:

$$V^* = \mathbb{E}\{\max\{X_1, V^*\}\} - c = \int_{-\infty}^{V^*} V^* \mathrm{d}F(x) + \int_{V^*}^{\infty} x \mathrm{d}F(x) - c, \tag{C.36}$$

where $F(x)$ is the distribution function of X_i. By transforming (C.36), we obtain

$$\int_{V^*}^{\infty} (x - V^*)\mathrm{d}F(x) = c \qquad \text{or} \quad \mathbb{E}[(X - V^*)^+] = c. \tag{C.37}$$

Obviously, there exists a unique solution for V^* for any $c > 0$.

Now, consider the case with recall, and define $Y_0 = Y_\infty = -\infty$ and $Y_n = M_n - nc$, where $M_n = \max\{X_1, \ldots, X_n\}$. Suppose that at some stage you have observed $M_n = m$ and it is optimal to continue. At the next stage, it is still optimal to continue due to the invariance of the problem in time. Hence, an observation from an earlier stage will not need to be recalled according to the principle of optimality. Therefore, the rule obtained in (C.35) is optimal for both the problems with and without recall.

Index

Energy Harvesting Wireless Communications, First Edition.
Chuan Huang, Sheng Zhou, Jie Xu, Zhisheng Niu, Rui Zhang and Shuguang Cui.

d

e

l

m